LARGE-SCALE DYNAMICAL PROCESSES IN THE ATMOSPHERE

LARGE-SCALE DYNAMICAL PROCESSES IN THE ATMOSPHERE

Edited by

BRIAN HOSKINS and ROBERT PEARCE

Department of Meteorology, University of Reading, England

1983

ACADEMIC PRESS

A Subsidiary of Harcourt Brace Jovanovich, Publishers

London   New York
Paris   San Diego   San Francisco   São Paulo
Sydney   Tokyo   Toronto

ACADEMIC PRESS INC. (LONDON) LTD.
24/28 Oval Road
London NW1

United States Edition published by
ACADEMIC PRESS INC.
111 Fifth Avenue
New York, New York 10003

British Library Cataloguing in Publication Data

Large-scale dynamical processes in the atmosphere.
1. Dynamic meteorology
I. Hoskins, B.   II. Pearce, R.
551.5'153   QC880

ISBN 0-12-356680-0
LCCCN 82-73231

Typeset by Mid-County Press, London
and printed in Great Britain by
Thomson Litho Ltd., East Kilbride, Scotland

LIST OF CONTRIBUTORS

I. N. James   Department of Meteorology, University of Reading, 2 Earley Gate, Whiteknights, Reading RG6 2AU, UK

J. M. Wallace   Department of Atmospheric Sciences, University of Washington, Seattle, Washington 98195, USA

M. L. Blackmon   National Centre for Atmospheric Research, P.O. Box 3000, Boulder, Colorado 80307, USA

R. M. Dole   Department of Meteorology, Massachusetts Institute of Technology, Cambridge, Massachusetts 02139, USA

Ngar-cheung Lau   Geophysical Fluid Dynamics Laboratory/NOAA, Princeton University, P.O. Box 308, Princeton, New Jersey 08540, USA

I. M. Held   Geophysical Fluid Dynamics Laboratory/NOAA, Princeton University, P.O. Box 308, Princeton, New Jersey 08540, USA

B. J. Hoskins   Department of Meteorology, University of Reading, 2 Earley Gate, Whiteknights, Reading RG6 2AU, UK

E. O. Holopainen   Department of Meteorology, University of Helsinki, Hallituskatu 11–13, 00100 Helsinki 10, Finland

P. J. Webster   CSIRO Division of Atmospheric Physics, P.O. Box 77, Mordialloc, Victoria 3195, Australia

J. R. Holton   Department of Atmospheric Sciences, University of Washington, Seattle, Washington 98195, USA

D. L. T. Anderson   Department of Atmospheric Physics, University of Oxford, The Clarendon Laboratory, Parks Road, Oxford OX1 3PU, UK

L. Bengtsson   European Centre for Medium-range Weather Forecasts, Shinfield Park, Reading RG2 9AX, UK

A. J. Simmons   European Centre for Medium-range Weather Forecasts, Shinfield Park, Reading RG2 9AX, UK

C. E. Leith   National Centre for Atmospheric Research, P.O. Box 3000, Boulder, Colorado 80307, USA

To the memory of the late
P. A. Sheppard

PREFACE

The Third Scientific Assembly of the International Association for Meteorology and Atmospheric Physics (IAMAP), held in Hamburg in August, 1981, included, as one of its symposium topics, the Dynamics of the General Circulation of the Atmosphere. This formed part of the scientific programme organised by its dynamical commission—the International Commission for Dynamical Meteorology (ICDM). The sessions on this topic were divided into three parts emphasising, respectively, the mid-latitude troposphere, the tropics and vacillation phenomena, the first part being arranged as a separate meeting, preceding the main assembly, at the University of Reading, England. This part was also supported by the Royal Society, the Royal Meteorological Society and the Natural Environment Research Council (NERC).

The exceptional level of support given to the symposium reflects both its importance and the substantial advances made in the subject in recent years. It was clear to all who participated that these should be widely disseminated through the publication of the main contributions in book form, and this present volume has therefore been prepared. Although it is based on papers presented at the Symposia in Reading and Hamburg, it is not simply a symposium proceedings. The individual chapters have been written in such a way that the book may be read as an advanced text on large-scale atmospheric dynamics, and additional material, not presented at the symposium, has been included to lend continuity and coherence to the book as a whole. It should be readily intelligible to students familiar with standard texts in dynamical meteorology and to research workers in the field, and has been written with these specific groups of readers in mind.

It is a pleasure to acknowledge the support given to the symposium by the organising bodies already mentioned, without which this book would not have been produced. The editors also wish to acknowledge, with much gratitude, the willing collaboration of each individual author in making a special effort to write his contribution along lines suggested to them in the interest of the coherence of the book as a whole. Finally, special thanks are due to Mrs. Nan Spicer who typed much of the manuscript, Mrs. Valerie Daykin who drew many of the diagrams, and the staff of Academic Press whose collaboration was so much appreciated.

Reading
January 1983

B. J. Hoskins
R. P. Pearce

CONTENTS

List of Symbols

The following symbols are used consistently throughout the book. Any additional special symbols are defined in the text as they are introduced.

x, y	Distances east, north in Cartesian coordinates
z	Height above mean-sea-level
ϕ	Latitude
λ	Longitude
$d/dt, D/Dt$	Differentiation following a fluid element
$\mathbf{u}$	Three-dimensional velocity vector
$\mathbf{v}$	Horizontal component of velocity vector
u	Eastward velocity component (dx/dt)
v	Northward velocity component (dy/dt)
w	Upwards velocity component (dz/dt)
p	Pressure
ω	dp/dt
f	Coriolis parameter $2\Omega \sin \phi$, where Ω is the Earth's angular velocity
β	df/dy
ζ	Relative vorticity
ζ_a	Absolute vorticity $(\zeta + f)$
ζ_g	Geostrophic relative vorticity
ψ	Streamfunction
g	Acceleration due to gravity
R	Gas constant for dry air
c_p	Specific heat of dry air at constant pressure
κ	R/c_p
$\hat{R}$	$\dfrac{R}{p}\left(\dfrac{p}{p_0}\right)^{\kappa}$, p_0 a reference pressure
T	Temperature
θ	Potential temperature $\left[\theta = T\left(\dfrac{1000}{p}\right)^{\kappa} \text{ with } p \text{ in mb}\right]$
Φ	Geopotential
Z	Geopotential height
N	Brunt–Väisälä frequency $\left[= \left(\dfrac{g\partial\theta}{\theta\partial z}\right)^{1/2}\right]$

E   Eliassen–Palm flux vector. In quasi-geostrophic theory, using
                  pressure coordinates, $\mathbf{E} = (-[u^*v^*], f[v^*\theta^*]/\Theta_p)$

$\overline{(\ )}$ Time average

$(\ )'$ Deviation from time average

$[\ ]$ Zonal average

$(\ )^*$ Deviation from zonal average

$\{\ \}$ Vertical average

$\tilde{(\ )}$ (1) Fourier amplitude; or (2) Residual

Introduction

The past decade has been one of considerable advance in global atmospheric circulation studies both observational and theoretical. Extensive analyses of the atmospheric circulation have been published during this period and diagnostics may now be obtained routinely from current operational global analyses. On the theoretical side, analytical and numerical methods have been used to extend understanding in many areas, for example the behaviour of perturbations on a baroclinic flow and the propagation of Rossby waves on the sphere. Particularly notable in the last few years has been the way that the theoreticians and those whose strengths have been in manipulating observational data have intensified their collaboration in a determined attempt to understand atmospheric behaviour.

In this spirit, the primary aim of the volume is to present an account of the main features of the atmospheric circulation as revealed by analysis of observational data. These are interpreted as far as possible in terms of dynamical processes using both analytical and numerical modelling approaches. The analytical theory is based for the most part on the conservation of quasi-geostrophic potential vorticity. For convenience a brief summary of quasi-geostrophic theory is presented in the Appendix. The numerical models referred to vary in complexity from small extensions of analytical models up to general circulation models including representations of a wide range of physical processes. The emphasis is on advancing our understanding of the large-scale atmosphere through experimentation with an hierarchy of models. The major hope for progress in predicting weather and weather types for periods longer than a few days rests on increasing our understanding of atmospheric processes. Without it, predictions on these time-scales can have no firm foundation.

The first four chapters present results of analyses of global observational data. The first compares the global circulations for January and July in a particular year, 1980, with emphasis on eddy fluxes of heat and westerly momentum, particularly those associated with so-called 'storm tracks'. These

1

analyses serve to draw attention not only to the zonal mean characteristics of each hemisphere, but also to what is now recognized as a key aspect of the circulations in the two hemispheres, their zonal variability. It is this variability which comprises the main theme of Chapter 2. Using a sequence of several years' data, the three-dimensional structure of the climatological mean stationary waves in the two hemispheres is exhibited. In recent years it has become increasingly clear that the transient waves of mid-latitudes—the depressions and associated transient ridges of high pressure—are by no means the only, or even the dominant, structures associated with transience on time scales less than a season. The 'low frequency' atmospheric variability, manifested locally in such phenomena as changes in position of the upper tropospheric jet maxima on time scales of two weeks or more, has been found to contribute at least as much to the total variance of the departures from seasonal means as the synoptic eddies (even when the seasonal trends are removed). Different aspects of these phenomena are considered in Chapters 3 and 4. It is probably true to say that one of the highest priorities in studies of the general circulation is to advance knowledge of the structures associated with these phenomena and to identify the dynamical processes which they represent.

Chapter 5 presents the results of analysing the global climatology generated by a highly sophisticated general circulation model in a 15-year simulation. As well as providing an excellent example of the state of the art of numerical modelling it raises intriguing questions as to the likely origin of atmospheric interannual anomalies; no non-seasonal perturbations are included in the model simulation yet such anomalies, resembling those observed, still arise in the simulation.

Chapters 6 and 7 are theoretical. They attempt to present aspects of generally accepted theory as well as some new perspectives relevant to the observed mid-latitude phenomena described in other chapters. These should enable the reader to appreciate the extent to which the processes underlying these phenomena are at present understood and the nature of the problems which remain to be answered. Chapter 6 uses a sequence of models to investigate the stationary wave patterns associated with local forcing by mountains and heat sources. It also provides, for the first time, a rigorous explanation as to why barotropic models of the upper tropospheric flow give such realistic representations of the behaviour of the long waves without involving the full baroclinic dynamics. The results of experiments with general circulation models with and without the Earth's orography included and also one with a uniform underlying surface are then described and their dynamical implications discussed. The emphasis in Chapter 7 is on the theory of the short period (up to one or two weeks) transient (baroclinic) eddies and their role in the general circulation. This contains an attempt to summarize the role of these eddies in momentum, heat and vorticity budgets in the storm-track regions of high synoptic eddy intensity.

Chapter 8 contains a summary of the observational analyses that have been carried out in recent years on the transient eddies of the mid-latitudes where there is a much denser observational network than in lower latitudes, at least in the Northern Hemisphere. A fundamental question concerning these eddies is to what extent they play a dissipative role in the maintenance of the general circulation and to what extent they help to maintain it. Using some of the quasi-geostrophic theory developed in Chapter 7, this chapter considers the extent to which the fluxes of heat and momentum associated with transient eddies dominate their net influence on the zonal mean flow.

The discussion of the role of the mid-latitude transients is followed, in Chapter 9, by an observational and theoretical survey of the tropical atmosphere. This starts with a description of its observed structure from the planetary scale down to the sub-synoptic scale. Theoretical interpretation is based on a scale analysis and a discussion of the fundamental modes for low latitudes. The forcing of motion by diabatic heating and the possibilities of interactions between the tropics and extra-tropics are explored.

An account of the large-scale atmospheric circulation would be incomplete without a chapter embodying recent advances in stratospheric studies. In some respects there have been more substantial advances in this area in the last two decades than in tropospheric studies. Chapter 10 presents a review of the larger-scale stratospheric flow based on observations and its features are then discussed theoretically making use of techniques based on the so-called transformed Eulerian-mean equations. Topics such as stratospheric sudden warmings, the quasi-biennial oscillation and interannual variability in the extratropical stratosphere are considered.

Studies of the atmospheric general circulation are clearly an essential component of a scientific programme on long-term atmospheric behaviour and climate–indeed the most fundamentally important component. However, the atmospheric circulation is strongly coupled to the ocean circulation and it is not possible to understand fully atmospheric behaviour without also studying the oceans. It is with this essential link in mind that one chapter of this volume on this topic (Chapter 11) has been included. This presents a wide review of knowledge of the large-scale oceanic circulation with particular emphasis on coupling with the atmosphere. This is again an area where numerical modelling techniques provides the most promise for the extension of analytical theory.

The last two chapters, 12 and 13, deal with the topic which is perhaps the most difficult, but at the same time the most important application of atmospheric sciences, namely atmospheric prediction. Chapter 12 presents and discusses some results of extended (up to 10-day) predictions with a global operational model, identifies the main error sources and suggests the most promising avenues for model improvement. Chapter 13 starts with a discussion of the fundamental question as to the inherent predictability of the atmosphere and of the limitations which we may ultimately have to accept in attempting to forecast its behaviour. The implications of the possible existence of 'almost-equilibrium' states are considered. This is followed by a theoretical (statistical) analysis of the interpretation of numerical model outputs, including the optimal incorporation of climatological information and additional data as the model integration proceeds.

In a volume which covers such an important and wide-ranging area of science at a time when rapid advances are still being made, it seems fitting to the editors that they should reserve for themselves the prerogative of adding a few reflective comments. These are contained in a brief epilogue.

– 1 –

Some aspects of the global circulation of the atmosphere in January and July 1980

I. N. JAMES

1.1 Analysis and sources of data

The global circulation of the Earth's atmosphere encompasses motions on a wide range of spatial and temporal scales, and any attempt to summarize its state must involve averaging. Traditionally much of the study of the general circulation has relied on zonal averages; the compilation of Oort and Rasmusson (1971) is but one excellent example of many such studies. Recently, time averaging of fields at a particular level in the atmosphere has been used to re-emphasize the longitudinal inhomogeneity of the flow. In particular the publications of Wallace, Blackmon and Lau (see, for example, Blackmon, 1976; Blackmon *et al.*, 1977; Lau, 1978; Lau and Wallace, 1979) have had considerable influence. Chapters 2 and 3 of this volume describe some of their results.

This chapter comprises a selection of mean and eddy fields obtained by time averaging data for January and July 1980. Unlike the results of Wallace and his co-workers, which were based on ensembles of seasonal means from the National Meteorological Center analyses, these data do not represent a climatology. They do, however, yield an interesting comparison between the summer and winter seasons in the two hemispheres for one particular year (1980), and they show the changing nature of the eddy – mean flow interaction. The climatological atlas of Lau *et al.* (1981) provides a useful comparison for the Northern Hemisphere.

The data were extracted from archived analyses produced at the European Centre for Medium-range Weather Forecasts (ECMWF). The analyses cover both hemispheres at high resolution, and although the Southern Hemisphere fields are undoubtedly less accurate, a study of the circulation they represent is informative. The archived fields have been balanced using a normal mode

5

initialization procedure, so that a vertical velocity field is also available. Unfortunately the initialization has a deleterious effect on the tropical divergent wind field. In mid-latitudes the effect is much less serious. The other region where the analyses seem to be unreliable is Antarctica, a result perhaps both of limited observations and of the high orography.

For the present purposes, it has proved sufficient to extract the data on a 5° latitude–longitude grid once every 24 h. Considerably higher spatial and temporal resolution would be possible but would lead to much greater computing expense.

In partitioning the flow into a 'mean' and 'transient' part and then producing fields representing the mean effect of the transients, there is an implicit assumption that the time-averaged eddy correlations represent a statistically stable set. That is, it is assumed that the fluxes of heat, momentum and so on by the transient motion field must be the accumulated result of many individual 'events'. Evidently, this will not be the case if the averaging period is very short and so it is worth enquiring into the minimum useful averaging period for the atmosphere.

Figure 1.1 shows the northward temperature flux, $\overline{v'T'}$, at 700 mb, integrated over each hemisphere and plotted against the length of the averaging period. A total of nine separate months from the summer and winter seasons of 1980 and 1981 forms the basis of the plots. The vertical bars indicate the scatter between the various periods used. Evidently, the important motions that transport heat had periods of less than 10–15 days, for the temperature flux increased quickly as the averaging period approached these values. The inclusion of longer periods generally did little to raise the total temperature flux. The exception is the Northern Hemisphere winter which, in all the months considered, showed a further significant increase of temperature flux when the averaging period was extended from 15 to 20 days. Consequently, an averaging period of 15 days or longer will generally include most of the transient eddy temperature flux in each hemisphere. Whether the local temperature flux is statistically stable for such short periods requires further consideration.

The scatter between the months employed generally had a maximum value for periods of around 10 days and decreased as the period approached 30 days, indicating that, although there may be some variability in the period required to saturate the temperature flux (in other words, in the frequency of the most important temperature transporting transients), the final value of the temperature transport was fairly constant in either season. The exception was once again the Northern Hemisphere winter, for which the scatter increased uniformly up to periods of 30 days. The explanation appears to lie in the very different character of the 1980 and 1981 Northern Hemisphere winters. The latter was notable for the very persistent splitting of the upper tropospheric jet

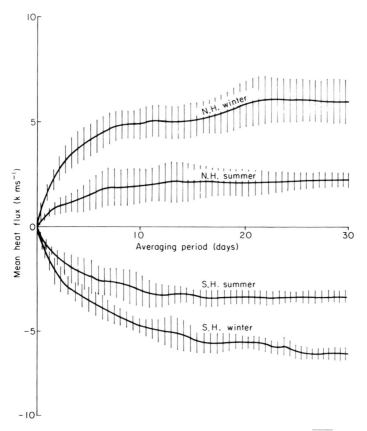

Fig. 1.1 Hemispherically averaged northward temperature flux, $\overline{v'T'}$, at 700 mb, plotted as a function of the length of the averaging period, for a number of summer and winter periods from December 1979 to July 1981. The vertical hatching indicates the standard deviation of the cases.

over the Atlantic and Pacific, whereas a relatively zonal pattern distinguished the 1980 winter. It would be interesting to confirm whether interannual variability is indeed greatest for the Northern Hemisphere winter season.

The total transient temperature flux in the Southern Hemisphere winter was comparable to that in the Northern Hemisphere winter. In the summer season, the Southern Hemisphere transports were much larger. Thus the seasonal variation of temperature flux in the Southern Hemisphere was a good deal less than in the Northern Hemisphere. In fact, the effect was even more pronounced than this diagram indicates, since the steady waves transported nearly as much heat as the transients in the Northern Hemisphere winter but were of less importance in the Southern Hemisphere.

If a comparable diagram for the northward momentum transport $\overline{u'v'}$ is produced, the saturation of the fluxes for periods in excess of 10 days is not nearly so marked. To some extent the difficulty arises because the momentum transport is not unidirectional in each hemisphere, so that the hemispheric average of $\overline{u'v'}$ involves a good deal of cancellation (evaluating the root mean square value of $\overline{u'v'}$ in each hemisphere would remove the problem). But there is a more fundamental issue, namely that low-frequency transients, with periods in excess of 10 days, made a large contribution to the momentum transports, but rather less to the temperature transports. Indeed Blackmon *et al.* (1977) comment on the difficulty of obtaining a statistically stable climatology for the low-frequency momentum transports even using their ensemble of 11 Northern Hemisphere winters. It is likely that momentum fluxes obtained from short period averaging will be much harder to interpret than heat fluxes.

Although Fig. 1.1 demonstrates that an averaging period of 15–30 days does yield sensible statistics (at least of the temperature flux) in a hemispheric average, it cannot confirm that this is so locally. The Hovmöller plot in Fig. 1.2 represents an attempt to consider the nature of the transients locally. It shows the transient meridional wind, v', averaged between 45°N and 55°N, for the period 1–31 January 1980. Zeros in the plot indicate either troughs or ridges in the transient wind field. The mean meridional flow, $\bar{v}$, for the same latitude is also shown.

The diagram shows that the most intense eddy activity was especially concentrated into two regions, more or less coinciding with the storm tracks described by Blackmon (1976). The first stretched across the Pacific, from about 150°E to the North American coast, and the other extended from about 90°W eastwards across the Atlantic. The western end of each track was characterized by short scale, mobile disturbances, whereas the eastern end displayed rather lower frequency, nearly stationary transients. It will emerge that the bulk of the heat transport was accomplished by the high-frequency transients on the earlier part of the storm track; Fig. 2 shows that during the 31-day period, typically five or six separate systems traversed these locations. Towards the eastern end of the storm tracks, or away from the storm track regions, the number of systems was a good deal less. Transient eddy transports calculated in these regions cannot be regarded as being so statistically reliable.

In the following sections, the mean state of the atmosphere in the selected months will be described first (Section 1.2). Plots of the heat fluxes will follow in Section 1.3 and those of the momentum fluxes in Section 1.4.

1.2 The time-averaged flow

Figure 1.3 shows the total mean wind speed $(\bar{u}^2 + \bar{v}^2)^{1/2}$ at 250 mb (which is generally only just below the jet cores) during the months of January and July

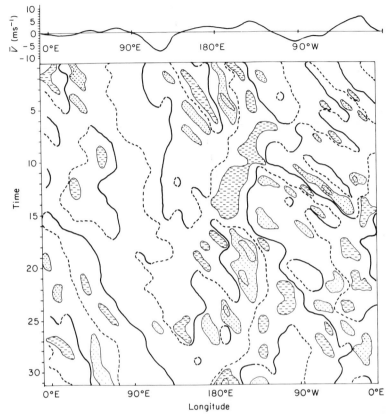

Fig. 1.2. Hovmöller plot of v' at 700 mb for the period 1–31 January 1980, averaged from 45°N to 55°N. Contour interval is 5 m s^{-1}; dots indicate northward winds, dashes southward. Troughs are delineated by a heavy solid line, and ridges by a dashed line. The graph at the top shows $\bar{v}$ for the period.

1980. The Northern Hemisphere January flow was remarkably similar to the climatological average reported by Blackmon *et al.* (1977). The most notable features were the very strong jet (up to 65 m s^{-1}) over the east coast of Asia and the weaker but broader jet over North America. A third jet extended over North Africa and across the Middle East; in this case it merged into the Asian jet, but it is more usual to observe a separate maximum over North Africa. The circulation had changed dramatically by July 1980. Wind speeds rarely exceeded 25 m s^{-1}, and the jet was more broken; the most notable jet regions are over North America and the West Atlantic and over Central Asia, a pattern which is reasonably similar to the climatological flow.

The seasonal variation of the Southern Hemisphere jets was less marked; the maximum wind speeds increased from somewhat over 35 m s^{-1} in January

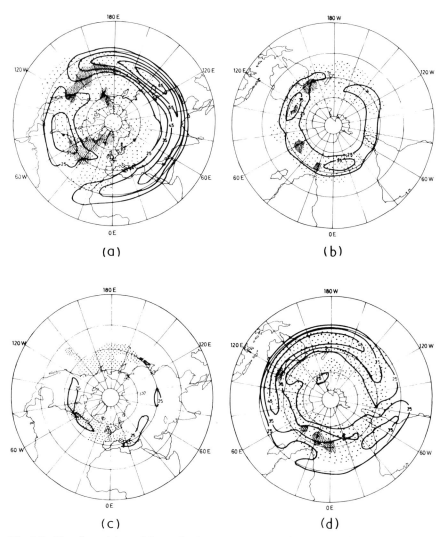

Fig. 1.3. The disposition of the major jets and storm tracks at 250 mb during January and July 1980. The contours show mean wind speeds, $(\bar{u}^2 + \bar{v}^2)^{1/2}$, greater than 25 m s^{-1}. The light dotting shows transient eddy kinetic energy, $\frac{1}{2}(u'^2 + v'^2)$, greater than 150 m^2 s^{-2} and the heavy dotting values greater than 300 m^2 s^{-2}. Lines of latitude and longitude are drawn every 20°. (a) January 1980, Northern Hemisphere; (b) January 1980, Southern Hemisphere; (c) July 1980, Northern Hemisphere; (d) July 1980, Southern Hemisphere.

to less than 50 m s^{-1} in July. The extent of the jets was considerably greater and it is evident that the total kinetic energy of the flow was a lot larger in winter. The jets were fairly continuous, with only one major break. The break was located over South America in July and near New Zealand in January.

The general level of eddy activity in the two periods is also indicated in Fig. 1.3, by shading regions of large transient eddy kinetic energy, $(\overline{u'^2} + \overline{v'^2})/2$. In the Northern Hemisphere, regions of large eddy kinetic energy were associated with the major Pacific and Atlantic storm tracks which, despite the considerable reorganization of the jet structure between the two periods, were located in much the same places. In the January data large values of eddy kinetic energy were also seen at high latitudes, over Alaska and over Greenland. These maxima were in fact associated with low frequency transients in the data, and were of quite a different character from the storm track maxima. Eddy activity in the Southern Hemisphere was more uniformly distributed, with generally higher values predominating in the eastern part of the hemisphere, from the Greenwich meridian to Australia. This asymmetry was even more marked when filtering was employed to select only the higher-frequency transients, and confirmed results based on the FGGE data (Physick, 1981).

Some indication of the three-dimensional structure of the flow is given in Fig. 1.4, which consists of cross-sections of zonal velocity and potential temperature, $[\bar{u}]$ and $[\bar{\theta}]$. The seasonal cycle is shown in both hemispheres as a poleward shift of around $10°$ in the location of the jet centre from winter to summer. The strength of the jets also varied by a factor of 2 in the Northern Hemisphere, but much less in the Southern. The monsoon was reflected in the much stronger tropical easterlies in July.

The zonally averaged baroclinity is clearly shown by the slope of the isentropes in these sections. Since the most unstable baroclinic disturbances have their maximum amplitude near the surface (see, for example, Simmons and Hoskins, 1978), the slope of the isentropes as they intersect the surface is of greatest relevance in explaining the intensity of baroclinic instability. Clearly, the surface baroclinity had its maximum value in the Northern Hemisphere in January, and it fell to very low values in July. In contrast, the change of surface baroclinity in the Southern Hemisphere was much less apparent. In the storm track latitudes in both hemispheres, the baroclinity generally declined away from the surface. On the tropical flanks of the main jets, however, maximum baroclinity was sometimes reached in the mid-troposphere; in such circumstances the flow may become unstable to internal baroclinic disturbances, with small amplitudes near the surface. The nature of such disturbances is relatively unknown.

In relating the level of eddy activity to some baroclinic instability parameter, it is probably not useful to confine attention to the zonally averaged fields. As Fig. 1.3 clearly shows, the eddy activity in the Northern Hemisphere is

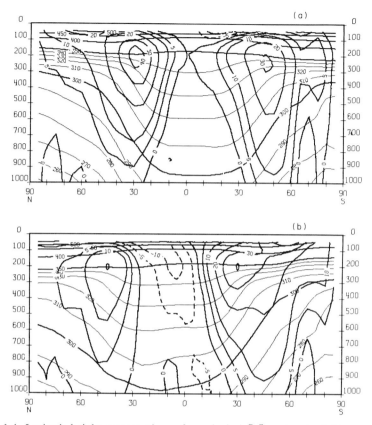

Fig. 1.4. Latitude height cross sections of zonal wind, $[\bar{u}]$, and potential temperature, $[\theta]$. Dashed contours show easterlies. The contour values for u are $\pm 5\,\mathrm{m\,s^{-1}}$, $\pm 10\,\mathrm{m\,s^{-1}}$ and every $10\,\mathrm{m\,s^{-1}}$ thereafter; for θ the contour interval is 10K to 350K and 50K thereafter. (a) January 1980; (b) July 1980.

localized, and the baroclinity at the start of the storm tracks is of greater importance than the zonally averaged value. The definition of 'storm tracks' in the Southern Hemisphere is less clear, but eddy activity is nevertheless localized to some degree.

1.3 Temperature fluxes

Heat is transported poleward in the atmosphere both by the transient disturbances and also by waves comprising the steady pattern, i.e., the stationary waves (considered in Chapter 2). Accordingly, Figs. 1.5 and 1.6 show zonally averaged cross-sections both of the transient eddy temperature

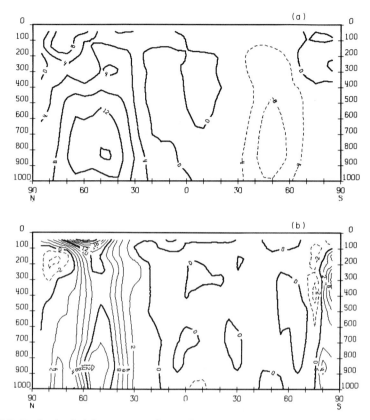

Fig. 1.5. Latitude–height cross-sections of the zonally averaged northward eddy-temperature fluxes for January 1980. Dashed contours indicate southward fluxes. (a) Transient eddy temperature flux, $[\overline{v'T'}]$—contour interval 4 K m s^{-1}; (b) steady eddy temperature flux, $[\bar{v}^*\bar{T}^*]$—contour interval 2 K m s^{-1}.

flux, $[\overline{v'T'}]$, and of the steady temperature flux, $[\bar{v}^*\bar{T}^*]$, for the periods January and July 1980, respectively. In the Northern Hemisphere winter, the steady temperature fluxes were nearly as strong as those due to transients. The steady flux was perhaps more restricted in latitude, but was more uniform in the vertical than the transient. The very large values of the steady temperature flux in the Northern Hemisphere stratosphere in January 1980 were notable. By contrast, the Southern Hemisphere temperature flux was dominated by the transient component which was rather more than half as large as that in the Northern Hemisphere. The only steady temperature fluxes were over Antarctica and cannot be regarded as reliable.

The Southern Hemisphere temperature fluxes were considerably larger in July. There was a steady component but it was only about half the size of the

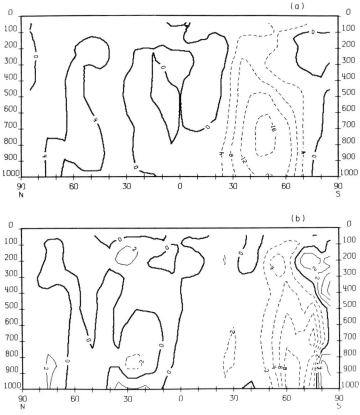

Fig. 1.6. As Fig. 1.5, but for July 1980.

transient component. Stratospheric heat fluxes are small and the steady component in particular was negligible. Of course, the Southern Hemisphere stratosphere is not well observed.

The July temperature fluxes in the Northern Hemisphere were generally small, and somewhat haphazard in appearance. The systematic poleward flux was achieved mainly by the transients and had a curious double maximum, with a second maximum at 250 mb.

Evidently, the temperature flux maximum was generally near 850 mb. Unfortunately, this pressure level intersects considerable areas of the Earth's surface, and so the 700 mb surface has been selected for plotting horizontal fields representative of the lower troposphere. This level is only below the surface over a significant area on the Tibetan plateau. Figures 1.7 and 1.8 therefore show the mean temperature field $\bar{T}$ and temperature fluxes $\overline{v'T'}$ at 700 mb for January and July 1980.

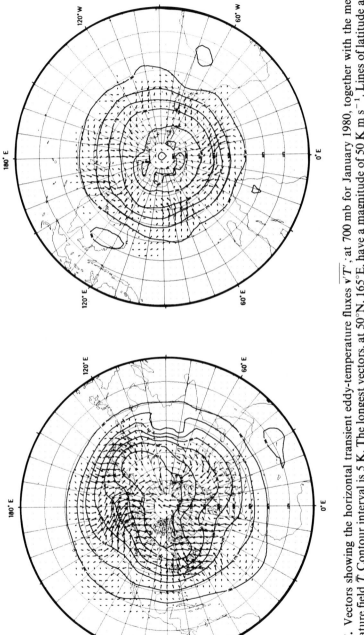

Fig. 1.7. Vectors showing the horizontal transient eddy-temperature fluxes $\overline{v'T'}$, at 700 mb for January 1980, together with the mean temperature field $\overline{T}$. Contour interval is 5 K. The longest vectors, at 50°N, 165°E, have a magnitude of 50 K m s⁻¹. Lines of latitude and longitude are drawn every 10°.

16

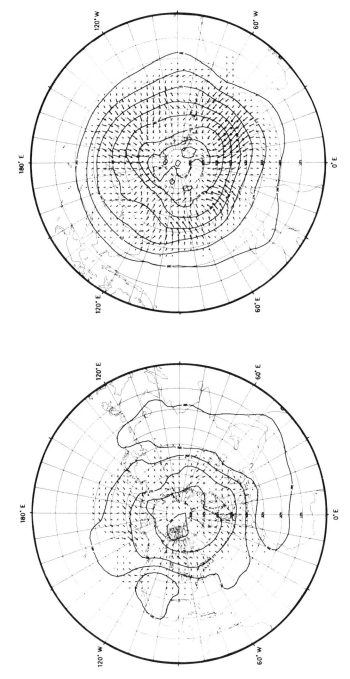

Fig. 1.8. As Fig. 1.7, but for July 1980. Vectors are scaled as in Fig. 1.7.

In both winter hemispheres, the transient temperature flux was a largely irrotational vector, pointing roughly down the temperature gradient. Filtering shows that the small rotational part was mainly associated with the lower frequency disturbances. Integrated over the hemisphere, as in Fig. 1.1, their contribution to $\overline{v'T'}$ was nearly zero. Clearly, the most intense temperature fluxes were localized, and indeed were associated with the major storm tracks. There was little obvious correlation with the strength of the local temperature gradient. Rather the correlation, at least in the Northern Hemisphere, is generally more with the surface baroclinity, emphasizing the baroclinic nature of the heat-transporting disturbances.

The divergent part of the temperature flux was rather reduced compared with the rotational part in the summer hemispheres. In the Northern Hemisphere in July 1980, in particular, the heat fluxes were very small, and were dominated by low-frequency, rotational fluxes. As one would expect, there was less change in the Southern Hemisphere, but the regions of strong poleward temperature flux were more limited in extent.

If the storm track eddies are baroclinic in origin, then one would expect upward as well as horizontal temperature fluxes to be prominent features of the storm tracks. As Figs. 1.9 and 1.10 reveal, this was indeed the case. Large upward values of the temperature flux, $\overline{\omega'T'}$, corresponded to large values of $\overline{v'T'}$ in both periods and in both hemispheres. Downward temperature flux (i.e., positive $\overline{\omega'T'}$) was very unusual and was confined to high latitudes or to the neighbourhood of steep topography. In the few cases where positive $\overline{\omega'T'}$ occurred, $\overline{v'T'}$ was reversed, and pointed up the temperature gradient.

1.4 Momentum fluxes

As was pointed out in Section 1.1, the determination of momentum fluxes from only 30 days of data is much more unreliable than that of temperature fluxes. Accordingly, the results contained in this section are much noisier and less structured than those in Section 1.3.

However, the roles of the eddies in forcing the zonally averaged flow are clearly seen in the cross-sections of $[\overline{u'v'}]$ and $[\bar{u}^*\bar{v}^*]$ in Figs. 1.11 and 1.12. The momentum fluxes reached their maximum values at 300–200 mb, slightly below the jet maxima, and the momentum converged into latitudes well poleward of the jets. In both periods and both hemispheres, the steady component of the momentum flux was rather unimportant compared with the transient, but, as might be expected, reached its largest values in the January Northern Hemisphere. In the Northern Hemisphere, the January and July $[\bar{u}^*\bar{v}^*]$ and $[\overline{u'v'}]$ fields were typical of the climatological patterns reported by Lau et al. (1981). Considerable variation is possible, however; as pointed out in

18

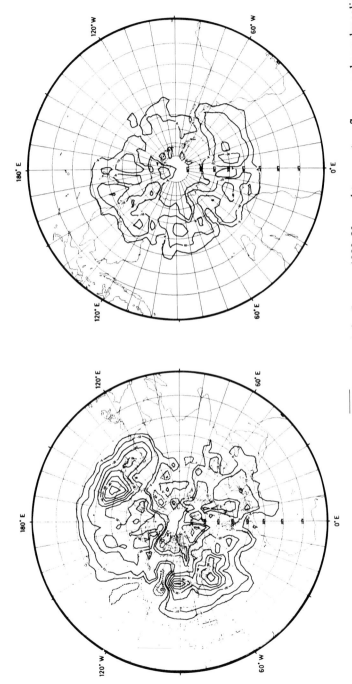

Fig. 1.9. Vertical transient eddy-temperature fluxes, $\overline{\omega'T'}$, at 700 mb for January 1980. Upward temperature fluxes are shown by solid contours (interval 10 K Pa s^{-1}) and downward temperature fluxes by dashed contours (interval 5 K Pa s^{-1}). For clarity, the zero contour is omitted; instead the -5 K Pa s^{-1} contour has been drawn.

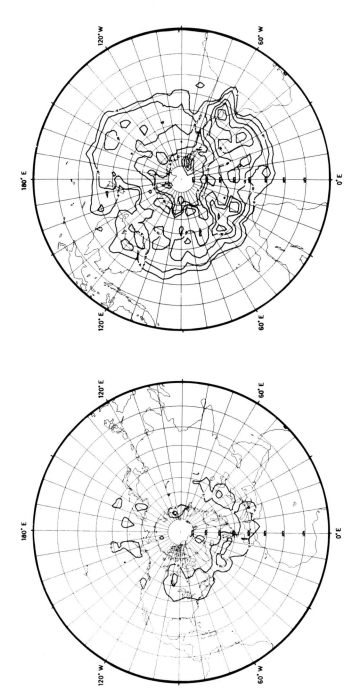

Fig. 1.10. As Fig. 1.9, but for July 1980.

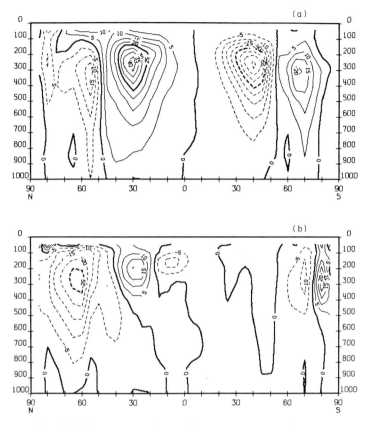

Fig. 1.11. Latitude–height cross-sections of the zonally averaged northward momentum fluxes for January 1980. Contour interval is 5 m² s⁻²; negative (southward) fluxes are indicated by dashed contours (a) Transient eddy momentum flux; (b) Steady eddy momentum flux.

Section 1.2, January 1981 exhibited a much more meridional mean flow field than January 1980, and accordingly the values of $[\bar{u}^*\bar{v}^*]$ were a good deal larger.

The horizontal field of momentum flux is plotted at 250 mb, close to the level of the jet maximum and momentum transport maximum, in Fig. 1.13 (January 1980) and Fig. 1.14 (July 1980). As well as vectors of $\overline{\mathbf{v}'u'}$, the $\bar{u}$ fields are also plotted. The most striking feature of the plots is the dominance of the zonal component, $\overline{u'^2}$, over the northward component, $\overline{u'v'}$, of the momentum flux. It is inevitable that the self-correlation $\overline{u'^2}$ will exceed $\overline{u'v'}$. What is less apparent *a priori* is that the momentum flux divergence should obtain a larger contribution from zonal variations in $\overline{u'^2}$ than from meridional variations of $\overline{u'v'}$. But evidently this was the case. When considering the zonally averaged

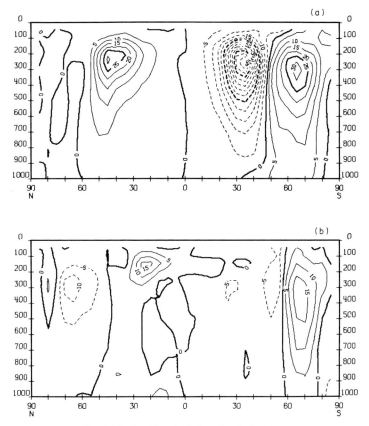

Fig. 1.12. As Fig. 1.11, but for July 1980.

state, the $\overline{u'^2}$ contribution to the acceleration will average to zero. Locally, however, in the region of jet entrances and exits, such zonal transports of momentum can be important (this will be discussed in Section 7.4). Many examples can be discerned in the Figures.

If, following Lau (1979), the idealized 'life-cycle' calculations of baroclinic wave evolution of Simmons and Hoskins (1978) are regarded as a model of events along a storm track, one would expect larger values of $\overline{u'v'}$ to be concentrated along the storm tracks, and especially towards their eastern ends. The two Northern Hemisphere storm tracks did exhibit such an effect in January 1980, but it is more clearly discerned in the Southern Hemisphere data, especially that for July 1980. The strong poleward momentum fluxes which developed in the Eastern Indian Ocean, towards the end of the major storm track, are very obvious.

22

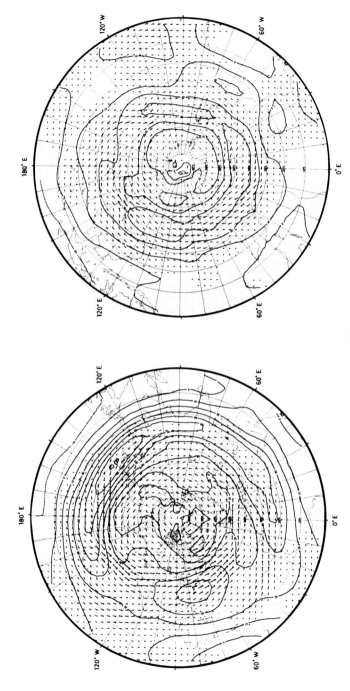

Fog. 1.13. Vectors showing the horizontal eddy momentum fluxes, $\overline{v'u'}$, at 250 mb, together with the mean zonal velocity $\bar{u}$ for January 1980. Contour interval is 10 m s^{-1}. The longest vectors, at 35°N, 140°W, have magnitudes of 450 m^2 s^{-2}.

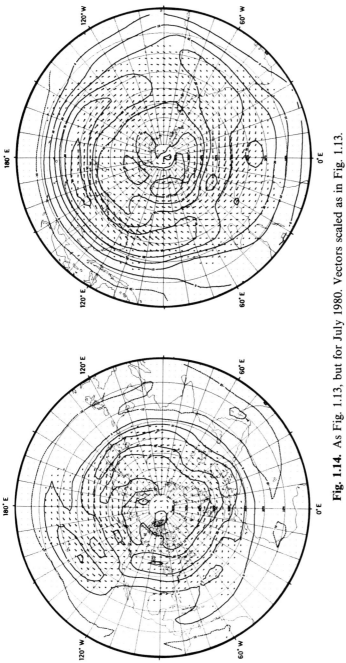

Fig. 1.14. As Fig. 1.13, but for July 1980. Vectors scaled as in Fig. 1.13.

In the case of the temperature flux, the association of large fluxes with the storm tracks clearly dominated the fields. The momentum fluxes did show some association with the storm tracks, but the relationship was confused by the large momentum fluxes in other regions. Once more, time filtering proves instructive, and shows that the higher frequency disturbances led to convergence of momentum into the storm tracks. The lower frequency transients are responsible particularly for large zonal transports of momentum, and indeed their contribution generally dominated the fields shown in Figs. 1.13 and 1.14.

1.5 Summary of results

The ECMWF archive, which contains high-resolution, 6-hourly analyses for many levels in both hemispheres, will undoubtedly become an important data set in the study of the global circulation as it increases in length. At present just two years of data are available and only case studies, such as those described here, can be undertaken. The agreement in so many respects with the climatology of the NMC analyses is an encouraging endorsement of both data sets. Little verification of the Southern Hemisphere results is possible, but many interesting features can be seen. In time, the ECMWF data promises to make a detailed climatology for the entire globe possible.

The most striking result of the comparison of January and July 1980 is perhaps the markedly less pronounced seasonal cycle in the Southern Hemisphere. Presumably this can be attributed to the greater preponderance of ocean in that hemisphere. Although a 'storm track'—defined on the basis of localized high-frequency transient eddy activity—can be identified in the Southern Hemisphere, it differs in important respects from the Pacific and Atlantic storm tracks. It appears less well defined and constant in its position than its Northern Hemisphere counterparts, and generally builds up in the mid-South Atlantic, rather than near a coast-line. Perhaps the reduced orographic forcing is important; it may be that the wandering of the South Atlantic storm track can be related to variations of sea surface temperatures at these latitudes.

Clearly the Northern Hemisphere storm tracks can be interpreted in terms of a 'life-cycle' model of developing baroclinic waves, as Lau (1979) has suggested (see Section 7.43 for a further discussion). Such a model accounts for the observed distribution of eddy kinetic energy, temperature flux, and, to a degree, of momentum flux. But it is not the complete picture. The low-frequency fluctuations in the strength and position of the major jets represent very large momentum and sometimes temperature changes that cannot be simply accounted for as the result of developing or decaying baroclinic eddies moving along the storm tracks. Reading some measure of order into these low frequency transients is a major, and difficult, objective of many current studies and is considered in Chapters 3–6.

Acknowledgements

This work was carried out as part of a collaborative project with the UK Meteorological Office; I am grateful to them for affording computing facilities at ECMWF. I particularly wish to thank Dr Sarah Raper of the Meteorological Office, Dr Tony Hollingsworth and his colleagues at ECMWF, and Dr Glenn White of the University of Reading for their advice and help. Finally, I thank Professors Robert Pearce and Brian Hoskins for their enthusiastic obstetric efforts on behalf of the project.

References

BLACKMON, M. L. (1976). A climatological spectral study of the 500 mb geopotential height of the Northern Hemisphere. *J. atmos. Sci.*, **33**, 1607–1623.

BLACKMON, M. L., WALLACE, J. M., LAU, N.-C. and MULLEN, S. L. (1977). An observational study of the Northern Hemisphere wintertime circulation. *J. atmos. Sci.*, **34**, 1040–1053.

LAU, N.-C. (1978). On the three-dimensional structure of the observed transient eddy statistics of the Northern Hemisphere wintertime circulation. *J. atmos. Sci.*, **35**, 1900–1923.

LAU, N.-C. (1979). The structure and energetics of transient disturbances in the Northern Hemisphere wintertime circulation. *J. atmos. Sci.*, **36**, 982–995.

LAU, N.-C. and WALLACE, J. M. (1979). On the distribution of horizontal transports by transient eddies in the Northern Hemisphere wintertime circulation. *J. atmos. Sci.*, **36**, 1844–1861.

LAU, N.-C., WHITE, G. H. and JENNE, R. L. (1981). Circulation statistics for the extra-tropical northern hemisphere based on N.M.C. analyses. N.C.A.R. Technical Note NCAR/TN–171 +STR.

OORT, A. H. and RASMUSSON, E. M. (1971). Atmospheric Circulation Statistics. N.O.A.A. Prof. Paper 5, U.S. Dept. of Commerce.

PHYSICK, W. L. (1981). Winter depression tracks and climatological jet streams in the Southern Hemisphere during the FGGE year. *Q. Jl R. met Soc.*, **107**, 383–398.

SIMMONS, A. J. and HOSKINS, B. J. (1978). The life cycles of some nonlinear baroclinic waves. *J. atmos. Sci.*, **35**, 414–432.

– 2 –

The climatological mean stationary waves: observational evidence

J. M. WALLACE

2.1 Introduction

One of the most important and most elusive goals of general circulation research in the period since World War II has been a thorough understanding of the structure and maintenance of the climatological mean circulation, including the deviations from zonal symmetry, which are commonly referred to as the stationary waves. (In many of the papers written by V. P. Starr and collaborators, the term 'standing eddies' is used.) An ability faithfully to reproduce the essential features of these circulations in numerical models and to interpret the results clearly and concisely, in terms of basic dynamical principles, would constitute an impressive demonstration of the validity of the dynamical principles and physical parameterizations used in numerical weather prediction and climate modelling, and it would provide a solid foundation upon which to build a theoretical framework for understanding and perhaps, to some extent, forecasting interannual climate variability.

The latest generation of general circulation models (GCM's) has been successful in simulating many aspects of the observed structure of the stationary waves (see, for example, Blackmon and Lau, 1980). However, the fundamental dynamical interpretation of the GCM results has proven more difficult than anticipated, partially because of the numerous feedback mechanisms that exist within the model code (and presumably within the atmosphere itself), and partially for want of a theoretical framework to serve as a basis for organizing them.

Much of the diagnosis and interpretation of observations and GCM simulations of the stationary waves has been carried out in the context of time-averaged budgets of quantities such as heat, vorticity, and kinetic energy. This 'balance requirement' approach, pioneered by Starr and Rossby in the late

1940s, has contributed much to our understanding of the zonally averaged general circulation. However, it is becoming increasing evident that, in the absence of theoretical guidance, the knowledge of how various budgets are satisfied is of limited value in elucidating cause and effect relationships.

Until quite recently most of the theoretical work on stationary waves forced by orography and diabatic heating was based on highly simplified models with beta-plane geometry, so that opportunities for comparison between theory and observations were rather limited. The more recent work described in Chapter 6, which incorporates spherical geometry and a realistic treatment of vertical structure, provides a much more comprehensive theoretical framework for interpreting observations and GCM simulations.

This chapter is devoted to a review of some of the more essential characteristics of the three-dimensional structure of the stationary waves in the extratropics of both hemispheres during both winter and summer seasons. Most of the theoretical interpretation of these results will be reserved for Chapter 6.

2.2 Meridional structure of the Northern Hemisphere stationary waves

Figure 2.1 shows time–latitude sections of zonal and meridional kinetic energy $[\bar{u}^{*2}]$ and $[\bar{v}^{*2}]$ associated with the climatological mean stationary waves at the 200 mb level. Poleward fluxes of zonal momentum by the stationary waves are denoted by the arrows in Fig. 2.1(a). These are based on an analysis of many years of data.

The dominance of the zonal component $\bar{u}^*$ in the stationary wave kinetic energy is a reflection of the fact that the meridional scale of the waves is shorter than the zonal scale. The meridional structure of the waves is strongly evident in the sections themselves.

In winter, the strongest concentration of zonal stationary wave kinetic energy $[\bar{u}^{*2}]$ is centred near 35°N, at the latitude where strong jetstreams near the east coast of Asia and North America alternate with much weaker westerlies over Europe and western North America (see also Fig. 1.3a). The large variance of zonal wind at this latitude is accompanied by a large poleward flux of zonal momentum $[\bar{u}^*\bar{v}^*]$.

The zonal wind perturbations associated with the stationary waves extend deep into the tropics, with a secondary maximum in $[\bar{u}^{*2}]$ near 10°N. This low-latitude variance maximum is characterized by a southward flux of westerly momentum, which is in agreement with the orientation of the major ridges and troughs in Sadler's (1975) climatological mean streamline field. The meridional wind component $\bar{v}^*$ (Fig. 2.1b) exhibits amplitude maxima near 60°N and 25°N, separated by a distinct minimum that lies just to the north of

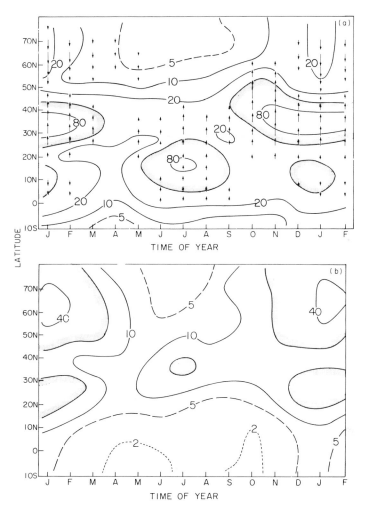

Fig. 2.1. Time–latitude sections of Northern Hemisphere climatological mean stationary wave kinetic energy at the 200 mb level based on data presented in the atlas of Oort and Rasmusson (1971). (a) Zonal wind component $[\bar{u}^{*2}]$; and (b) meridional wind component $[\bar{v}^{*2}]$, in units of m^2 s^{-2}; logarithmic contour interval. Arrows in (a' denote the direction and relative magnitude of the meridional flux of zonal momentum by the stationary waves, based on the same data source. All the major features in this figure appear in independent analyses by Newell *et al.* (1974) and the extratropical features have been confirmed in analyses of NMC data by Lau (1979) and White (1982a).

the climatological mean jetstream. The latitudinal profile of the amplitude of geopotential height $\bar{Z}^*$ exhibits a qualitatively similar structure.

The summertime regime is characterized by a strong maximum of $[\bar{u}^{*2}]$ centred near 15°N at the latitudes of the mid-ocean upper tropospheric troughs (e.g., see Fig. 2.22a). The corresponding maxima in $[\bar{v}^{*2}]$ (Fig. 2.1b), $[\bar{Z}^{*2}]$ and $[\bar{T}^{*2}]$ (not shown) are located further poleward, near 35°N. Poleward of 40°N the summertime stationary wave kinetic energy is much less than the corresponding wintertime values. However, even at these higher latitudes the stationary waves possess a distinct structure that exerts an important influence upon summertime climate. The wintertime and summertime regimes are quite distinct and separated by transition seasons in which the stationary waves are relatively weaker.

In the following two sections, we will consider in further detail the three-dimensional structure of the Northern Hemisphere stationary waves in the wintertime and summertime regimes.

2.3 Northern Hemisphere wintertime stationary waves

As a basis for a more detailed description of the Northern Hemisphere wintertime stationary waves, the following information is presented in the figures in this section:

1. the Northern Hemisphere wintertime stationary wave geopotential height field $\bar{Z}^*$ at the 200 mb level (Fig. 2.2),

2. longitude–height sections of stationary wave geopotential height and temperature perturbations along 60°N, 40°N, and 25°N (Figs. 2.3 and 2.4),

3. the climatological mean 500 mb height and sea-level pressure maps for the winter months (Figs. 2.5 and 2.6).

Sources for the figures are given in the captions, together with references to other works that show comparable results based on other data sets. All the major features discussed in this section are highly reproducible in independent data sets based on different sets of winters and different analysis procedures. They appear in individual winters with only relatively minor year to year variations in structure.

The stationary wave 200 mb height field, shown in Fig. 2.2, is characterized by distinct high and low latitude regimes, as was already remarked upon in the previous section. The transition between these regimes is not a true node in the stationary wave pattern; amplitudes of $\bar{v}^*$ and $\bar{Z}^*$ are about two-thirds as large at 35°N as at 25°N and 60°N.

The transition latitude is characterized by a significant poleward flux of zonal momentum (Fig. 2.1a) which, according to simple wave theory discussed in Chapters 7 and 10, is indicative of an equatorward Eliassen–Palm (EP) flux from the high latitude regime to the low latitude one. Hence it would appear that the waves in the low latitude regime may be at least partially forced from

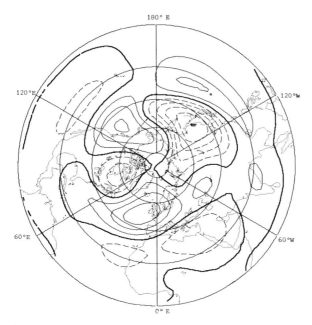

Fig. 2.2. Northern Hemisphere climatological mean January distribution of stationary wave geopotential height at the 200 mb level based on the atlas of Crutcher and Meserve (1970), digitized, spatially and temporally smoothed and archived on magnetic tape in the NCAR data library. Contour interval 60 m; the zero contour is thickened; positive contours are solid and negative ones are dashed. For a comparable analysis based on a different data set, see Holopainen (1970). Lines of latitude and longitude are drawn every 20° and 60°, respectively.

higher latitudes. Further implications of the observed latitudinal structure of the stationary waves are discussed in Section 6.3.

The vertical structure of the waves (Figs. 2.3 and 2.4) is rather complicated. There is some evidence of a westward tilt of the geopotential field with height, particularly at higher latitudes where the waves extend upward into the lower stratosphere. This westward tilt is reflected in the stationary wave heat fluxes, which are poleward at all levels, with maxima near 850 and 200 mb (Oort and Rasmusson, 1971; Lau, 1979). However, it would be an over-simplification to describe the vertical structure purely in terms of vertically propagating wave modes. The stationary wave amplitudes in the geopotential height field increase markedly with height up to the tropopause level, which is indicative of an equivalent barotropic component in the vertical structure (i.e., no phase tilt with height—see Appendix Section A3). This equivalent barotropic component is reflected in the small longitudinal phase difference between $\bar{Z}^*$ and $\bar{T}^*$ which is evident from a careful comparison of Figs. 2.3 and 2.4, and in the relatively low correlation coefficients between $\bar{v}^*$ and $\bar{T}^*$ (generally below 0.5

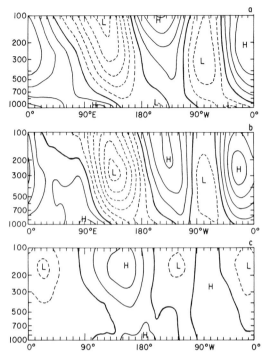

Fig. 2.3. Longitude–height cross-sections of stationary-wave geopotential height $\bar{Z}^*$ for the winter season, derived from 11 years of NMC operational analyses, adapted from Lau (1979). (a) 60°N; (b) 45°N; (c) 25°N. Contour interval 50 m; the zero contour is thickened.

between 700 mb and 300 mb: Oort and Rasmusson, 1971, Table F10). Above the tropopause there is evidence of vertical trapping, particularly at the lower latitudes, with amplitudes of the geopotential height fluctuations decreasing with height and $\bar{Z}^*$ and $\bar{T}^*$ having a marked out-of-phase component. Only in the 60°N section is the vertically propagating component comparable with the equivalent barotropic or 'trapped' component at these levels. In the low latitude regime, exemplified by the sections for 25°N, there is little if any evidence of vertical phase propagation, and the large wave amplitudes appear to be trapped in the upper troposphere. Several of these features are explained theoretically in Chapter 6.

The vertical structure at levels below 700 mb deserves special emphasis. In this layer the poleward heat fluxes by the stationary waves are particularly large and the correlation coefficient between $\bar{v}^*$ and $\bar{T}^*$ reaches values on the order of $+0.8$. Note also that at the Earth's surface, most of the maxima and minima in the temperature field lie roughly a quarter wavelength to the west of the corresponding features in the geopotential field, and the latter show a

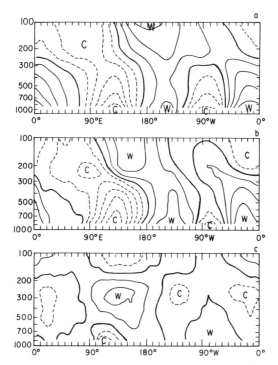

Fig. 2.4. As in Fig. 2.3 but for the temperature. Contour interval 2 C. For comparison with a similar analysis based on data from the U.K. Meteorological Office, see Saltzman and Rao (1963).

pronounced westward tilt with height. It will subsequently be shown that this peculiar low-level structure has no counterpart in the summertime stationary waves in the Northern Hemisphere, or in the Southern Hemisphere stationary waves during either season.

Figures 2.5 and 2.6 show the 500 mb height field and sea-level pressure field, respectively. The 500 mb height pattern shows northwesterly flow over the Himalayas and Rockies, with prominent downstream troughs that lie on the poleward flanks of the climatological mean jetstreams. The related vertical velocity pattern (e.g., see Lau, 1979, Figs. 7–9; or White, 1982b, Figs. 7–9) shows an even more obvious relation to the two major mountain ranges. The apparent relationship between the middle and upper tropospheric geopotential height pattern and the major mountain ranges is borne out in general circulation modelling experiments such as that of Manabe and Terpstra (1974) and the one described in Section 6.6 of this book, which indicate that mountains play the primary role in determining the longitudes of the major ridges and troughs in the wintertime Northern Hemisphere 500 mb height pattern.

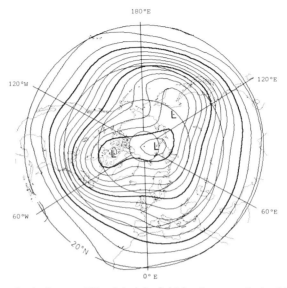

Fig. 2.5. Climatological mean 500 mb height field for January, derived from the same data source as Fig. 2.2. Contour interval 60 m; the 5100, 5400, and 5700 m contours are thickened. For comparison with the corresponding field derived from NMC operational analyses, see Blackmon (1976) or Lau *et al.* (1981). The outer latitude circle is 20°N.

The sea-level pressure field, on the other hand, bears the mark of thermal influences. Low pressure covers the high latitude oceans, and the isobars seem to be crowded along the coastlines. (Less highly smoothed analyses show evidence of troughs or centres of low pressure over bodies of warm water such as the Barents, Norwegian, Mediterranean and Black Seas, and even over ice-covered waters such as Baffin and Hudson Bays and the Great Lakes. These small-scale features are largely confined to the layer below 850 mb, where they have a 'warm core' equivalent barotropic structure. Their existence complicates the interpretation of longitude–height sections such as Fig. 2.3 at low levels.) A similar situation exists in the Antarctic during winter. The same GCM experiments cited above indicate that a realistic climatological mean sea-level pressure distribution can be produced by thermal influences at the lower boundary, even in the absence of mountain ranges (Fig. 6.24).

Hence it seems likely that orographic forcing is the dominant factor in determining the positions of the major ridges and troughs in the Northern Hemisphere wintertime stationary waves at the jetstream level, whereas thermal forcing makes an important contribution to maintaining the surface lows over the high latitude oceans. To the extent that thermal forcing is dominant at low levels, it would appear that the strong westward tilt of the stationary waves with height in the lower troposphere is at least partially

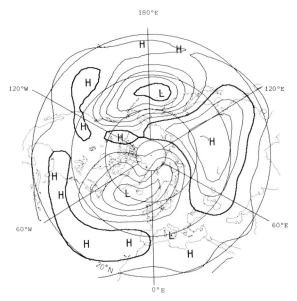

Fig. 2.6. Climatological mean sea-level pressure field for January, derived from the same data source as Fig. 2.2. Contour interval 4 mb; the 1000 and 1020 mb contours are thickened. For comparison with the corresponding field derived from NMC data, see Blackmon *et al.* (1977) or Lau *et al.* (1981). The outer latitudinal circle is 20°N.

fortuitous. For example, suppose that the Rockies were located along the east coast of North America instead of the west coast, so that the orographically induced upper level trough lay over the warm North Atlantic ocean, instead of over eastern Canada. In this situation, the stationary waves in the geopotential height field might exhibit little, if any, westward tilt with height. For certain hypothetical (not necessarily geophysically realistic) combinations of terrain and sea-surface temperature distributions it is even conceivable that the stationary waves might tilt eastward with height in the lower troposphere as they do during summer.

If the vertical tilt of the stationary waves in the lower troposphere is dictated by the arrangements of mountain ranges, coastlines, and warm and cold ocean currents on the underlying surface, rather than by fundamental physical principles governing the maintenance of forced stationary waves in a shear flow, then similar considerations apply to the energetics. Holopainen (1970) showed that the kinetic energy of the existing waves is maintained primarily by a baroclinic conversion from zonal available potential energy to stationary wave available potential energy, and thence to stationary wave kinetic energy. The former conversion is intimately related to the westward tilt of the waves with increasing height; hence, if there were no tilt in the lower troposphere

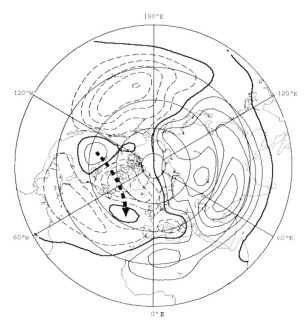

Fig. 2.7. Climatological mean July distribution of stationary wave geopotential height at the 200 mb level. Data source, and plotting conventions as in Fig. 2.2, except that the contour interval is 30 m. For comparable distributions based on different data sets, see Holopainen (1970) and White (1982a) or Lau *et al.* (1981). The arrow shows the wavetrain discussed in the text. The outer latitude circle is the equator.

there would be much less conversion and the energetics could conceivably be quite different.

2.4 Northern Hemisphere summertime stationary waves

The discussion in this section is based on the following observations results:

1. the stationary wave 200 mb geopotential height field (Fig. 2.7),

2. vertical cross-section of stationary wave geopotential height and temperature fields along 62.5°N, 45°N, and 30°N (Figs. 2.8 and 2.9, respectively),

3. sea-level pressure and 500 mb height maps for July (Figs. 2.10 and 2.11, respectively).

The summertime stationary waves attain their maximum amplitude in the geopotential height and temperature fields near 30°N, where dominant features in the 200 mb height field are the Tibetan anticyclone and the two

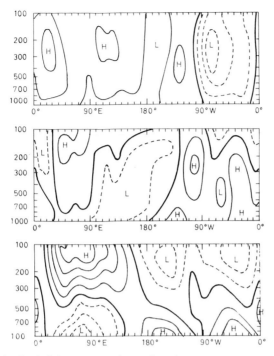

Fig. 2.8. Longitude–height cross-sections of stationary wave geopotential height $\bar{Z}^*$ for the summer season, derived from 12 years of NMC operational analyses, adapted from White (1982a). (a) 60°N; (b) 45°N; (c) 25°N. Contour interval 30 m. The zero contour is thickened.

mid-ocean troughs. Secondary features include the extension of the Atlantic trough into the Mediterranean and the anticyclones over North Africa and the Rockies. All the above features are clearly evident in Fig. 2.7 and they all appear in the sea-level pressure field with reversed polarity (Fig. 2.10). Hence these features have a vertical structure similar to that of many weather systems in the tropics which are associated with deep cumulus convection.

The dominance of tropical-type features in the summertime stationary waves is consistent with Holopainen's (1970) results concerning the summertime energetics, which show a strong generation of available potential energy by diabatic heating gradients associated with thermal contrasts between the warm continents and the cooler oceans, and a subsequent conversion to kinetic energy by thermally direct circulations with ascent over the eastern continents and sinking over the oceans. The same vertical structure is evident in the vertical cross-sections for 30°N in Figs. 2.8 and 2.9, with strong thermal contrasts at the 300 mb level and a phase reversal between upper and lower tropospheric geopotential height fields. Hence, it is evident that

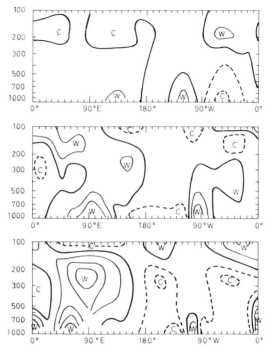

Fig. 2.9. As in Fig. 2.8 but for the temperature. Contour interval 2 C. For comparison with a similar analysis based on data from the U.K. Meteorological Office, see Saltzman and Rao (1963).

monsoon circulations account for the prominent summertime maxima in stationary wave kinetic energy in Fig. 2.1 and the energy conversions that maintain them.

Poleward of the summertime mean jetstream the vertical structure of the summertime standing waves is more complicated. A careful comparison of Figs. 2.7 and 2.10 suggests that vestiges of a tropical-type structure extend to quite high latitudes over the Pacific and Eurasian sectors, with major features in the $\bar{Z}^*$ field appearing with opposite phase at the 1000 and 200 mb levels. However, the heavy rainfall rates that characterize the rising branch of deep tropical circulations do not extend to these high latitudes and the node in the vertical profile of geopotential height occurs at much lower levels than in the tropics. Some of the high-latitude features, such as the prominent low over eastern Canada, exhibit an equivalent barotropic structure, with maximum amplitude at the tropopause, and geopotential height in phase at all levels.

White (1982a) has suggested that some of the features at higher latitudes might be a remote response to the monsoon circulations. The cold upper level low over eastern Canada and the ridge over the North Atlantic, which are the

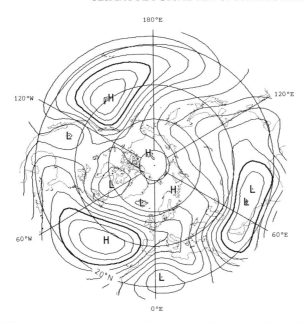

Fig. 2.10. Climatological mean sea-level pressure field for July, derived from the same data source as Fig. 2.2. Contour interval 2 mb; the 1000 and 1020 mb contours are thickened. For comparison with the corresponding field derived from NMC operational analyses, see White (1982a) or Lau *et al.* (1981)). The outer latitude circle is 20°N.

features in Fig. 2.7 that are perhaps most clearly distinct from the monsoons, have the kind of orientation suggestive of a wavetrain propagating along a 'great circle route' downstream from the ridge along the eastern slope of the Rockies, reminiscent of the forced steady-state solutions described by Hoskins *et al.* (1979) and Hoskins and Karoly (1981).

One of the more puzzling aspects of the observed summertime stationary wave structure is the prevailing tendency for an eastward slope with height in the lower troposphere, particularly at latitudes around 50°N. This eastward slope is evident in Fig. 2.8(b) and it is reflected in the observed equatorial fluxes of sensible heat by the stationary waves (see, for example, Fig. 1.6b). Although the fluxes themselves are small, Oort and Rasmusson (1971) report negative spatial correlations between $\bar{v}^*$ and $\bar{T}^*$ ranging as high as -0.60 at 50°N, 900 mb for the month of July.

The possible role of mountains in forcing the summertime stationary waves has not received much emphasis, perhaps because of the absence of a westward tilt with height or an upward EP flux which is usually associated with forcing from below. However, as we have noted, there is reason to question whether the vertical structure of the waves is a reliable indicator of the source of

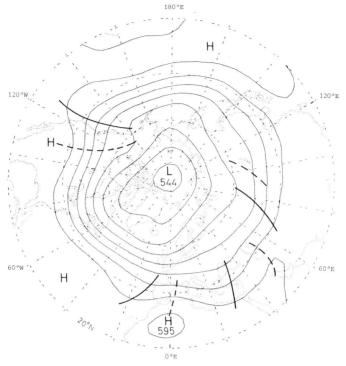

Fig. 2.11. Climatological mean 500 mb height for July, derived from NMC analyses. Contour interval 60 m. Short wavelength features discussed in the text are marked with heavy lines; solid for troughs and dashed for ridges. The same features are evident in the unsmoothed analyses of Crutcher and Meserve (1970), but they are strongly attenuated by the smoothing applied to the digitized data. The outer latitude circle is 20°N.

excitation in situations where both thermal and orographic forcing might be simultaneously present.

A more generally applicable indicator of the importance of orographic forcing is the sea-level pressure gradient across the major mountain ranges. In the presence of a westerly zonally averaged flow, a pressure decrease from west to east across a mountain range is indicative of a destruction of zonal kinetic energy and an upward boundary flux of stationary wave kinetic energy. It is evident from Fig. 2.10 that sea-level pressure decreases from west to east across every mountain range in the hemisphere, within the belt of upper level westerlies. The gradient is conspicuously large across the western ranges of the Rockies.

Perhaps the strongest evidence that orographic influences play a significant role in forcing the summertime stationary waves is the presence of ridges in the

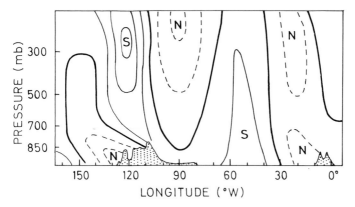

Fig. 2.12. Partial longitude–height cross-section for the meridional wind component $\bar{v}$ during summer along 45°N, adapted from van Loon *et al.* (1972a). Contour interval 2 m s^{-1}; the zero contour is thickened.

upper level flow over the Rockies, the Atlas Mountains of Northwest Africa, the Alps and the Caucasus, with prominent short-wave troughs to the west of each range. These features are evident in the 500 mb height field (Fig. 2.11) and in the geopotential height fields for all levels between 200 mb and 700 mb. The distance from trough to ridge is on the order of 1500 km; far shorter than the half-wavelength of the corresponding wintertime features.

On the basis of thermal wind considerations, a surface anticyclone coexisting with an upper level trough on the west side of a mountain range implies a highly baroclinic low level flow, with colder air lying to the west of the range. If a mountain range lies along the west side of a continent, atmospheric cold advection, wind driven ocean currents, upwelling, and stratus clouds can conspire to maintain a cold pool of low-level air just offshore, thus enhancing the thermal contrast across the range. These conditions are typified by the flow regime along the west coast of North America, from California to southern Alaska, during mid-summer. A zonal cross-section of meridional wind component along 40°N through this region (Fig. 2.12) shows shallow northerly flow associated with the cold, offshore anticyclone extending up to about the 800 mb level with southerly flow aloft, extending from the shortwave trough just off the coast to the ridge over the Rockies. The low-level baroclinicity is, in some sense, a small-scale, poleward extension of the monsoon, with cold air over the ocean and warm air over land. However, the related vertical structure is shallower than in the planetary-scale monsoons with the node in the geopotential height and wind fields lying near 800 mb rather than near 500 mb, and there is an absence of the heavy rainfall characteristic of the subtropical monsoon. Analogous, but weaker structures seem to exist along the west coasts of all the continents during summertime, at middle latitudes.

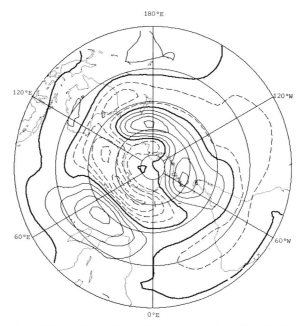

Fig. 2.13. Southern Hemisphere climatological mean July distribution of stationary wave geopotential height $\bar{Z}^*$ on the 200 mb level, based on the atlas of Taljaard *et al.* (1969) digitized, spatially and temporally smoothed, and archived on magnetic tape in the NCAR data library. Contour interval 30 m. The zero contour is thickened; positive contours are solid and negative ones are dashed. For comparison with the 500 mb $\bar{Z}^*$ field derived from Australian operational analyses, see Trenberth (1980). The outer latitude circle is the equator.

2.5 Southern Hemisphere stationary waves

The Southern Hemisphere general circulation exhibits a less pronounced seasonal variation than that of the Northern Hemisphere and the stationary waves do likewise (see Fig. 1.3). Nevertheless, there is still enough of a seasonal cycle to make it worth considering the winter and summer seasons separately. Also the study of the forcing of the seasonal cycle in the Southern Hemisphere, with its absence of large mountain plateaux, is of fundamental importance for both hemispheres.

The discussion of the wintertime stationary waves is based on the following figures:

1. the stationary wave 200 mb height and sea-level pressure fields for July (Figs. 2.13 and 2.14, respectively), and

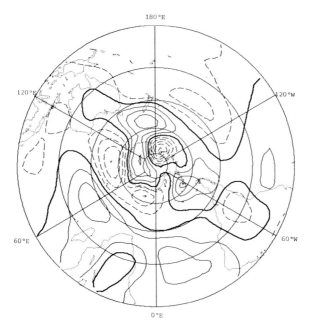

Fig. 2.14. As in Fig. 2.13, but for the sea-level pressure. Contour interval 2 mb; the zero contour is thickened.

2. vertical cross-sections of stationary wave geopotential height and temperature along 60°S, 45°S and 25°S (Figs. 2.15 and 2.16, respectively) for
   June–August 1981. The maps and cross-sections are from entirely different
   data sources and they cover different periods of record, but it has been
   verified that they are consistent with regard to the major climatological
   features discussed here.

The wintertime stationary waves in the 200 mb height field (Fig. 2.13) show the
same tendency for low- and high-latitude regimes as their Northern Hemisphere counterparts. The node that separates the two regimes is located near
45°S. The high-latitude regime is dominated by a broad ridge across the
Pacific sector and a trough in the Indian Ocean sector. Amplitudes of these
features approach 120 m, which is roughly half that of their Northern
Hemisphere counterparts. It is apparent from a comparison of Figs. 2.13 and
2.14 and from the cross-sections that this predominantly zonal wavenumber 1
pattern has an approximately equivalent barotropic structure in the troposphere with only a modest increase of amplitude with height. van Loon and
Jenne (1972) have pointed out that the geopotential height and temperature
fields in this wave are essentially in phase with the underlying sea-surface

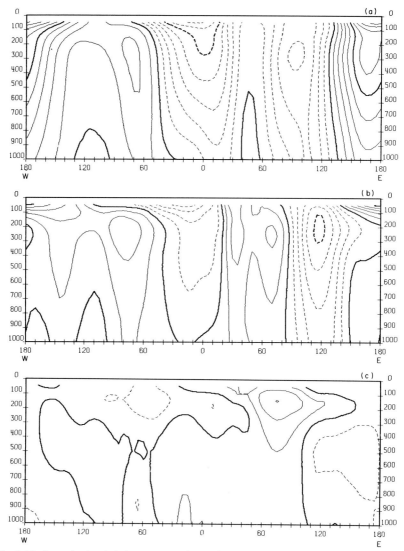

Fig. 2.15. Longitude–height cross-sections of stationary wave geopotential height $\bar{Z}^*$ for the winter season (June–August) 1981, based on ECMWF operational analyses. (a) 60°S; (b) 45°S; (c) 25°C. Contour interval 30 m. The zero contour is thickened.

temperature pattern, but which is cause and which is effect is not obvious from the observations alone. Modelling results by Grose and Hoskins (1979) indicate that the asymmetry of the Antarctic continent about the South Pole could produce a strong zonal wavenumber 1 response with the observed phase at latitudes near 55–60°S. The sea-level pressure pattern in Fig. 2.14 shows

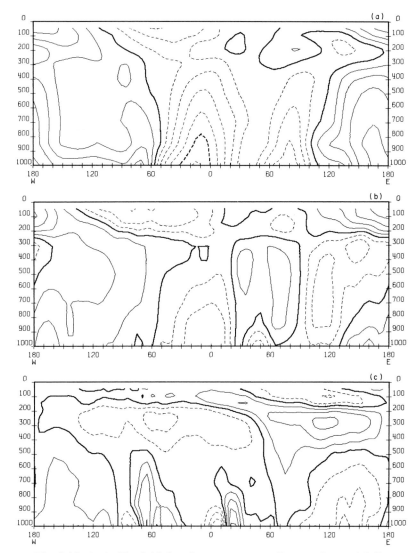

Fig. 2.16. As in Fig. 2.15, but for temperature. Contour interval 1 C.

evidence of shallow thermal lows in the Ross and Weddell Seas, analogous to features in the Northern Hemisphere polar seas during wintertime. (The pronounced ring of low sea-level pressure surrounding Antarctica does not show up in Fig. 2.14, which shows only departures from zonal averages.)

The strongest feature at lower latitudes is the ridge just to the east of South Africa, which also displays a nearly equivalent barotropic structure, though perhaps with some slight westward tilt with height. The remaining features lie

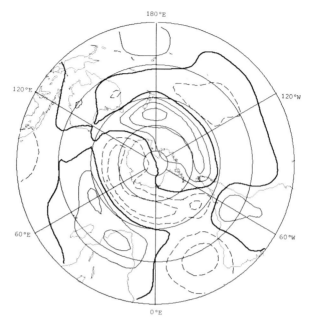

Fig. 2.17. As in Fig. 2.13, but for January.

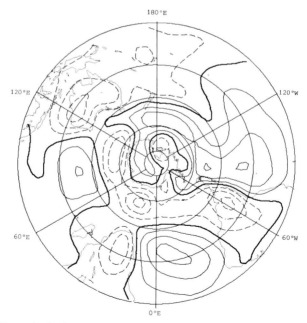

Fig. 2.18. Climatological mean January distribution of stationary wave sea-level pressure. Data source and plotting conventions as in Fig. 2.13. Contour interval 2 mb.

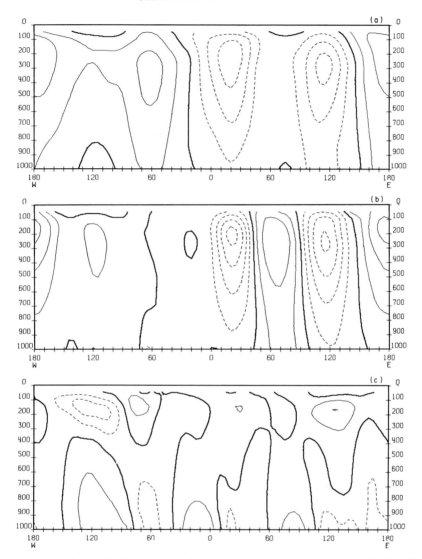

Fig. 2.19. Longitude–height cross-sections of stationary wave geopotential height $\bar{Z}^*$ for the summer season December 1981–February 1982, based on ECMWF operational analyses. (a) 60°S; (b) 45°S; (c) 25°S. Contour interval 30 m. The zero contour is thickened.

at tropical and subtropical latitudes and have a vertical structure more typical of tropical systems, with warm lows over the continents at low levels.

The corresponding figures for the summer season are shown in Figs. 2.17–2.20. The stationary wave patterns may seem similar, at first glance, to their wintertime counterparts, but they show marked differences that in some

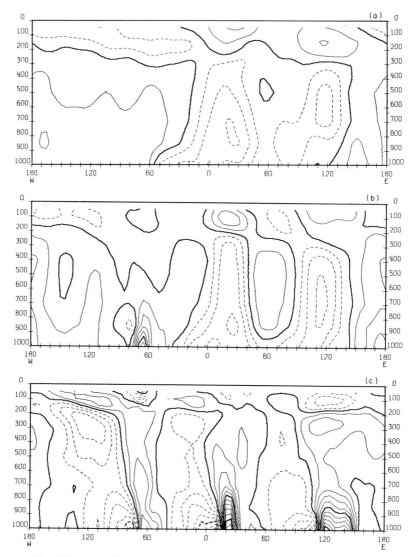

Fig. 2.20. As in Fig. 2.19, but for temperature. Contour interval 1 C.

respects mirror the seasonal changes in the Northern Hemisphere stationary waves. The low-latitude features increase in amplitude from winter to summer and assume a more well-defined 'tropical' structure with 'heat lows' in the sea-level pressure pattern over the subtropical continents overlain by ridges at the 200 mb level, and the subtropical oceanic anticyclones at sea-level overlain by 200 mb troughs. As in the Northern Hemisphere, the east–west sea-level pressure gradients tend to be strongest along the west coasts of the continents, particularly where mountain ranges are present. These low-latitude features, which are quite obviously related to the summertime monsoon, extend poleward to about 40°S. The high-latitude regime is little changed from winter except for the fact that the shallow surface lows over the polar seas are considerably weaker.

2.6 The tropical stationary waves

Detailed discussion of the tropical circulations is reserved for Chapter 9. However, for the sake of completeness, the tropical wind fields at the 200 mb and 850 mb levels for December–February 1980–81 and June–August 1981 are presented in Figs. 2.21 and 2.22, respectively. In nearly all respects, these distributions are representative of the long-term climatological means. At the higher latitudes they are qualitatively consistent with the geopotential height distributions shown in previous figures and with the distributions of $[\bar{u}^{*2}]$ and $[\bar{v}^{*2}]$ shown in Fig. 2.1. In interpreting these figures, the reader may find it helpful to refer to Fig. 9.2, which gives some indication of the distribution of diabatic heating which drives the tropical circulation patterns.

2.7 Summary and conclusions

On the basis of the discussion in the previous sections it is possible to ascribe the dominant features associated with the observed stationary waves to specific mechanisms which are quite well understood, and whose importance is widely acknowledged:

1. orographic forcing by the Himalayas, Rockies and Antarctica which determines the positions of the major features in the upper level geopotential height field during wintertime (and the high latitude pattern of the Southern Hemisphere during summer as well), and

2. thermal forcing associated with land–sea temperature contrasts which account for the monsoonal circulations of the summer hemisphere equatorward of the jetstream, as well as an assortment of shallow features at high latitudes of the winter hemisphere.

In addition, the observations would appear to suggest that some of the more subtle features in the stationary wave patterns may be associated with other mechanisms, namely

50

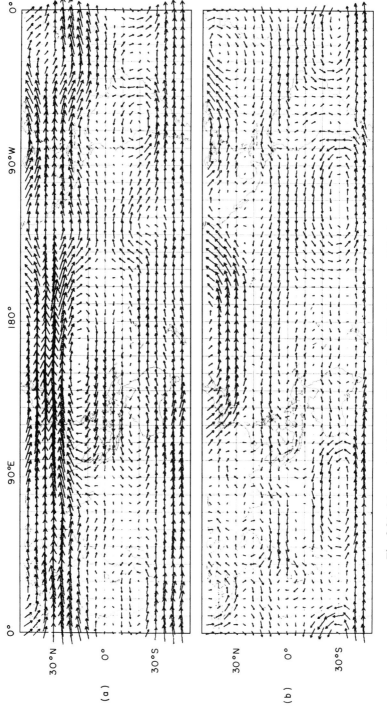

Fig. 2.21. Mean wind field for December 1980–February 1981, based on ECMWF operational analyses. (a) 200 mb; (b) 850 mb. From White (1982c).

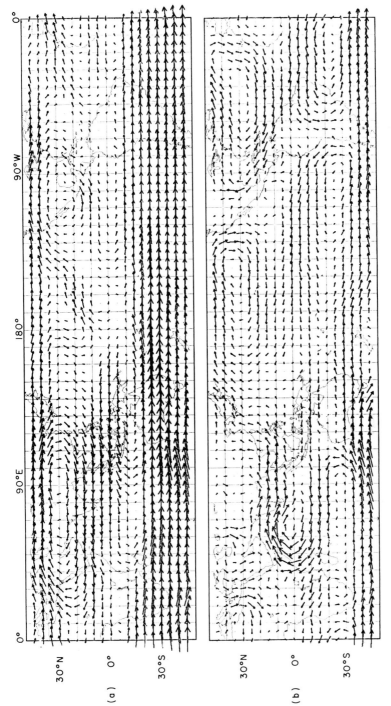

Fig. 2.22. Mean wind field for June–August 1981, based on ECMWF operational analyses. (a) 200 mb; (b) 850 mb. From White (1982c).

1. remote forcing via Rossby wave propagation, and

2. orographic (or possibly combined orographic and thermal) forcing of short waves, which is best exemplified by the wave structure in the vicinity of mountain ranges along the west coasts of continents during summer.

On the basis of observations alone it is not possible to infer whether thermal forcing makes a substantial contribution to the wintertime stationary wave pattern at levels above the lower troposphere. However, the GCM simulation described in Section 6.6 appears to indicate that thermal forcing is capable of producing a deep, vertically propagating zonal wavenumber 1 response at high latitudes of the Northern Hemisphere during wintertime.

Acknowledgements

I would like to thank Glenn White for supplying the figures based on the ECMWF analyses and for helpful suggestions. Part of the work was sponsored by the US National Science Foundation, Climate Dynamics Research Section under Grant ATM-81-06099. ECMWF facilities were used in the preparation of the figures.

References

BLACKMON, M. L. (1976). A climatological spectral study of the 500 mb geopotential height of the Northern Hemisphere. *J. atmos. Sci.*, **33**, 1607–1623.

BLACKMON, M. L. and LAU, N.-C. (1980). Regional characteristics of the Northern Hemisphere circulation: a comparison of a simulation of a GFDL general circulation model with observations. *J. atmos. Sci.*, **37**, 497–514.

BLACKMON, M. L., WALLACE, J. M., LAU, N.-C. and MULLEN, S. L. (1977). An observational study of the Northern Hemisphere wintertime circulation. *J. atmos. Sci.*, **34**, 1040–1053.

CRUTCHER, H. L. and MESERVE, J. M. (1970). *Selected level heights, temperatures and dew points for the Northern Hemisphere.* NAVAIR Atlas 50-IC-52, Chief Naval Operations, Washington, D.C. Available through Sup't of Documents, U.S. Gov't Printing Office, Washington, D.C.

GROSE, W. L. and HOSKINS, B. J. (1979). On the influence of orography on the large-scale atmospheric flow. *J. atmos. Sci.*, **36**, 223–234.

HOLOPAINEN, E. O. (1970). An observational study of the energy balance of the stationary disturbances in the atmosphere. *Qt. Jl R. met. Soc.*, **96**, 626–644.

HOSKINS, B. J. and KAROLY, D. (1981). The steady linear response of a spherical atmosphere to thermal and orographic forcing. *J. atmos. Sci.*, **38**, 1179–1196.

HOSKINS, B. J., SIMMONS, A. J. and ANDREWS, D. G. (1979). Energy dispersion in a barotropic atmosphere. *Qt. Jl R. met. Soc.*, **103**, 553–567.

LAU, N.-C. (1979). The observed structure of tropospheric stationary waves and local balances of vorticity and heat. *J. atmos. Sci.*, **36**, 996–1016.

LAU, N.-C., WHITE, G. H. and JENNE, R. L. (1981). *Circulation statistics for the extratropical Northern Hemisphere based on NMC analyses.* NCAR Tech. Note 171+STR, April 1981. Available from the National Center for Atmospheric Research, Boulder, Colorado.

MANABE, S. and TERPSTRA, T. B. (1974). The effects of mountains on the general circulation of the atmosphere as identified by numerical experiments. *J. atmos. Sci.*, **31**, 3–42.

NEWELL, R. E., KIDSON, J. W., VINCENT, D. G. AND BOER, G. J. (1972), (1974). *The general circulation of the tropical atmosphere and interactions with extratropical latitudes*, **1**, 258 pp.; **2**, 371 pp. The M.I.T. Press, Cambridge, Mass.

OORT, A. H. and RASMUSSON, E. M. (1971). *Atmospheric circulation statistics*. NOAA Prof. Pap. No. 5, U.S. Dept. of Commerce, Washington, D.C., 323 pp.

SADLER, J. C. (1975). *The upper tropospheric circulation over the global tropics*. Univ. of Hawaii, Dept. of Meteor., Tech. Rept. UHMET 75-05, 35 pp.

SALTZMAN, B. and RAO, M. S. (1963). A diagnostic study of the mean state of the atmosphere. *J. atmos. Sci.*, **20**, 414–432.

TALJAARD, J. J., VAN LOON, H., CRUTCHER, H. L. and JENNE, R. L. (1969). Climate of the upper air: Southern Hemisphere, **1**, *Temperatures, dew points and heights at selected pressure levels*. NAVAIR Atlas 50-IC-55. Chief Naval Operations, Washington, D.C., 135 pp. Available through Sup't of Documents, U.S. Gov't Printing Office, Washington, D.C.

TRENBERTH, K. E. (1980). Planetary waves at 500 mb in the Southern Hemisphere. *Mon. Wea. Rev.*, **108**, 1378–1389.

VAN LOON, H. and JENNE, R. L. (1972). The zonal harmonic standing waves in the Southern Hemisphere. *J. Geophys. Res.*, **77**, 992–1003.

VAN LOON, H., JENNE, R. L. and LABITZKE, K. (1972a). Climatology of the stratosphere in the Northern Hemisphere, Part 2. Geostrophic winds at 100, 50, 30 and 10 mb. *Meteorologische Abhandlungen*, **100**, No. 5, 54–55.

VAN LOON, H., TALJAARD, J. J., SASAMORI, T., LONDON, J., HOYT, D. V., LABITZKE, K. and NEWTON, C. W. (1972b). *Meteorology of the Southern Hemisphere*. Meteor. Monographs, **13**. Amer. Meteor. Soc., 263 pp.

WHITE, G. H. (1982a). An observational study of the Northern Hemisphere extratropical summertime general circulation. *J. atmos. Sci.*, **38**, 28–40.

WHITE, G. H. (1982b). Estimates of the seasonal mean vertical velocity fields of the extratropical Northern Hemisphere. *Mon. Weath. Rev.*, in press.

WHITE, G. H. (1982c). *The global circulation of the atmosphere December 1980–November 1981 based upon ECMWF analyses*. Technical note, Department of Meteorology, University of Reading.

– 3 –

Observations of low-frequency atmospheric variability

J. M. WALLACE and M. L. BLACKMON

3.1 Introduction

In contrast to the features on daily weather charts, whose three-dimensional structure and time evolution is already quite well understood both from the standpoint of pattern recognition and theoretical interpretation, the features that appear on weekly or monthly averaged charts are still something of an enigma. The patterns themselves can be described only in rather vague terms, and their time variation is so chaotic that it is difficult to distinguish between sequences of time-averaged charts arranged in forward and reverse chronological order (Jenne et al., 1972).

The purpose of this chapter is to document some of the characteristics of atmospheric fluctuations with time scales on the order of a week or longer, drawing upon the results of several recent observational studies that build upon the earlier works of G. T. Walker, H. Flohn, J. Namias, W. H. Klein, J. Sawyer and many others. For the sake of brevity, the discussion will be largely confined to extratropical latitudes of the Northern Hemisphere, during the winter season. Discussion of the more familiar day-to-day atmospheric variability associated with baroclinic disturbances will be taken up in Chapters 7 and 8.

In Section 3.2 it will be shown that there exist strong regional contrasts in the level of intensity of low-frequency fluctuations that are distinct from those associated with the higher-frequency baroclinic disturbances. Section 3.3 deals with the horizontal structure of the low-frequency fluctuations, as manifested in patterns based on the simultaneous correlations between the geopotential height at a given location with those at other gridpoints throughout the hemisphere. Such correlations, sometimes referred to as 'teleconnections', have been intensively investigated in the earlier works cited above. The distinguishing characteristic of the more recent work described herein is the interpretation of the results in terms of two-dimensional Rossby wave dispersion on a sphere. This same theoretical framework provides a partial interpretation of the time evolution of the low-frequency fluctuations, as discussed in Section 3.4.

Much of the existing literature on low-frequency atmospheric variability deals with it in a spectral context, as a function of frequency and zonal wavenumber. In Section 3.5 it will be shown that the signature of two-dimensional Rossby wave dispersion on the sphere is evident even in the behaviour of the individual ultra-longwave components in zonal wavenumber space.

In Section 3.6, the nature and causes of low-frequency atmospheric variability will be discussed in the context of the observations, with specific reference to:

1. external forcing from the lower boundary by anomalies in sea-surface temperature or soil moisture,

2. vacillation phenomena,

3. the possibility of two or more preferred climatic states that might coexist in the presence of the same external forcing,

4. the development of long-lived circulation features such as gyres, modons and solitons,

5. wave interactions,

6. forcing by high-frequency transients.

The extent to which the observed low-frequency variability can be attributed to one or more of these mechanisms has obvious implications upon the prospects for improved long-range weather prediction.

3.2 Geographical distribution of low-frequency fluctuations

Figure 3.1 shows the geographical distribution of the temporal variance of 500 mb height during the winter season for four different data sets: (a) unfiltered daily data; (b) band-pass filtered data which emphasize the time scales associated with baroclinic waves; (c) low-pass filtered data which emphasize fluctuation with periods longer than 10 days; and (d) monthly mean data (approximated here by using successive 30-day means). Further information on the data sets is provided in the figure caption. Frequency responses of the filters used for (b) and (c) are shown in Fig. 3.2.

The distribution in Fig. 3.1(b) is characterized by elongated maxima over the oceans near 45°N which are believed to be associated with the 'storm tracks', where baroclinic wave activity is most intense (Blackmon et al., 1977; see also Figs. 1.9 and 1.10). Figures 3.1(a), (c) and (d) show a contrasting pattern, with less elongated maxima located over the north Atlantic and Pacific Oceans and over the Siberian arctic. The locations of high variance on these maps correspond closely to the regions of frequent blocking identified by Rex (1950) and documented more recently by Knox (1981). (The term 'blocking', as it is conventionally used by synoptic meteorologists, refers to the development of strong, long-lived anticyclones that extend through the depth of the troposphere with an equivalent barotropic structure. A 'block' or 'blocking ridge' appears as a closed anticyclone at lower levels and as a strong ridge in the westerlies in the upper troposphere so that at that longitude the jetstream lies far poleward of its usual position. Short-wavelength baroclinic disturbances tend to follow the jetstream in its poleward excursion; hence these features temporarily 'block' the normal eastward progression of weather systems in middle latitudes.) The similarity of the patterns in Figs. 3.1(c) and (d) (apart from differences in magnitude) suggests that the geographical distribution of amplitude is not strongly frequency dependent at time scales longer than a week. The fact that the geographical distribution of variance in the unfiltered daily data closely resembles the 'low-frequency signature' in Figs. 3.1(c) and (d) indicates that the frequency spectrum of geopotential height fluctuations is sufficiently red that the geographical distribution of temporal variance is largely determined by variability within the low-frequency domain rather than by baroclinic disturbances.

This interpretation is supported by the fact that the low-pass filtered variances (Fig. 3.1c) are considerably larger than the band-pass filtered variances (Fig. 3.1b). In a similar manner, the vertical structure of the geopotential height fluctuations is dominated by a geographically dependent 'low-frequency signature' that is distinct from that associated with baroclinic waves (Blackmon et al., 1979).

Hence, the observations show evidence of a 'low-frequency regime', with characteristic time scales of a week or longer, in which the temporal variance and vertical structure of geopotential height perturbations show well-defined geographical distributions that are distinguishable from those associated with baroclinic waves. These distinctions are even clearer in the horizontal structure, which is described in the following section.

3.3 Horizontal structure of the low-frequency fluctuations

Figure 3.3 shows the simultaneous correlation between 500 mb height at 55°N, 20°W (denoted by the small circle in Fig. 3.1a) and 500 mb height at

58

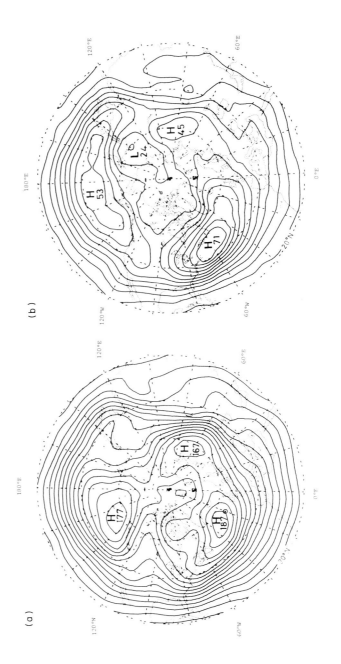

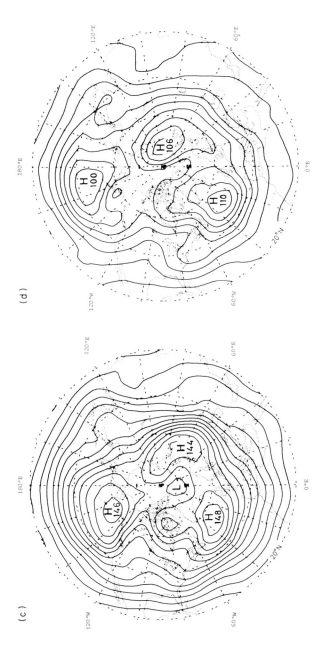

Fig. 3.1. Variance of 500 mb height based on NMC analyses for 18 winter seasons (1962–63 to 1979–80, inclusive) where winter is defined as the 90 days beginning 1 December. The contribution from the climatological mean annual cycle has been removed. (a) Unfiltered twice daily data; contour interval 10 m. (b) Band-pass filtered data emphasizing fluctuations in the 2.5–6 day period range; contour interval 5 m. (c) Low-pass filtered data emphasizing fluctuations with periods longer than 10 days; contour interval 10 m. (d) 30-day mean data; contour interval 10 m. Lines of latitude and longitude are drawn every 20°, the outer latitude circle being 20°N.

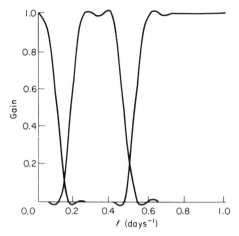

Fig. 3.2. Frequency response curves (in terms of amplitudes) for the filters used in Figs. 3.1 and 3.3(b),(c).

every other gridpoint, again for (*a*) unfiltered daily data, (*b*) band-pass filtered data, (*c*) low-pass filtered data, and (*d*) monthly mean data where the filters are the same as those used in the previous section. (Correlation patterns for this gridpoint are also presented by Sawyer, 1970.) A similar pattern is evident in (*a*), (*c*) and (*d*): the base gridpoint shows negative correlations with a broad belt across the subtropical Atlantic and North Africa and weaker positive correlations with a region in Siberia northwest of Lake Baikal. The fact that this correlation pattern is strongest in the monthly mean data (*d*), substantially weaker in the unfiltered data (*a*), and completely absent in the band-pass filtered data (*b*), clearly indicates that it is associated with the same low-frequency regime described in the previous section. (The band-pass filtered data exhibit an entirely different pattern which is suggestive of wavelike fluctuations in the upper level flow with wavelengths on the order of 4000 km. For a further discussion of these patterns see Blackmon *et al.*, 1983a, b.) Similar patterns are observed for other gridpoints, but not all gridpoints show such strong correlations with distant regions. Figure 3.4 shows a summary of the strongest correlation patterns as revealed by a recent analysis of monthly mean data. All these patterns have been subjected to the analysis described in Fig. 3.3, and the relationships within the resulting sets of patterns prove to be very similar to those described above.

Recent theoretical work on Rossby wave dispersion on a sphere, described in Chapter 6, provides a basis for interpreting these correlation patterns, which resemble trains of waves with ray paths oriented along 'great circle routes' emanating from the tropics, turning eastward to become tangent to a latitude circle, and curving back into the tropics. Such wavetrains constitute the atmosphere's planetary-scale response to any sustained, geographically localized forcing.

Modelling results indicate that the characteristic time scale required for a two-dimensional Rossby wave pattern to be set up in response to the initiation of a steady localized forcing is of the order of a week. (The term *two-dimensional Rossby wave* will be used to distinguish the wavetrains described here from the one-dimensional concept of Rossby waves with a prescribed meridional structure propagating along latitude circles.) During this 'set up time' the nodes and antinodes of the emerging wave pattern remain geographically fixed, whereas wave action disperses along the ray path, causing new centres to develop and amplify at locations successively more remote from the forcing (Hoskins *et al.*, 1977). Hence it seems quite reasonable that such wavetrains should play a role in shaping the correlation patterns for geopotential height fluctuations with characteristic time scales on the order of a week or longer, but they should be much less important for the higher frequency fluctuations which survive the bandpass filter used in Fig. 3.3(b).

Wallace and Gutzler (1981) have shown that the wavelike structures associated with these correlation patterns are much more clearly defined at the 500 mb level than at the Earth's surface and that, to a first approximation, their structure may be regarded as equivalent barotropic with amplitude increasing with height. Hoskins and Karoly (1981) have obtained a similar vertical structure in orographically and thermally forced Rossby wave trains in a five-layer primitive equation model.

The strong geographical dependence of the correlation patterns in Fig. 3.4 is of considerable interest. There is mounting evidence that the existence of the climatological mean stationary waves described in Chapter 2 is responsible for much of this structure, Kalnay-Rivas and Merkine (1981) showed that the downstream response to a given vorticity source in a barotropic flow in a channel is sharply enhanced by the presence of an orographically forced wavetrain in certain favoured positions relative to the source. Webster and Holton (1982) examined Rossby wave dispersion in a two-layer model in the presence of a zonally asymmetric background flow and found that propagation out of and through the tropics is strongly enhanced if the forced wavetrain encounters westerlies in the range of longitudes in which it disperses through the tropics. Some of these results are discussed in Section 9.5. Branstator (1983) considered barotropic Rossby wave dispersion through a number of idealized background flows and found that critical line absorption, reflection from regions of weak gradients of potential vorticity and the 'waveguiding' action of strong westerly jets can all play a role in organizing the wavetrains. Simmons (1982) and Branstator (1983) have both conducted experiments with a barotropic model, linearized about the climatological mean 300 mb Northern Hemisphere wintertime flow pattern. The results seem to indicate that large responses in the north-central Pacific can be excited by

62

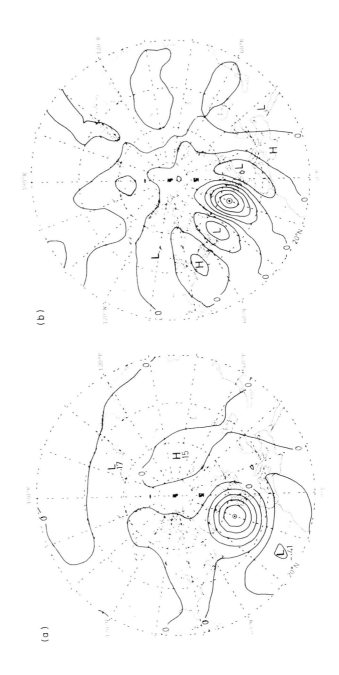

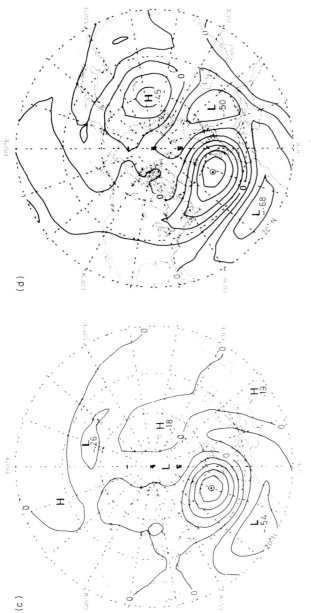

Fig. 3.3. Simultaneous temporal correlations between 500 mb height fluctuations at 55°N, 20°W and other Northern Hemisphere locations during wintertime, based on the same data set as Fig. 3.1 ; contour interval 0.2. (a) Unfiltered twice daily data. (b) Band-pass filtered data emphasizing fluctuations with periods longer filtered data emphasizing fluctuations in the 2.5–6 day period range. (c) Low-pass filtered data emphasizing fluctuations with periods longer than 10 days. (d) Monthly mean data. For further details, see Blackmon *et al.* (1983a).

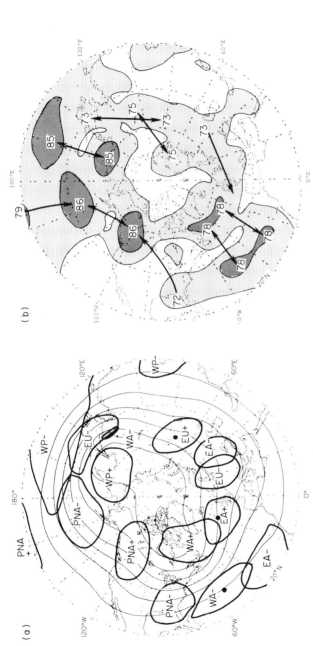

Fig. 3.4. Depiction of the predominant simultaneous correlation patterns in winter-time monthly mean 500 mb height data. (a) Areas enclosed by heavy lines represent centres of the five strongest patterns identified in the correlation statistics: PNA—the Pacific/North American pattern; WA—the West Atlantic pattern; EA—the East Atlantic pattern; EU—the Eurasian pattern; and WP—the West Pacific pattern. The plus and minus signs denote the sense of the correlations within each pattern; centres of like (unlike) sign are positively (negatively) correlated with one another. The lighter contours show the wintertime mean 500 mb height pattern. (b) Shading denotes regions that exhibit strong negative correlations with distant gridpoints (light shading, stronger than −0.6; and heavy shading, stronger than −0.75). Arrows indicate the region with which they are strongly negatively correlated. For example, gridpoints over the southeastern United States show strong negative correlations (up to −0.72) with gridpoints in western Canada and those gridpoints show even stronger negative correlations (up to −0.86) with gridpoints in the Central Pacific. After Wallace and Gutzler (1981). For confirmation of these results based upon methods of factor analysis, see Horel (1981). Small circles in (a) denote gridpoints for which lag correlation statistics are shown in the following three figures

forcing from a range of locations in both the tropics and in the extratropics. Further work by these authors (personal communication) suggests that the Pacific/North American pattern and possibly the East Atlantic pattern as well (see Fig. 3.4a) may be associated with the normal modes of the barotropic vorticity equation, linearized about the climatological mean 300 mb wintertime flow pattern. Lau (1981) has shown that similar teleconnection patterns emerge in extended integrations of a general circulation model which simulates the observed stationary waves but has no time-dependent external forcing apart from the annual cycle (see Section 5.3). Evidence presented by Held in Section 6.7 indicates that the corresponding one-point correlation patterns are much weaker in GCM simulations without stationary waves.

3.4 Time variation of the low-frequency patterns

Correlation patterns analogous to those shown in Fig. 3.3, but for various lag times should be capable of revealing preferred patterns of time variation, if they exist. Results of such an analysis are dependent upon the manner in which the data are time averaged or low-pass filtered before the correlations are computed. If the data represent successive time averages for a rather long time period such as monthly means, or if they are severely low-pass filtered, it is observed that the lagged correlation patterns are merely a damped version of the corresponding simultaneous correlation patterns (i.e., the shapes of the patterns are qualitatively similar to those for zero lag time, but the correlations approach zero with increasing lag time). This result is understandable in terms of the theoretical framework alluded to in the previous section. If the averaging time (or the high-frequency cutoff of the low-pass filter) is long in comparison to the 'setup time' for the two-dimensional Rossby wave patterns, then the growth and decay of the wavetrains cannot be resolved, and the only information that can be derived from the analysis is the rate at which the correlation pattern decays with increasing lag time, which gives some indication of the frequency spectrum of the forcing. More informative patterns emerge when one considers lag correlation statistics for shorter averaging times.

Figures 3.5–3.7 show examples of lag correlation patterns based upon twice daily 500 mb height data. The base gridpoints for these patterns are denoted by the black circles in Fig. 3.4(a), from which it can be seen that all three of them correspond to 'centres of action' for well-defined correlation patterns in the monthly mean data. (The gridpoint used for Fig. 3.5 is the same one that was used for generating Fig. 3.3.)

When the gridpoint near 55°N, 20°W, is correlated with every gridpoint in the hemisphere two days earlier (Fig. 3.5a) and two days later (Fig. 3.5b), the resulting patterns are rather similar, but the latter shows a substantially

66

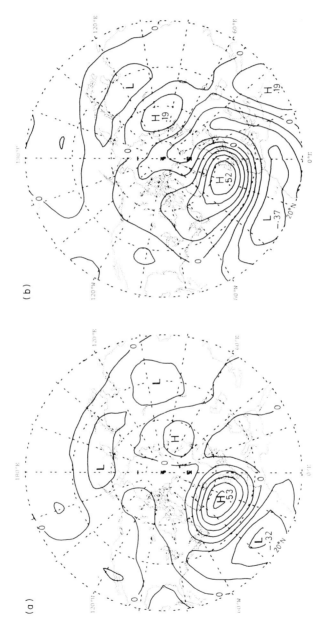

Fig. 3.5. Lag correlations between 500 mb height at (55°N, 20°W) and 500 mb height throughout the hemisphere (a) 2 days earlier and (b) 2 days later, based on the unfiltered daily data set described in the caption of Fig. 3.1. Contour interval 0.1.

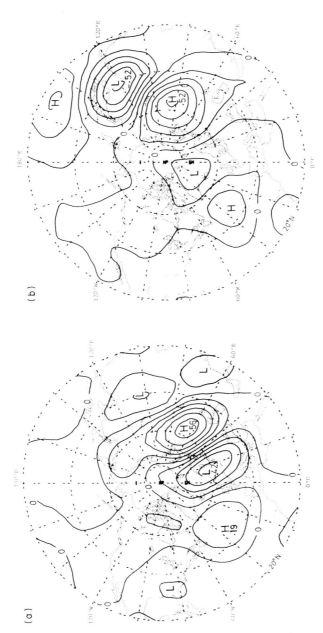

Fig. 3.6. As in Fig. 3.5, but for the gridpoint (55°N, 75°E).

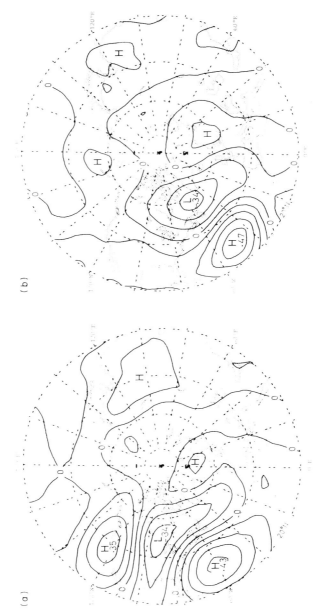

Fig. 3.7. As in Fig. 3.5, but for the gridpoint (30°N, 75°W).

stronger pattern reminiscent of Fig. 3.3(a),(c),(d) downstream of the base gridpoint. The Siberian gridpoint near (55°N, 75°E) shows a stronger contrast between the lag correlation patterns for two days earlier and two days later. The former (Fig. 3.6a) is characterized by relatively strong upstream centres of action over Scandinavia and the North Atlantic, whereas the latter (Fig. 3.6b) shows a strong downstream centre over Manchuria. Hence, the two patterns are indicative of waves propagating along a broad trajectory across the Eurasian continent. There is some evidence of eastward phase propagation in the daily data, as evidenced by the eastward shift of the positive centre over Siberia from (a) to (b), but the time evolution is dominated by the intensification of the downstream centres and the weakening of the upstream centres, which is consistent with the notion of Rossby wave dispersion. The correlation pattern in the Pacific/North American sector (not shown) exhibits a similar pattern of time variation.

Not all the lag correlation patterns are as simple as the above examples. Figure 3.7 shows the minus and plus two-day lag correlations for the gridpoint (30°N, 55°W) near Bermuda. The former (a) is dominated by a wavetrain approaching along a 'great circle' path across North America and the latter is suggestive of energy dispersing away from the region along a different ray path directed poleward and then eastward toward Europe. Lag correlation statistics for certain gridpoints near the Asian coast exhibit a similar behaviour.

Patterns for additional gridpoints and corroborative evidence based on low-pass, time filtered 500 mb height data are presented in Blackmon *et al.* (1983b). Lau (1981) has shown evidence of similar behaviour in lag correlation statistics generated from the output of a general circulation model. Hence, the interpretation of low-frequency variability in terms of Rossby way dispersion on a sphere appears to be a useful one, for descriptive purposes at least. However, it should be stressed that the spatial correlations used to define these patterns are not very strong; hence it is not surprising that the time variation appears to be rather chaotic in individual realizations, despite the existence of a certain amount of structure in the spatial correlations.

The bandpass filtered data exhibit a completely different kind of time variation from that described above, with advection of individual centres of high and low correlation by a 'steering flow' which resembles the 700 mb wind field (Blackmon *et al.*, 1983b). This behaviour is clearly revealed in the time-lapse movie of time-filtered 500 mb height fields by Jenne *et al.* (1972).

3.5 A spectral view of low-frequency variability

From the spatial correlation statistics presented in Sections 3.2 and 3.3 one obtains the distinct impression that the horizontal structure of low-frequency variability is wavelike in two dimensions (i.e., when viewed in terms of wavetrains propagating along 'great circle' routes), but local in one dimension, in the sense that these wavetrains occupy only a small portion of the latitude circle to which they are tangent at their turning points. Hence, for example, they would not appear as waves on traditional 'Hovmöller diagrams'. Furthermore, there is little indication of any systematic phase propagation during the development and decay of these wavetrains, which can be viewed as stationary waves, with fixed nodes and antinodes. When interpreting these results, it should be borne in mind that the analysis approach used in those sections is designed to emphasize structures associated with large local 500 mb height anomalies: it does not necessarily present a balanced view of low-frequency variability as a whole.

We must consider the possibility that the structure and time variation of the patterns associated with the ultra-longwave components may be considerably different from those described in the preceding sections. For example, there is widespread evidence of a prevailing tendency for westward phase propagation or retrogression of both the wavenumber 1 and wavenumber 2 components. For a discussion of the structure and interpretation of these westward propagating ultra-longwaves, the reader is referred to a recent review article by Madden (1979).

For the purposes of the present discussion, it is of interest to examine whether there is any relationship between the ultra-longwave components on individual circles and the two-dimensional Rossby wave patterns described in the previous section. For this purpose, we have prepared a set of correlation maps, based upon the amplitude and phase of zonal wavenumbers 1 and 2 at 50°N, as computed from successive 5-day mean 500 mb height charts for 30 winter seasons. Further information concerning the data set and the compositing procedure is provided in the figure caption.

Figure 3.8(a) shows the pattern of temporal correlation between the cosine coefficient of zonal wavenumber 1 for 500 mb height on the 50°N latitude circle (referred to the Greenwich meridian) and local gridpoint values of 500 mb height throughout the hemisphere at the same time. If the ultra-longwaves behave as distinct entities in the low-frequency transient fluctuations, one should expect such correlation maps to be dominated by simple geometric patterns of the appropriate zonal wavenumber. Figure 3.8 does indeed show a strong zonal wavenumber 1 component in the vicinity of 50°N, with a maximum in the vicinity of the Greenwich meridian and a minimum in the vicinity of the dateline. However, it is evident that higher zonal wavenumbers are present, even at 50°N, as reflected in the westward displacement of the centre of highest positive correlations relative to the Greenwich meridian. In the Atlantic and Eurasian sectors this correlation pattern shows some of the features of the 'East Atlantic Pattern' documented by Sawyer (1970) and Wallace and Gutzler (1981) and shown in Figs. 3.3(a),(c),(d) and Fig. 3.4(a).

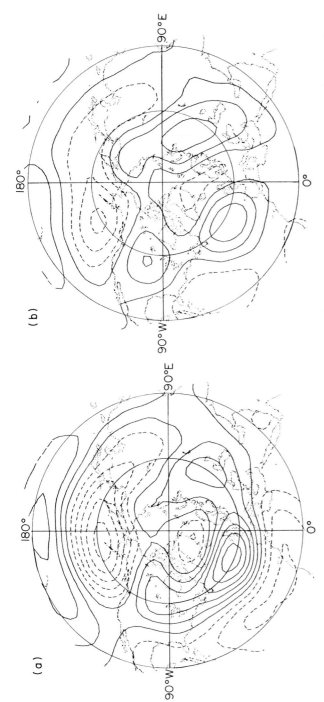

Fig. 3.8. (a) Simultaneous correlation between the cosine coefficient of zonal wavenumber 1 on the 50°N latitude circle (referred to the Greenwich meridian) and the geopotential height at each individual gridpoint throughout the hemisphere, based on 500 mb height data for consecutive five-day means (pentads). The data set consists of NMC analyses for 30 winter seasons (1948–49 to 1979–80, omitting 1959–60 and 1960–61), each consisting of the 30 pentads beginning 2 November. The climatological mean annual cycle was removed from the data before the analysis was performed. Contour interval 0.1. The inner latitude circle is 50°N. (b) As in (a) but, in this case, the same cosine coefficients are correlated with 500 mb height at each individual gridpoint throughout the hemisphere for the previous pentad. Contour interval 0.1. For further details, see Wallace and Hsu (1983a).

Figure 3.8(b) shows the pattern when the same cosine coefficient is correlated with 500 mb height throughout the hemisphere 5 days earlier. If the ultra-longwaves behave as distinct, zonally propagating entities in the low frequency fluctuations, then one should expect the same simple, geometric wavenumber 1 pattern to be dominant in such lag correlation maps, but it should be rotated with respect to the pole by an amount consistent with the preferred frequency and direction of zonal propagation; for example, if zonal wavenumber 1 propagates westward with a period on the order of 3 weeks, the pattern in Fig. 3.8(b) should be rotated counterclockwise about the pole by about 90° relative to that in Fig. 3.8(a). There is, perhaps, some evidence of westward propagation of the ultra-longwave features in Fig. 3.8, but the dominant features in Fig. 3.8(b) are rearranged in such a way as to line up along 'great circle' routes. This rearrangement has resulted in a breaking up of the zonal wavenumber 1 pattern along 50°N. With liberal imagination one can see features of the 'Eurasian' and 'Pacific/North American' patterns described in Fig. 3.4(a).

Figure 3.9 shows another example, based on the cosine component of zonal wavenumber 2 on 50°N. As expected, the zonal wavenumber 2 signature is quite evident in the simultaneous correlation pattern (Fig. 3.9a) with maxima near the Greenwich meridian and the dateline and minima near 90°W and 90°E. However, there are substantial departures from geometrical symmetry with respect to zonal wavenumber 2; over the western part of the hemisphere the signature of the 'Pacific/North American pattern' (Fig. 3.4a) is evident. In the lag correlation pattern for 5 days later (Fig. 3.9b) the zonal wavenumber 2 pattern is greatly attenuated, but the geographically localized 'Pacific/North American pattern' persists.

Analogous charts have been prepared for other phases of zonal wavenumbers 1 and 2 and the results were found to be qualitatively similar to those presented above. (Further examples will be presented in a forthcoming paper by Wallace and Hsu, 1983a.) Higher wavenumber features were always found to be present in combination with the ultra-long waves and there was always evidence of two-dimensional Rossby wave patterns; particularly in the lag correlations for the pentads 5 days earlier and 5 days later. Hence, in comparison with the two-dimensional Rossby wave signatures considered in the previous two sections, the individual Fourier wave components have only an ephemeral existence in the observational statistics, even when the analysis is especially designed to isolate them.

How is one to reconcile the above results with the conclusions of other studies based on analysis of Fourier wave components on latitude circles, which emphasize the zonal propagation characteristics of the ultra-long waves? Perhaps this apparent contradiction can be resolved by noting that the zonally propagating component of the ultra-long waves accounts for only about 10% of the variance associated with them (e.g., in Table 5a of Blackmon,

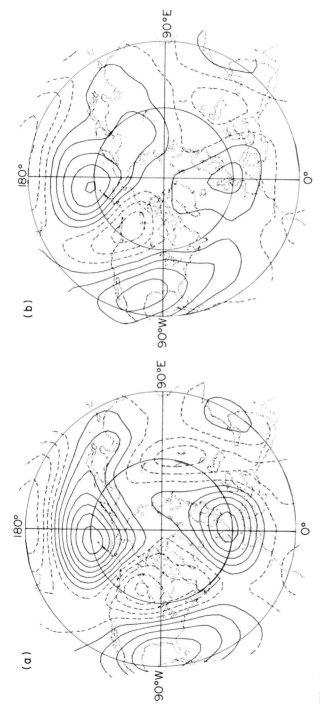

Fig. 3.9. (a) As in Fig. 3.8, but for the cosine coefficient of zonal wavenumber 2 on the 50°N latitude circle (referred to the Greenwich meridian). (b) The lag correlation map for the same cosine coefficient correlated with 500 mb height at each individual gridpoint for the subsequent pentad.

1976, the normalized quadrature spectrum for zonal wavenumbers $m = 1$ and 2 reaches values of about -0.3 for the gravest two or three modes in two-dimensional wavenumber n for the low-frequency fluctuations in wintertime 500 mb height. The square of the normalized quadrature spectrum $(-0.3)^2$ gives the fraction of the variance of these modes that can be attributed to the presence of a regular, westward propagating component.) Hence it is possible to detect this westward propagating component only with rather specialized forms of analysis, which are based upon the implicit working hypothesis that everything of interest in the ultra-long waves can be described in terms of a superposition of Fourier modes propagating along latitude circles (e.g., see Deland, 1964; Pratt, 1976; Hayashi, 1979). Although this spectral approach has been fruitful for studies of stratospheric dynamics (see Chapter 10) it is evidently of more limited usefulness for the Northern Hemisphere troposphere where so much of the low-frequency variance appears to be associated with two-dimensional Rossby wave dispersion.

3.6 Causes of low-frequency variability—observational evaluation of theories

Although the notion of Rossby wave propagation is helpful in interpreting the structure and time evolution of low-frequency atmospheric variability, it should perhaps be emphasized that it is not, in and of itself, the cause of the low-frequency variability. For example, none of the evidence presented in the previous sections would be sufficient grounds for rejecting what might be regarded as a 'null hypothesis' that low-frequency variability is nothing more than a response to stochastic forcing by higher-frequency disturbances (analogous to 'Brownian motion') as suggested by Leith (1973) and Hasselmann (1976). Therefore, if one is to understand the nature and causes of low-frequency variability of the atmospheric general circulation it is essential to go beyond the kinds of observational analyses described in the previous sections of this chapter and to focus upon mechanisms.

(a) External forcing

Interpretation of low-frequency atmospheric variability as a response to slowly varying thermal forcing from the oceans and cryosphere is a recurrent theme in the literature on interannual climate variability (see, for example, Walker and Bliss, 1932, and Namias and Cayan, 1981). However, it is only rather recently that the importance of this mechanism is becoming widely accepted. Stimulated by the observational studies which Bjerknes (1966, 1969) and others carried out in the late 1960s, Rowntree (1972, 1979), Julian and Chervin (1978) and Keshavamurty (1982) have performed general circulation

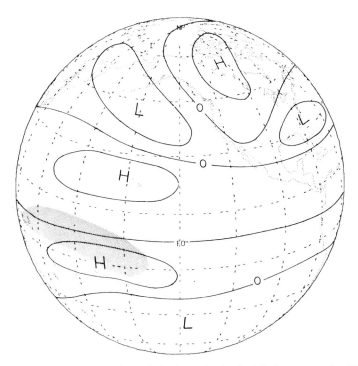

Fig. 3.10. Schematic illustration of the hypothesized global pattern of middle and upper tropospheric geopotential height anomalies (solid lines) during a Northern Hemisphere winter which falls within an episode of warm sea-surface temperatures in the equatorial Pacific. The shading indicates regions of enhanced rainfall. After Horel and Wallace (1981).

modelling (GCM) experiments designed to determine the nature of the atmospheric response to sea-surface temperature anomalies in the tropical oceans. The results appear to indicate that such anomalies can produce a substantial response in the large-scale circulation in extratropical latitudes of the winter hemisphere.

More recent observational studies by Trenberth (1976), Barnett (1977), van Loon and Madden (1981) and Horel and Wallace (1981) have added credibility and detail to Bjerknes' description of the global atmospheric response to warm sea-surface temperature anomalies in the equatorial Pacific during the Northern Hemisphere winter. The composite picture, shown schematically in Fig. 3.10, is suggestive of a train of two-dimensional Rossby waves emanating from an equatorial heat source. A more detailed theoretical interpretation of this structure is given in Gill (1980), Opsteegh and van den Dool (1980), Hoskins and Karoly (1981), Webster (1981) and in Section 6.5 and 9.5. Very similar patterns have been obtained in a recent series of GCM experiments by Shukla and Wallace (1983).

The equatorial Pacific sea-surface temperature anomalies undergo a distinctive pattern of time evolution described by Rasmusson and Carpenter (1982), which provides a basis for long-range prediction out to several seasons in advance.

There is also evidence of a link between sea-surface temperature anomalies and interannual climatic variability in the Atlantic sector. Hastenrath and Heller (1977) have shown that drought in northeast Brazil tends to occur during seasons when sea-surface temperatures are above normal in the tropical Atlantic north of the equator and below normal south of the equator. Moura and Shukla (1981) have successfully simulated the pattern of anomalous rainfall in a series of GCM experiments with realistic sea-surface temperature anomalies. There is also evidence based on the work of Rowntree (1976b) that sea-surface temperature anomalies in the subtropical north Atlantic during wintertime may give rise to circulation anomalies at higher latitudes over the Atlantic sector and Europe.

There are numerous references in the observational literature to extratropical sea-surface temperature anomalies as a causal factor in interannual climate variability (Nicholls, 1980; Namias and Cayan, 1981), but there is little in the ways of supporting evidence based on GCM simulations. Davis (1976) has presented statistical evidence that indicates that in extratropical latitudes, it is the climate anomalies that tend to force the sea-surface temperature anomalies. Furthermore, there are theoretical grounds for expecting extratropical sea-surface temperature anomalies to have only a rather small feedback upon the atmospheric circulation (Rowntree, 1976a; Webster, 1981, 1982; see also Chapter 6).

Sea-surface temperature anomalies are not the only form of external forcing that might be a significant causal factor in short-term climate variability. During summertime, variations in soil moisture can produce large changes in the surface heat and moisture budgets over the land masses. Recent GCM simulations by Walker and Rowntree (1977), Shukla and Mintz (1982) and others indicate that these effects could have a significant impact upon the large-scale summertime circulation.

It should be emphasized that external forcing accounts for only a small fraction of the low-frequency variability of the atmospheric circulation in extratropical latitudes. For example, according to estimates of Horel and Wallace (1981), sea-surface temperature fluctuations in the equatorial Pacific account for less than half the variance of seasonal mean 500 mb height, even over the most sensitive regions of North America, and they account for less than one-fourth of the variance of monthly mean 500 mb height in these regions. Furthermore, it is evident from GCM simulations (see, for example, Manabe and Hahn, 1981, and Chapter 5 of this book) that the extratropical atmosphere is capable of a considerable amount of variability, even at very low frequencies, even in the absence of external forcing. Therefore it is necessary to consider other sources of low-frequency variability which are internal to the atmosphere.

(b) Vacillation phenomena

In the early literature on long-range weather forecasting (e.g., Namias, 1950) there was widespread emphasis on the notion of an 'index cycle', consisting of quasi-periodic fluctuations between a 'high index' state of the general circulation, with strong westerlies and weak waves or eddies, and a 'low index' state with weak westerlies and strong waves. It was widely believed that this index cycle behaviour might be a manifestation of some sort of vacillation phenomenon analogous to those that occur, under certain controlled conditions, in a rotating annulus.

There is ample evidence that the winter hemispheric circulation does undergo large fluctuations in the partition between zonal and eddy forms of kinetic energy (K_Z and K_E, respectively) and available potential energy (A_Z and A_E, respectively). Because of the constraint of thermal wind balance, K_Z and A_Z tend to vary in phase with one another, and similarly for K_E and A_E. Furthermore, it is observed that the zonal and eddy forms of energy vary out of phase with one another, as in the hypothesized index cycle, and that certain of the processes that represent conversions between the zonal and eddy components of the energy vary in quadrature with the energies themselves, in agreement with energy budget considerations (Winston and Krueger, 1961; Webster and Keller, 1975; McGuirk and Reiter, 1976). The characteristic period of these index cycle fluctuations is of the order of a few weeks, and the eddy energy varies by up to a factor of two between the high and low index phases.

Webster and Keller (1975) and McGuirk and Reiter (1976) have found evidence of spectral peaks in time series of energetics parameters but they account for only a very small fraction of the low-frequency variability. Hence it appears that the index cycle will be of marginal utility in long range forecasting. Nevertheless, it is of interest to determine whether these index cycle fluctuations can be identified with one or more distinct forms of vacillation.

In order to contrast the circulation during high and low index regimes, Wallace and Hsu (1983b) use an eddy index E, defined as the hemispherically integrated eddy variance of 500 mb height, Z^{*2}, as defined in Section 2.2 (the integration is carried out for the region poleward of 20°N). A time series of E was generated on the basis of consecutive 5-day mean 500 mb height maps for 30 winter seasons. Figure 3.11 shows the composite anomaly map for the 46 pentads with the highest value of E, adjusted for the annual cycle. A

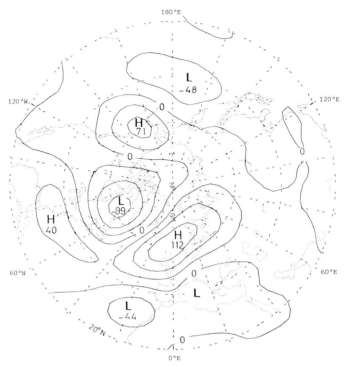

Fig. 3.11. 500 mb height anomalies for the 46 pentads in which the index E (a measure of the intensity of low-frequency eddy activity integrated over the hemisphere) showed the largest positive deviations from the seasonally adjusted climatological mean value. Contour interval 30 m.

comparison with Fig. 2.2 reveals that in these 'low index' situations, the transient fluctuations (whose time-averaged distribution is represented by the composite anomaly chart) tend to be in phase with the climatological mean stationary waves. This tendency for reinforcement of the stationary waves is observed for all the major extrema in $\bar{Z}^*$.

The extent to which this tendency for reinforcement of the climatological mean stationary waves during low-index situations contributes to the high eddy energy levels is revealed by the statistics summarized in the bar graph (Fig. 3.12). The left-hand column shows the contributions of the climatological mean stationary waves (3700 m^2) and transient eddies (7700 m^2) to the wintertime mean value of E $(11\,400 \text{ m}^2)$. The right-hand column shows the corresponding statistics for the low index composite. Here the 'enhanced stationary wave contribution' (8000 m^2) represents the value of E derived from the composite map for the 46 low index pentads, and the total $(17\,800 \text{ m}^2)$ represents the mean value of E computed from the 46 maps for the individual

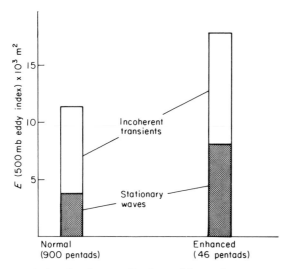

Fig. 3.12. Bar graph showing the contributions of the stationary waves and incoherent transient eddies to the index E for climatological mean conditions and for low-index situations. See text and Wallace and Hsu (1983b) for further explanation.

pentads in the composite. The remaining 9800 m² can be regarded as the contribution from incoherent transients (i.e., time-dependent fluctuations which are averaged out in the compositing procedure). Hence, the low index composite is characterized by a value of E some 6400 m² (17 800–11 400 m²) above the climatological mean, of which 4300 m² (8000–3700 m²) can be regarded as a consequence of the enhancement of the stationary waves.

Hence it appears that an important (and perhaps the dominant) component of 'index cycle' type fluctuations in bulk energetics parameters for the Northern Hemisphere winter tropospheric circulation is the alternating constructive and destructive interference between the climatological mean stationary waves and the low frequency transients, a process altogether different from those observed in conventional experiments with an azimuthually symmetric rotating annulus. This phenomenon is presumably of less importance in the Southern Hemisphere, where the stationary waves are much weaker. It is interesting to note that Madden (1975) found evidence of an analogous index cycle phenomenon in stratospheric energetics parameters, related to alternating constructive and destructive interference between the stationary waves and the zonally propagating ultra-long waves.

It would seem that the observed out of phase relation between the zonal and eddy energies might be nothing more than a reflection of energy balance constraints; in the absence of strong diabatic heating gradients or frictional dissipation, the eddy energy can increase (either by the transient disturbances coming more into phase with the stationary waves in the hemispheric average, or by an increase in the amplitudes of the incoherent part of the transient eddies) only at the expense of the energy associated with the zonally symmetric circulation. Such compensation would not be inconsistent with the null hypothesis that low frequency variability is nothing more than a response to stochastic forcing by higher frequency disturbances. Thus, at this point, the case in favour of a more organized vacillation phenomenon rests largely upon the observational evidence of spectral peaks reported by Webster and Keller (1975) and McGuirk and Reiter (1976).

(c) Transition between different climatic regimes

It has been postulated that some of the observed low-frequency atmospheric variability may be related to the existence of two or more distinct 'weather regimes' that are compatible with the same external forcing, each with its own zonally averaged flow and stationary wave configuration and its own pattern of transient variability. Under certain conditions it might be possible for the hemispheric circulation to undergo rather abrupt transitions from one weather regime to another (Lorenz, 1968; Charney and Devore, 1979). An observational study lending some support to this hypothesis is described in Chapter 4. Some theoretical evidence in support of this possibility is discussed in Chapter 7.

The existence of bi- or multi-modal frequency distributions in general circulation statistics would constitute a convincing observational demonstration of the existence of multiple weather regimes. White (1980) has examined the frequency distribution of 500 and 1000 mb height over the Northern Hemisphere for both winter and summer data. He found substantial deviations from normality, including rather box-like distributions similar to the sample shown in Fig. 3.13, which occur in regions of frequent blocking over the eastern oceans. Such distributions could be regarded as evidence of bimodality. However, it is also quite possible that they are a reflection of a normal distribution which is truncated near the extremes, perhaps because of dynamical constraints that restrict the observed distribution of geopotential height to within a certain range. It is possible that the signature of multiple weather regimes would show up more distinctly in multivariate frequency distributions analogous to the phase diagrams used in theoretical papers on the subject. However, in the absence of rather specific theoretical guidance, it is difficult to know how to select, from among the myriad of possibilities, the most revealing representations of the state of the general circulation in multidimensional phase space.

An alternative approach to the problem of identifying multiple weather regimes is to classify persistent circulation patterns in terms of the equilibrium

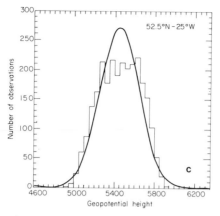

Fig. 3.13. Frequency distribution of wintertime 500 mb height at a gridpoint near (55°N, 20°W), based on 11 years of twice daily NMC analyses. Winter is defined as the 120-day season beginning 15 November. After White (1980).

solutions derived from simplified models that describe the planetary wave response to the observed orographic and/or thermal forcing. The results of Charney *et al.* (1981), based on this approach, show some intriguing qualitative similarities between observed and theoretically derived wave profiles in the vicinity of 45°N. However, their barotropic model, with its highly truncated treatment of the latitudinal structure is probably too idealized to warrant the kind of detailed comparison that might serve as conclusive evidence of the existence of multiple weather regimes in the atmosphere.

In the absence of more detailed theoretical guidance, perhaps it is justifiable to use a more subjective approach in the tradition of the Grosswetterlagen of Baur (1951) for identifying the kinds of circulation patterns that might be associated with preferred weather regimes (if, indeed, the notion is a valid one). In view of the prominence of orographic forcing in present theoretical thinking about multiple-equilibrium states (e.g., Charney and Devore, 1979), it would seem logical to emphasize pressure patterns in the vicinity of the major mountain ranges as a basis for classifying circulation regimes, and since the large-scale circulation exhibits a good deal more variability near the Rockies than near the Himalayas it seems reasonable, at least as a starting point, to focus on circulation patterns in their part of the hemisphere.

Figure 3.14 shows the 500 mb and 1000 mb height patterns averaged over a period of 24 consecutive days beginning with 31 December 1980. The corresponding patterns for the remaining 76 days of the 1980–81 winter (1–30 December 1980, plus 24 January to 10 March 1981, inclusive) are shown in Fig. 3.15. Throughout the former period a strong ridge is centred over the Rockies

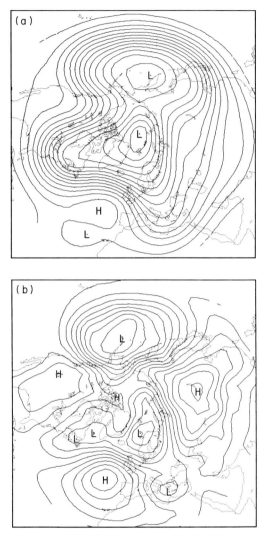

Fig. 3.14. Composite geopotential height patterns for the 24-day period beginning 31 December 1980, based upon daily analyses made at ECMWF. (a) 500 mb height, contour internal 60 m; and (b) 1000 mb height, contour interval 30 m.

at both levels. This ridge exhibited only a weak westward tilt with height between the 1000 and 500 mb levels and it was characterized by a strong, warm westerly flow over the Canadian Rockies at all levels. The troughs over the central Pacific and over eastern North America were unusually pronounced during this 24-day period. Even more distant stationary wave features such as the east Atlantic ridge, the trough over the eastern Mediterranean, the

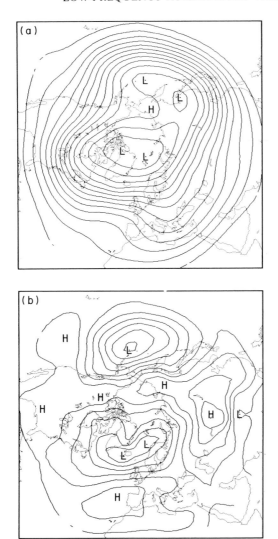

Fig. 3.15. As in Fig. 3.14, but for the other 76 days of the 1980–81 winter (1–30 December 1981 and 24 January–10 March, inclusive).

Siberian high and the Aleutian low were more pronounced than in the climatological mean. The other 76 days of the winter saw a wider variety of circulation patterns, the average of which was very similar to the long-term wintertime climatology (compare Fig. 3.15 and Figs. 2.5 and 2.6).

Extended periods with circulation patterns similar to those in Fig. 3.14 over the Pacific and North American sectors of the hemisphere have occurred during previous winters, the most notable case being the winter of 1976–77. It should perhaps be emphasized that during these earlier episodes, conditions in other parts of the hemisphere have varied widely; for example, geopotential heights have sometimes been below normal over the eastern Atlantic. Extended episodes with similar circulation anomalies have also been produced in general circulation model simulations, without the aid of external forcing (see Chapter 5). Thus it would appear that this circulation type (which is obviously related to the Pacific/North American correlation pattern described in Fig. 3.4a) possibly represents one of the atmosphere's preferred weather regimes or, to be more precise: one of the atmosphere's preferred adaptations to the Rockies.

There appears to be at least one other way in which the atmosphere can adapt to the Rockies for periods of a week or longer. Figure 3.16 shows composite 500 and 1000 mb flow patterns for the 22 days: 1–9 and 18–24 December 1980 and 5–10 February 1981, inclusive. In this situation the ridge at upper levels is located over the Gulf of Alaska so that the flow over the northern Rockies is parallel to the range rather than perpendicular to it, as in Fig. 3.14a). There have been numerous periods during previous winters when this northerly flow over the Rockies extended further southward into the western United States. It can be seen that the ridge in this parallel flow regime is characterized by a distinct split in the upper level westerlies whereas the one in the perpendicular flow regime considered previously is characterized by a strong, high-latitude zonal flow. The low-level pattern (Fig. 3.16b) is also very different, with a cold anticyclone over the northern Rockies and predominantly easterly flow across the range, as opposed to westerly flow in the perpendicular flow regime. The vertical structure is much more baroclinic than in the perpendicular flow regime, with almost a $90°$ phase shift between the surface anticyclone over the Yukon and the 500 mb ridge over the Gulf of Alaska. The geopotential height anomalies associated with this parallel flow regime do not resemble any of the correlation patterns shown in Fig. 3.4(a). These two circulation regimes bear certain similarities to multiple weather regimes derived by Reinhold and Pierrehumbert (1982) in numerical experiments with a two-layer model. (Specifically, they resemble the '$30°$ ridge' and '$90°$ ridge' equilibrium states. The latter corresponds to one of the two 'weather regimes' apparent in the model statistics. For further discussion of these experiments, see Chapter 7.)

Prolonged episodes of the two circulation types described above account for 46 of the 100 days of the 1980–81 winter (defined here as starting with 1 December). Hemispheric maps for the remaining 54 days include a few brief episodes of the same two circulation types, interspersed with an array of different patterns, including a week with no long-wave ridge in the vicinity of the Rockies at all. Hence it remains to be seen whether these perpendicular and parallel flow regimes represent preferred time-averaged states of the general

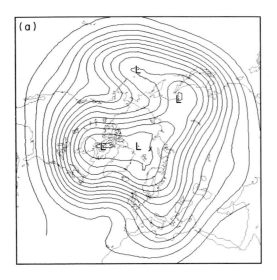

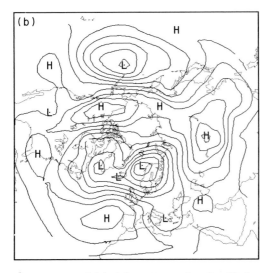

Fig. 3.16. Composite geopotential height patterns for the 22 days 1–9 and 18–24 December 1980 and 5–10 February 1981, inclusive, based on daily analyses made at ECMWF. (a) 500 mb height, contour interval 60 m; (b) 1000 mb height, contour interval 30 m.

circulation or whether they are merely a pair of contrasting samples drawn from the multi-dimensional continuum of possible atmospheric states.

(d) Gyres, solitons and modons

Under certain conditions, portions of the hemispheric circulation are capable of assuming configurations in which the flow is equivalent barotropic and the absolute vorticity contours closely parallel the streamlines so that the potential vorticity advection and the geopotential height tendencies are small. In polar latitudes where the beta-effect is small and the mean zonal flow and the baroclinicity both tend to be weak, closed cyclonic and anticyclonic gyres frequently assume such a configuration that, once established, tends to be rather long lived. Such systems probably account for the high 1-, 2- and 3-day lag correlations observed over the Arctic for geopotential height and the large fractional contribution of low frequency fluctuations to the total variance in that region (e.g., see Blackmon, 1976, Fig. 6; Blackmon and Lee, 1983).

In middle latitudes, more specialized conditions are required in order to produce such stable, long-lived flow configurations. Such conditions are most often realized over the eastern oceans where the climatological mean zonal flow and the baroclinicity are weakest and the amplitudes of the transients are large enough so that closed circulations or gyres are often observed, even within the belt of strongest westerlies. At these latitudes such long-lived features often have a dipole component; a high latitude anticyclone being paired with a lower latitude cyclone. The dynamical significance of these dipole structures, which have been likened to idealized flow configurations called solitons and modons, is discussed in Section 13.3. Examples of two such patterns, one which dominated the eastern Atlantic circulation from 1–5 December 1980 and another which developed just east of the dateline on 6 December and persisted for a week, are shown in Fig. 3.17.

(e) Wave interactions

In a series of numerical time integrations, Egger (1978) has shown that for flow confined to a mid-latitude beta-plane channel, interactions between orographically forced waves and slow-moving transient planetary waves are capable of producing rather long-lived, large amplitude features that resemble blocks. Loesch (1974) has considered the role of weakly non-linear wave interactions involving resonant triads in relation to blocking. If such mechanisms are to be of use in interpreting blocking or low-frequency variability in general, then one should expect to see distinctive signatures in one- or two-dimensional planetary wave spectra.

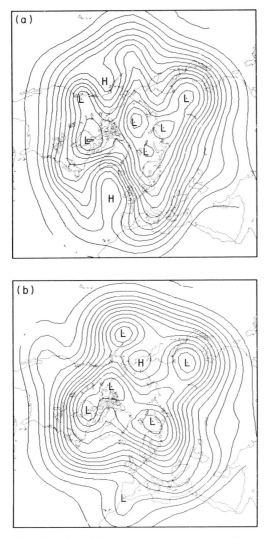

Fig. 3.17. Composite 500 mb height patterns based on daily analyses made at ECMWF: (a) 1–5 December 1980; (b) 6–12 December 1980. Contour interval 60 m.

The spectral signatures of blocking episodes have been considered by a number of different investigators over the past few years, but relatively little has been published on the subject. Colucci *et al.* (1981) documented two blocking episodes that displayed a spectral behaviour qualitatively consistent with Loesch's mechanism. However, the examination of larger numbers of episodes by R. Dole (1978, personal communication) and D. L. Hartmann

Fig. 3.18. Time-averaged 500 mb height field for the Southern Hemisphere based on ECMWF operational analyses for the period 6 July–4 August 1981. Contour interval 60 m.

(1980, personal communication) failed to reveal any simple patterns of time evolution that were common to a large number of events. In most cases the spectral description of blocking episodes proved to be more complex and more variable from case than the more conventional synoptic description. These predominantly negative results are consistent with the evidence presented in the previous section, which indicates that the structure of low-frequency variability cannot be concisely represented in terms of the superposition of the conventional zonal or spherical harmonics. This same evidence would appear to argue against any mechanism for generating low-frequency variability in which periodic boundary conditions in the longitude domain play a crucial role.

Lest we overstate the case against mechanisms involving wave interaction, it should perhaps be noted that remarkable simple spectral signatures are sometimes observed in low-frequency fluctuations in the Southern Hemisphere. An example from the 1981 winter is shown in Fig. 3.18. It portrays a zonal wavenumber 3 pattern that persisted throughout much of July. Note that the low-latitude flow during this period had a strong zonal wavenumber 6 component that appears to be phase-locked with the higher latitude pattern. The superposition of the wavenumber 3 and 6 patterns results in split flow at the longitude of New Zealand.

(f) Forcing by high-frequency transients

The notion that forcing by high-frequency transients, such as baroclinic waves, might contribute to the observed low-frequency variability has been discussed not only in statistical context as in the works of Leith (1973), Madden (1976) and Hasselmann (1976), but also in a dynamical context. Green (1977) has suggested that the high-frequency transients might act in a systematic manner, through their fluxes of vorticity, to sustain certain existing low-frequency circulation anomalies such as blocking configurations. This interpretation draws some support from results of a GCM experiment reported by Mahlman (1979) which indicate that blocking anticyclones tend to re-intensify in response to the approach of decaying cyclones from upstream which are preceded by strong poleward flows of tropical air, low in potential vorticity.

Observational results of Green (1977), Savijarvi (1977) and Edmon (1980) all support the idea that the fluxes of vorticity due to the high-frequency transients are in the sense as to sustain the low-frequency anomalies. However, the latter two authors found that the associated heat fluxes tend to act in the opposite sense, so as to produce damping. The relative importance of the two effects can be compared by computing the transport of potential vorticity by the high-frequency transients, as has been done by Savijarvi (1978), whose results indicate that the damping produced by the heat fluxes tends to predominate. For a further discussion of the role of the high-frequency transients in relation to the stationary waves and the low-frequency transients, the reader is referred to Chapters 7 and 8.

It should perhaps be noted that conclusive observational evidence to the effect that high-frequency transient eddy fluxes act in a sense as to maintain circulation anomalies would not necessarily mean that other mechanisms, such as those described above, are unimportant. To cite a possible analogy: radial transports by spiral rain bands undoubtedly contribute to the angular momentum budget of a developing hurricane (Palmen and Riehl, 1957; Pfeffer, 1958), yet these transports are not generally regarded as the primary cause of hurricanes, and they certainly do not alone determine when and where hurricanes form. In a similar manner, it is possible that high-frequency transients might contribute to the potential vorticity budgets of certain types of low-frequency circulation anomalies, but other processes might be more influential in determining when and where the anomalies develop, and in anchoring them in certain favoured geographical positions. Hence, the forcing of low-frequency variability by high-frequency transients, if it indeed occurs, need not be an entirely random process.

From the foregoing discussion it is evident that several different dynamical mechanisms contribute to the observed low-frequency variability of the extratropical tropospheric circulation. On the seasonal time scale, external forcing has been shown to be of some importance, and on the time scale of a week, quasi-stable flow configurations similar to gyres and modons very likely play a role. Transitions between weather regimes and forcing by baroclinic waves could contribute to low-frequency variability on a wide range of time scales, but further observational evidence is needed in order to demonstrate their importance. In view of the evidence presented in the previous section, it would appear that mechanisms such as vacillation and non-linear wave interaction, which imply that the low-frequency transients exhibit a wavelike behaviour in longitude, are more difficult to reconcile with existing observations for the Northern Hemisphere troposphere.

3.7 Concluding remarks

Perhaps the most important message in this chapter is the demonstration that a more coherent observational picture of low-frequency variability of the tropospheric circulation emerges when one considers its full two-dimensional structure in the space domain (at any given level), and not just its one-dimensional structure along latitude circles, as in Fourier decompositions in terms of zonal wavenumber. The two-dimensional structure has greater dynamical significance because Rossby wave dispersion in a barotropic atmosphere is inherently a two-dimensional phenomenon. (In the stratosphere where only the ultra-long waves are present and large amplitude disturbances are confined to the higher latitudes, a one-dimensional treatment may be more justifiable).

The contrasting horizontal structure of the high- and low-frequency geopotential height fluctuations (Fig. 3.3) has implications on a wide range of general circulation statistics. For example, the high-frequency fluctuations with their relatively long meridional scales and shorter zonal scales (as evidenced by the elongated ellipses in Fig. 3.3b) have about twice as much kinetic energy in the meridional wind component as in the zonal wind component (Blackmon et al., 1977), whereas the low-frequency fluctuations, with their characteristic north/south dipole structures, elongated in the zonal direction, tend to be anisotropic in the opposite sense, particularly at lower latitudes. Furthermore, it seems reasonable to expect that patterns of momentum and heat fluxes for the two frequency domains might be rather different, with the pattern of storm tracks determining the distributions for the higher frequencies and the prevailing orientation of the two-dimensional Rossby wave patterns determining the distributions for the lower frequencies. The view of the two-frequency domains as being characterized by basically different dynamical processes has influenced the format of this book, with the low frequencies being emphasized in Chapters 3–6 and the higher frequencies in Chapters 7 and 8.

In view of the foregoing discussion, it seems quite possible that terms such as 'blocking', which are widely used in the literature to describe low-frequency variability, do not refer to a single, well-defined phenomenon resulting from a single causal mechanism but, rather, to a collection of different low-frequency phenomena that have similar structures but different causes. For example, an anticyclonic circulation anomaly over western Canada in the mean for a winter season might well be a result of different dynamical processes than one that persists continuously for a three-week period, though both might be loosely termed 'blocking ridges'. Furthermore, it would appear that 'blocking ridges' over the Rockies and the mid-oceans might be maintained by different dynamical processes. The discussion of causal mechanisms in the previous section is intended as a tentative approach toward categorizing low-frequency variability in terms of underlying causes rather than structures. The list of causal mechanisms presented there may not be complete, and it may contain entries that are irrelevant, but at least it might serve some purpose as a basis for discussion that might lead to a clarification of these issues.

Acknowledgement

We would like to thank H.-H. Hsu, Y.-H. Lee and M. Revell who prepared some of the figures and made them available for early publication in this paper. Part of the work was sponsored by the US National Science Foundation under Grant ATM-81-06099. ECMWF facilities were used in the preparation of some of the figures.

References

BARNETT, T. P. (1977). An attempt to verify some theories of El Niño. *J. phys. Oceanogr.*, 7, 633–647.
BAUR, F. (1951). Extended range weather forecasting. *Compendium of Meteorology, Amer. met. Soc.*, 814–833.
BJERKNES, J. (1966). A possible response of the atmospheric Hadley circulation to equatorial anomalies of ocean temperature. *Tellus*, 18, 820–829.
BJERKNES, J. (1969). Atmospheric teleconnections from the equatorial Pacific. *Mon. Weath. Rev.*, 97, 162–172.
BLACKMON, M. L. (1976). A climatological spectral study of the 500 mb geopotential height of the Northern Hemisphere. *J. atmos. Sci.*, 33, 1607–1623.
BLACKMON, M. L., WALLACE, J. M., LAU, N.-C. and MULLEN, S. L. (1977). An observational study of the Northern Hemisphere wintertime circulation. *J. atmos. Sci.*, 34, 1040–1053.
BLACKMON, M. L., MADDEN, R. A., WALLACE, J. M. and GUTZLER, D. S. (1979). Geographical variations in the vertical structure of geopotential height fluctuations. *J. atmos. Sci.*, 36, 2450–2466.

BLACKMON, M. L., LEE, Y.-H. and WALLACE, J. M. (1983a). Horizontal structure of 500 mb height fluctuations with short, intermediate and long time scales. *J. atmos. Sci.*, in press.

BLACKMON, M. L., LEE, Y.-H., WALLACE, J. M. and HSU, H.-H. (1983b). Time variation of 500 mb height fluctuations with short, intermediate and long time scales. Submitted to *J. atmos. Sci.*

BRANSTATOR, G. (1983). Horizontal energy propagation in a barotropic atmosphere with meridional and zonal structure. Submitted to *J. atmos. Sci.*

CHARNEY, J. G. and DEVORE, J. G. (1979). Multiple flow equilibria in the atmosphere and blocking. *J. atmos. Sci.*, **36**, 1205–1216.

CHARNEY, J. G., SHUKLA, J. and MO, K. C. (1981). Comparison of a barotropic blocking theory with observation. *J. atmos. Sci.*, **38**, 762–779.

COLUCCI, S. J., LOESCH, A. Z. and BOSART, L. F. (1981). Spectral evolution of a blocking episode and comparison with wave interaction theory. *J. atmos. Sci.*, **38**, 2092–2111.

DAVIS, R. E. (1976). Predictability of sea surface temperature and sea level pressure anomalies over the North Pacific Ocean. *J. phys. Oceanogr.*, **6**, 249–266.

DELAND, R. J. (1964). Travelling planetary waves. *Tellus*, **16**, 271–273.

EDMON, H. J. JR. (1980). A study of the general circulation over the Northern Hemisphere during the winters 1976–77 and 1977–78. *Mon. Weath. Rev.*, **108**, 1538–1553.

EGGER, J. (1978). Dynamics of blocking highs. *J. atmos. Sci.*, **35**, 1788–1801.

GILL, A. E. (1980). Some simple solutions for a heat-induced tropical circulation. *Q. Jl R. met. Soc.*, **106**, 447–462.

GREEN, J. S. A. (1977). The weather during July 1976: some dynamical considerations of the drought. *Weather*, **32**, 120–125.

HASSELMANN, K. (1976). Stochastic climate models. Part I Theory. *Tellus*, **6**, 473–485.

HASTENRATH, S. and HELLER, L. (1977). On the modes of tropical circulation and climate anomalies. *Qt. Jl R. met. Soc.*, **103**, 77–92.

HAYASHI, Y. (1979). A generalized method of resolving transient disturbances into standing and travelling waves by space–time spectral analysis. *J. atmos. Sci.*, **36**, 1017–1029.

HOREL, J. D. (1981). A rotated principal component analysis of the Northern Hemisphere 500 mb height field. *Mon. Weath. Rev.*, **109**, 2080–2092.

HOREL, J. D. and WALLACE, J. M. (1981). Planetary scale atmospheric phenomena associated with the Southern Oscillation. *Mon. Weath. Rev.*, **109**, 813–829.

HOSKINS, B. J. and KAROLY, D. (1981). The steady linear response of a spherical atmosphere to thermal and orographic forcing. *J. atmos. Sci.*, **38**, 1179–1196.

HOSKINS, B. J., SIMMONS, A. J. and ANDREWS, D. G. (1977). Energy dispersion in a barotropic atmosphere. *Qt. Jl R. met. Soc.*, **103**, 553–567.

JENNE, R. L., WALLACE, J. M., YOUNG, J. A. and KRAUS, E. B. (1972). Daily 500 mb heights and long period fluctuations. NCAR motion picture film J4. (Available at cost on 16 mm film, C/O Roy L. Jenne, NCAR, Boulder, Colo. 80307.)

JULIAN, P. R. and CHERVIN, R. M. (1978). A study of the Southern Oscillation and Walker Circulation phenomenon. *Mon. Weath. Rev.*, **106**, 1433–1451.

KALNAY-RIVAS, E. and MERKINE, L. O. (1981). A simple mechanism for blocking. *J. atmos. Sci.*, **38**, 2077–2091.

KESHAVAMURTY, R. N. (1982). Response of the atmosphere to sea-surface temperature anomalies over the equatorial Pacific and the teleconnections of the Southern Oscillation. *J. atmos. Sci.*, **39**, 1241–1259.

KNOX, John L. (1981). Atmospheric blocking in the Northern Hemisphere. Ph.D. Thesis, Department of Geography, University of British Columbia, Vancouver, B.C., V6T 1W5, 274 pp.

LAU, N.-C. (1981). A diagnostic study of recurrent meteorological anaomalies appearing in a 15-year simulation with a GFDL general circulation model. *Mon. Weath. Rev.*, **109**, 2287–2311.

LEITH, C. E. (1973). The standard error of time averaged estimates of climatic means. *J. appl. Meteor.*, **12**, 1066–1069.

LOESCH, A. Z. (1974). Resonant interactions between unstable and neutral baroclinic waves. *J. atmos. Sci.*, **31**, 1177–1201.

LORENZ, E. N. (1968). Climatic determinism. *Met. Monogr.*, **8**, No. 30, 1–3.

MADDEN, R. A. (1975). Oscillations in the winter stratosphere, 2, The role of horizontal heat transport and the interaction of transient and stationary planetary-scale waves. *Mon. Weath. Rev.*, **103**, 717–729.

MADDEN, R. A. (1976). Estimates of the natural variability of time averaged sea-level pressure. *Mon. Weath. Rev.*, **104**, 942–952.

MADDEN, R. A. (1979). Observations of large-scale travelling Rossby waves. *Rev. Geophys. and Space Phys.*, **17**, 1935–1949.

MAHLMAN, J. D. (1979). Structure and interpretation of blocking anticyclones as simulated in a GFDL general circulation model. *Proceedings of the (thirteenth) Stanstead Seminar.* Publication in Meteorology No. 123 (T. Warn, ed.), Dept. of Meteorology, McGill Univ., Montreal, Quebec, pp. 70–76.

MANABE, S. and HAHN, D. G. (1981). Simulation of atmospheric variability. *Mon. Weath. Rev.*, **109**, 2260–2286.

McGUIRK, J. P. and REITER, E. R. (1976). A vacillation in atmospheric energy parameters. *J. atmos. Sci.*, **33**, 2079–2093.

MOURA, A. D. and SHUKLA, J. (1981). On the dynamics of droughts in Northeast Brazil: Observations, theory, and numerical experiments with a general circulation model. *J. atmos. Sci.*, **38**, 2653–2675.

NAMIAS, J. (1950). The index cycle and its role in the general circulation. *J. Meteor.*, **7**, 130–139.

NAMIAS, J. and CAYAN, D. R. (1981). Large-scale air–sea interactions and short period climatic fluctuations. *Science*, **214**, 869–876.

NICHOLLS, N. (1980). Long range weather forecasting: value, status and prospects. *Rev. Geophys. and Space Phys.*, **18**, 771–788.

OPSTEEGH, J. D. and VAN DEN DOOL, H. M. (1980). Seasonal differences in the stationary response of a linearised primitive equation model: Prospects for long range forecasting? *J. atmos. Sci.*, **37**, 2169–2185.

PALMEN, E. and RIEHL, H. (1957). Budget of angular momentum and energy in tropical cyclones. *J. Meteor.*, **14**, 150–159.

PFEFFER, R. L. (1958). Concerning the mechanism of hurricanes. *J. Meteorol.*, **15**, 113–120.

PRATT, R. W. (1976). The interpretation of space–time spectral quantities. *J. atmos. Sci.*, **33**, 1060–1066.

RASMUSSON, E. M. and CARPENTER, T. (1982). Variations in tropical sea-surface temperatures and surface wind fields associated with the Southern Oscillation/El Niño. *Mon. Weath. Rev.*, **110**, 354–384.

REINHOLD, B. B. and PIERREHUMBERT, R. T. (1982). Dynamics of weather regimes: Quasi-stationary waves and blocking. *Mon. Weath. Rev.*, **110**, 1105–1145.

REITER, E. R. (1978). Long term wind variability in the tropical Pacific: Its possible causes and effects. *Mon. Weath. Rev.*, **106**, 324–330.

REX, D. F. (1950). Blocking action in the middle troposphere and its effect upon regional climate II. The climatology of blocking action. *Tellus*, **2**, 275–301.

ROWNTREE, P. R. (1972). The influence of tropical east Pacific Ocean temperatures on the atmosphere. *Qt. Jl R. met. Soc.*, **98**, 290–321.

ROWNTREE, P. R. (1976a). Tropical forcing of atmospheric motions in a numerical model. *Qt. Jl R. met. Soc.*, **102**, 583–605.

ROWNTREE, P. R. (1976b). Response of the atmosphere to a tropical Atlantic Ocean temperature anomaly. *Qt. Jl R. met. Soc.*, **102**, 607–625.

ROWNTREE, P. R. (1979). The effects of changes in ocean temperature on the atmosphere. *Dyn. Atmos. Oceans.*, **3**, 373–390.

SAVIJARVI, H. (1977). The interaction of the monthly mean flow and large-scale transient eddies in two different circulation types. Part II: Vorticity and temperature balance. *Geophysica*, **14**, 207–229.

SAVIJARVI, H. (1978). The interaction of the monthly mean flow and large-scale transient eddies in two different circulation types. Part III: Potential vorticity balance. *Geophysica*, **15**, 1–16.

SAWYER, J. S. (1970). Observational characteristics of fluctuations with a time scale of a month. *Qt. Jl R. met. Soc.*, **96**, 610–625.

SHUKLA, J. and MINTZ, Y. (1982). The influence of land-surface evaporation on the Earth's climate. *Science*, **215**, 1498–1501.

SHUKLA, J. and WALLACE, J. M. (1983). Numerical simulation of the atmospheric response to equatorial Pacific sea-surface temperature anomalies. In preparation.

SIMMONS, A. J. (1982). The forcing of stationary wave motion by tropical diabatic heating. *Qt. Jl R. met. Soc.*, **108**, 503–534.

TRENBERTH, K. E. (1976). Spatial and temporal variations of the Southern Oscillation. *Qt. Jl R. met. Soc.*, **102**, 639–653.

VAN LOON, H. and MADDEN, R. A. (1981). The Southern Oscillation. Part I: Global associations with pressure and temperature in the northern winter. *Mon. Weath. Rev.*, **109**, 1150–1162.

WALKER, G. T. and BLISS, E. W. (1932). World weather V. *Mem. R. meteor. Soc.*, **4**, 53–84.

WALKER, J. and ROWNTREE, P. R. (1977). The effect of soil moisture on the circulation and rainfall in a tropical model. *Qt. Jl R. met. Soc.*, **103**, 29–46.

WALLACE, J. M. and GUTZLER, D. S. (1981). Teleconnections in the geopotential height field during the Northern Hemisphere winter. *Mon. Weath. Rev.*, **109**, 785–812.

WALLACE, J. M. and HSU, H.-H. (1983a). Ultra-long waves and two dimensional Rossby waves. Submitted to *J. atmos. Sci.*

WALLACE, J. M. and HSU, H.-H. (1983b). Low frequency fluctuations in zonal indices and stationary wave configurations. Submitted to *J. atmos. Sci.*

WEBSTER, P. J. (1981). Mechanisms determining the atmospheric response to sea surface temperature anomalies. *J. atmos. Sci.*, **38**, 554–571.

WEBSTER, P. J. (1982). Seasonality in the local and remote atmospheric response to sea surface temperature anomalies. *J. atmos. Sci.*, **39**, 41–52.

WEBSTER, P. J. and HOLTON, J. R. (1982). Cross equatorial response to middle latitude forcing in a zonally varying basic state. *J. atmos. Sci.*, **39**, 722–733.

WEBSTER, P. J. and KELLER, J. L. (1975). Atmospheric variations: Vacillations and index cycles. *J. atmos. Sci.*, **32**, 1283–1300.

WHITE, G. H. (1980). Skewness, kurtosis and extreme values of Northern Hemisphere geopotential heights. *Mon. Weath. Rev.*, **108**, 1446–1455.

WINSTON, J. S. and KRUEGER, A. F. (1961). Some aspects of a cycle of available potential energy. *Mon. Weath. Rev.*, **89**, 307–318.

— 4 —

Persistent anomalies of the extratropical Northern Hemisphere wintertime circulation

RANDALL M. DOLE

4.1 Introduction

Observational meteorologists frequently suggest that there may be some recurrent flow anomalies that typically persist beyond the periods associated with synoptic-scale variability. The prototypical example cited is the phenomenon of blocking, often described as having considerable persistence, frequently for durations of more than two weeks (Rex, 1950; Sumner, 1954) and a tendency toward the recurrence of qualitatively similar flow patterns (Namias, 1947; Elliott and Smith, 1949; Rex, 1951). The descriptions provided in these studies are often highly provocative. Nevertheless, our current understanding of the characteristics and causes of recurrent persistent flow anomalies remains quite limited. A challenging problem facing meteorologists is to determine if these earlier observations can be placed within a more general and systematic framework.

This chapter summarizes results from an observational study of quasi-stationary persistent anomalies of the extratropical Northern Hemisphere wintertime circulation. For brevity we will focus on results at one level (500 mb). Questions on vertical structure, as well as on the relationship between the occurrence of persistent anomalies and changes in the location of, and activity along, storm paths are discussed in Dole (1981).

4.2 Data

The data used in this study are twice-daily (00Z and 12Z) National Meteorological Center (NMC) final analyses of the Northern Hemisphere

95

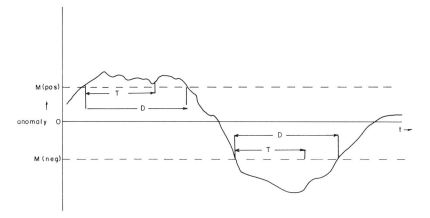

Fig. 4.1. Method for defining cases. See text for explanation.

500 mb geopotential heights for the 14 winter seasons from 1963–1964 to 1976–1977, inclusive. The winter season is defined as the 90-day period from 1 December to 28 February. Raw anomalies are defined as the departures of the analysed heights from the corresponding long-term seasonal trend values. The seasonal trend time series at a point is determined by a least-squares quadratic fit to the 14-winter mean time series for that point (e.g., the first value of the winter mean time series is the average of the 14 Dec. 1, 00Z values, the second value is the average of the Dec. 1, 12Z values, etc.). The raw height anomalies Z' are further normalized by a scale factor which is inversely proportional to the sine of latitude:

$$Z'_n = Z'(\sin 45°/\sin \phi). \tag{4.1}$$

Note that this normalization is similar to that used in obtaining a geostrophic stream function from height data. Henceforth, 'anomalies' refers to the latitude normalized anomalies.

4.3 Method for defining cases

The definition of a persistent anomaly used here is simple and intuitive: a persistent anomaly is defined at a point if the anomaly at that point exceeds a threshold value for a sufficient duration. The method, illustrated in Fig. 4.1, is as follows:

1. Specify a 'magnitude' criterion, M, and a duration criterion, T, where for positive anomaly cases $M \geqslant 0$ and for negative anomaly cases $M \leqslant 0$.

2. Define the occurrence of a persistent positive (negative) anomaly case at a particular gridpoint satisfying selection criteria (M, T) if the anomaly at that point remains equal to or greater (less) than M for at least T days.

3. Define the duration, D, for a positive (negative) case as the time from which the anomaly first becomes greater (less) than M to the time when the anomaly next becomes less (greater) than M at that point.

Note that these criteria act as lower bounds, so that all events which meet or exceed the threshold values are counted as persistent anomaly cases satisfying the specified selection criteria.

4.4 Geographic distribution

The number of persistent anomaly cases occurring over the 14 winter seasons was determined for each point on the 5° latitude by 5° longitude grid using the following values for the selection criteria: $M = \pm 0$ m, ± 50 m, ..., ± 250 m; $T = 5$ days, 10 days, ..., 25 days. Figure 4.2 presents the results of typical geographic distribution calculations for (a) positive anomaly cases (100 m, 10 days), (b) negative anomaly cases (-100 m, 10 days), and (c) the corresponding 'sum' distribution, obtained by adding the numbers of cases from the two previous distributions.

The three distributions shown use data that have been lightly low-pass filtered to remove brief (~ 1 day) interruptions by mobile transient disturbances. The total numbers of cases are increased by about 50% over the unfiltered values, but the regional characteristics are otherwise almost unchanged. We see that:

1. There are three primary regions for the occurrence of persistent anomalies: the North Pacific to the south of the Aleutians (PAC), the North Atlantic to the southeast of Greenland (ATL) and the northern Soviet Union, extending northeastward into the Arctic Ocean (NSU). These correspond to the three regions of high variance shown in Fig. 3.1.

2. For each region the maximum in the frequency of occurrence of positive anomalies is approximately co-located with, and has a comparable magnitude to, the corresponding maxima of negative anomalies.

3. There is considerable latitudinal variability in the number of cases, despite the anomaly normalization, with maxima occurring near 50N for ATL and PAC regions and near 60N for NSU.

4. The range in the number of cases is substantial: the three major regions each have in excess of twenty cases over the 14 winter seasons, whereas

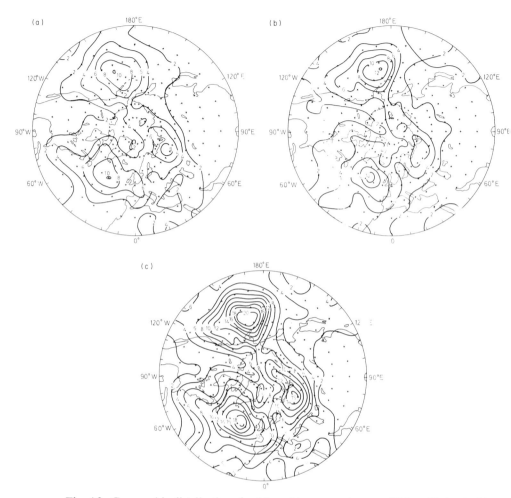

Fig. 4.2. Geographic distributions for (a) positive anomaly cases (100 m, 10 days); (b) negative anomaly cases (− 100 m, 10 days); and (c) the sum of the cases in (a) and (b). The outer circle is 20°N.

parts of Asia, the subtropics and central North America have less than two events satisfying these criteria over the period.

Qualitatively similar distributions are obtained if the raw anomalies are defined as departures from the season (rather than long-term) means, although the magnitudes of the maxima are slightly reduced (5%–25%).

Figure 4.3 illustrates regional variations in persistence. The number of events $y(n)$ exceeding the threshold criteria (M) for n or more consecutive days are shown on a semi-logarithmic diagram for:

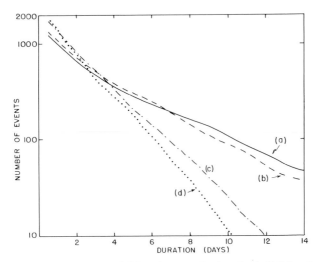

Fig. 4.3. The number of events satisfying the magnitude criteria (M) for durations of *n* days or longer for (a) positive events (50 m) in regions having high numbers of persistent anomalies; (b) as in (a) for negative events (− 50 m); (c) as in (a) for positive events (50 m) in a region experiencing few persistent events (North America); and, (d) as in (c) for negative events.

(a) positive events (50 m), for regions having high numbers of persistent anomalies, obtained by averaging the distributions of the three major persistent anomaly regions (the distribution for a region is determined by combining the distributions obtained at each of nine points in a 10° latitude by 20° longitude box centered at a point within the region. The combined distributions are qualitatively similar to, and have smaller sampling errors than, the corresponding distributions at individual points);

(b) negative events (− 50 m) for regions having high numbers of persistent anomalies, obtained similarly to (a): and, for comparison;

(c) positive events (50 m) for a region with relatively few persistent anomalies, centered over North America; and

(d) negative events (− 50 m) for the same region.

We notice that:

1. for durations beyond about 3 days, there are more cases in the persistent anomaly (PA) regions than in the non-persistent anomaly (NPA) regions. The ratio of the number of events in the two regions, $y(n)_{PA}/y(n)_{NPA}$, increases with increasing *n*.

2. the distributions for the NPA region form nearly straight lines, whereas the slopes for the PA distributions vary as a function of duration.

Since the conditional probability that an event which has lasted n days will last at least $n + 1$ days is given by:

$$P(n + 1 \mid n) = y(n + 1)/y(n) \qquad (4.2)$$

a linear relationship in this display indicates that the persistence probability P is independent of the duration n. Conversely, if the relationship is non-linear, the persistence probability varies with the duration. Estimates of anomaly 'half-lives' (the time required for half of the anomaly events to be terminated) can be obtained from this figure: for NPA, the half-life is about 1.5 days; for PA, the half-lives vary from about 1.5 days at very short durations to around 3.5 days after about $n = 6$ days, with slightly more positive than negative cases at long durations. Although the shift in time scales appears small, the consequences are highly significant: for durations of about 10 days or longer, roughly an order of magnitude more events are observed than would be predicted by assuming a single distribution following the rapid decay rate. The cumulative distributions in the PA regions can effectively be described as a sum of two simple decay processes whose decay time scales lie on either side of the comparable time scale for a region experiencing few persistent anomalies (Dole, 1981).

4.5 Horizontal structure

The previous results indicate that there are three primary regions having high numbers of persistent anomalies. Consequently, a large percentage of the Northern Hemisphere wintertime persistent anomalies can be studied by selecting cases from these regions. For brevity, we focus on cases from only one such region: the North Pacific to the south of the Aleutians (PAC).

Composite or 'grand mean' persistent anomaly maps were obtained by first constructing case mean maps for each of the cases defined at certain key points, then averaging the case means, taking each case as one realization. Key points are the points with the largest sums of positive and negative cases. Figures 4.4(a) and (b) show composite anomaly maps for 7 PAC positive anomaly cases (100 m, 10 days) and 7 PAC negative anomaly cases (-100 m, 10 days), respectively (key point 50°N, 165°W). Comparing these maps, we see that, in addition to the anomaly at the key point, there is a larger scale pattern evident which, to a first approximation, appears with opposite polarity in the two maps. The overall impression of this pattern is of a dominant centre near the key point, a centre of opposite sign located about 25° of latitude to the south of the dominant centre and a train of anomaly centres of alternating

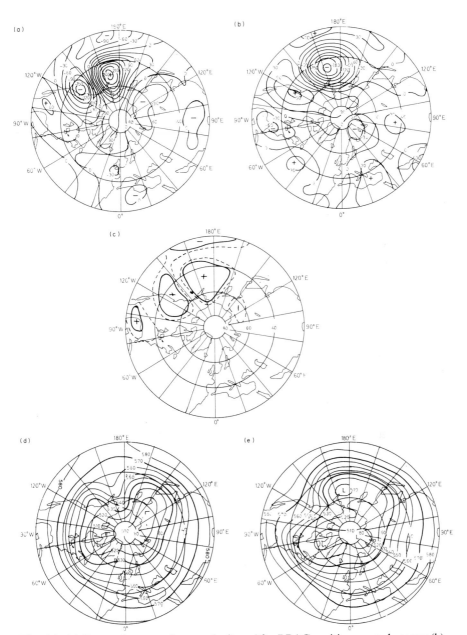

Fig. 4.4. (a) Composite anomaly maps (units: m) for 7 PAC positive anomaly cases; (b) as in (a) for 7 negative anomaly cases; (c) confidence levels for a two-sided t-test estimating the significance of the differences in means (a)–(b). Dashed at 95% confidence level, solid at 99% level; sign is the sign of the difference (a)–(b); (d) composite 500 mb heights (units: dam) for the positive cases; and (e) as in (d) for the negative cases.

signs and decreasing magnitudes extending downstream from the main centre. The statistical significances of the differences between means (null hypothesis of equal means), obtained by a two-sided t-test, are displayed in Fig. 4.4(c). The principal centres of the pattern are associated with statistically significant differences at the 99% confidence level.

Figures 4.4(d) and (e) present the corresponding composite of the heights for the positive and negative cases, respectively. The positive map displays a flow configuration typical of the central North Pacific blocking patterns described by White and Clark (1975). The principal features of the negative map are an intensified westerly flow across the central North Pacific, an amplified ridge along the west coast of North America and an enhanced trough in the East. It is interesting that the contrasting flows over the Pacific, which are often descrbed as the two extreme stages of the index cycle (Namias, 1950), appear in the present analyses as opposite phases of a single basic anomaly pattern. The particular flow configurations highlighted in Figs. 3.14(a) and 3.16(a) show many similarities with Figs. 4.4(e) and (d), respectively.

Concurrently with out study, Wallace and Gutzler (1981) have examined temporal correlation patterns between Northern Hemisphere gridpoints using 15 winters of monthly mean 500 mb NMC analyses. One of the outstanding correlation patterns that they describe (which they call the Pacific North America, or PNA, pattern) closely resembles our PAC pattern (see Fig. 3.4). As Wallace and Gutzler point out, this pattern resembles solutions obtained in simple linear models for forced stationary waves on a sphere (Hoskins et al., 1977); Grose and Hoskins, 1979; Hoskins and Karoly, 1981). Wallace and Gutzler also noted that this pattern appears in other studies on teleconnections (e.g., Martin 1953; O'Connor, 1969; Dickson and Namias, 1976).

It is interesting that although Wallace and Gutzler use selection and data analysis procedures substantially different from ours, the point that they choose for constructing teleconnection maps (45°N, 165°W) is remarkably close to our key point. Empirical orthogonal function (EOF) analysis of low-pass filtered height data (periods beyond 10 days) establishes that the points selected by Wallace and Gutzler and ourselves correspond to the main centres of the dominant regional patterns of low-frequency variability (Lorenz, 1956, and Davis, 1976, thoroughly discuss the theory and methods of EOF analysis). An EOF analysis of the 14 case mean patterns enables us to extract the single spatial pattern that best describes the patterns of both the positive and negative anomaly cases. The principal EOF pattern calculated from the case data alone (Fig. 4.5a) and the principal EOF pattern calculated from the 14 winter seasons of low-pass filtered data (Fig. 4.5b) are virtually identical. This suggests, and analysis of time behaviour (Dole, 1982) confirms, that the cases selected are predominantly enhancements of the primary regional pattern of low-frequency variability. Most (about 75%) of the low-frequency variance of

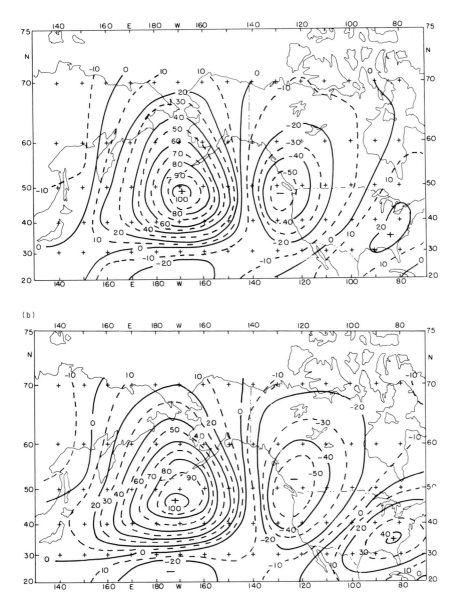

Fig. 4.5. (a) The first EOF of the PAC persistent anomaly cases; and (b) for the same region, the first EOF of the low-pass filtered anomaly data.

this pattern is contributed by within-season, rather than between-season (i.e., interannual) variability.

4.6 Time evolution

We now briefly consider the question of how persistent anomalies evolve in time, placing particular attention on identifying typical time scales and on isolating systematic propagation characteristics. We examine here only the development of PAC positive anomalies. We first study the 'slow' time evolution of the pattern by constructing composites from low-pass filtered data (retaining periods of 10 days or longer). We then examine corresponding composites constructed from unfiltered data around certain key times when the patterns rapidly evolve. Cases are selected by applying the usual procedure to low-pass filtered data. Composites are formed relative to the time when the anomaly first reaches the threshold value at the key point. This time is defined as day 0.

Figure 4.6 presents composite analyses of 15 PAC positive anomaly cases [selection criteria (100 m, 10 days)] at 2-day intervals from 4 days before onset (day −4) to 6 days after onset (day +6). There is little evidence in these analyses of a precursor until a few days before onset; indeed, at day −4, there are weak negative anomalies over the key region. By day −2 weak positive anomalies extend east–west over the entire mid-latitude North Pacific, with weak negative anomalies located to the north and south. This structure suggests that the associated anomalies in the wind field are primarily in the zonal component. At day 0 a single major positive centre becomes established over the key region. Starting at this time, anomaly centres form to the south of, and in sequence downstream from, the main centre. Intensification of the centres occurs with little evidence of phase propagation. By day +4 the PAC positive pattern is established.

The gross features of the development are strongly reminiscent of the behaviour seen in simple models of energy dispersion on a sphere away from a localized, transient (e.g., switch-on) source of vorticity (Hoskins *et al.*, 1977; Hoskins, 1978). The near simultaneity of development, the almost north–south orientation of the centres and the absence of tilts in the anomaly axes over the Pacific make interpretation of the development in this region more difficult; indeed, the pattern somewhat resembles a standing wave in the north–south direction.

Although data used in the above analysis were low-pass filtered, the main centre appeared to develop quite rapidly. We now examine similar analyses conducted on unfiltered data for further clues to the character of this rapid development. The starting dates are identical, so that the only difference from the previous analyses is the filtering procedure.

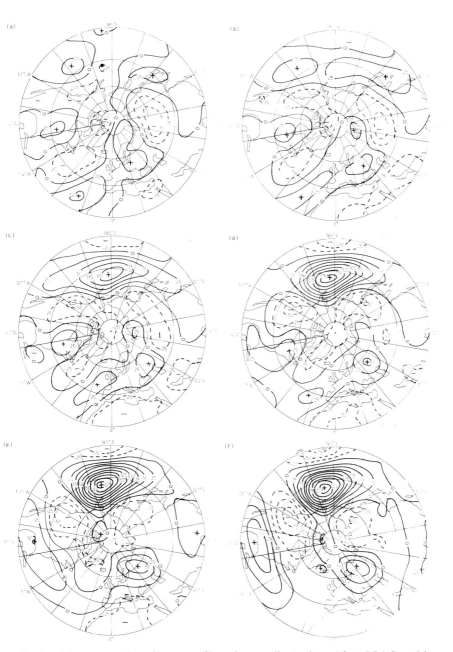

Fig. 4.6. Time composites of low-pass filtered anomalies (units : m) for 15 PAC positive cases at day (a) -4; (b) -2; (c) 0; (d) $+2$; (e) $+4$; and (f) $+6$.

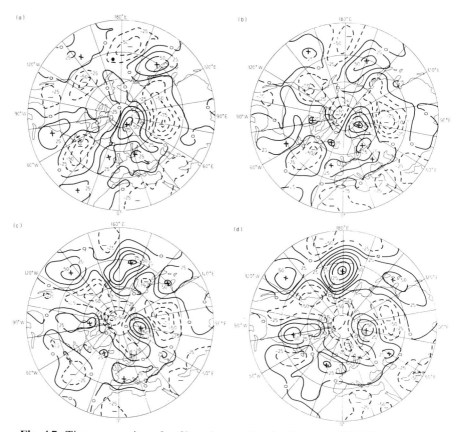

Fig. 4.7. Time composites of unfiltered anomalies for the same 15 PAC positive cases as in Fig. 4.6 at day (a) -3; (b) -2; (c) -1; and (d) 0.

Figure 4.7 displays the unfiltered composites at 1-day intervals from day -3 to day 0. The major differences in evolution are mainly associated with a positive anomaly centre located to the east of Japan at day -3. This centre propagates eastward and intensifies through the period, reaching the key region at day 0. In advance of this feature, a negative centre moves southeastward to the subtropical central North Pacific by day 0. This sequence of development suggests that the initial rapid growth of the main centre is primarily associated with the propagating, intensifying disturbance that originates in mid-latitudes near Japan. This disturbance slows up, but continues to intensify as it approaches the key region, thereafter appearing as a quasi-stationary negative vorticity centre over the mid-latitude North Pacific. The main centre over the subtropical North Pacific also apparently originates from a mid-latitude disturbance. Maps following day 0 (not shown) indicate that subsequent developments are qualitatively similar to those displayed in the low-pass analyses.

4.7 Discussion

It is interesting that some simple theoretical models do produce analogues to the persistent patterns we observe. As described in Section 6.2, the low-order barotropic channel model studied by Charney and DeVore (1979) exhibits two stable equilibrium states, one of which is characterized by an enhanced zonal flow and a relatively small wave amplitude and the second of which has a weak zonal flow and a highly amplified wave resembling blocking. The wave patterns of the two stable equilibrium states have nearly opposite phases. The high- and low-index equilibria correspond, respectively, to zonal flows above and slightly below the values for linear resonance of the forced wave. Charney and Strauss (1980) also obtain high- and low-index equilibria in a highly truncated, two-layer baroclinic model.

A behaviour suggestive of multiple equilibria would be the occurrence of multiple well-separated modes in suitably defined frequency distributions. To construct such a distribution, we have assumed that the first eigenvector in the EOF analyses discussed above adequately defines the spatial patterns of the hypothesized positive anomaly and negative anomaly quasi-equilibria. We then generated corresponding time series from the 14 winter seasons of twice-daily anomaly data. From these time series, histograms of the values of the first EOF time coefficient were constructed. Although the distributions for each of the three regions (not shown) are somewhat flatter than corresponding normal distributions, none shows a strongly bimodal character. This result holds for both unfiltered and filtered data. This suggests that, if the positive anomaly and negative anomaly patterns are associated with two quasi-equilibrium states, then either the means of the states are not well separated or the time spent between the two quasi-equilibrium states is not small compared to the time spent within the states. Alternatively, the forcing may vary sufficiently between seasons (or perhaps even within seasons) to alter the 'attractor set' and thus preclude simple identification in long-term statistics.

The persistent patterns that we have identified resemble solutions obtained in simple linear models of forced stationary waves on a sphere (Hoskins, 1978; Grose and Hoskins, 1979; Hoskins and Karoly, 1981). The temporal development of the patterns also qualitatively agrees with corresponding time-dependent models for energy dispersion away from a localized source of vorticity (Hoskins et al., 1977; Hoskins, 1978). As discussed in Section 3.5, Horel and Wallace (1981) have recently presented convincing observational evidence indicating some relationship between tropical Pacific sea-surface temperature anomalies and the sign of the North Pacific anomaly pattern.

Their work is supported by results obtained from general circulation models and simple theoretical studies on the atmosphere's response to changes in external forcing (particularly to tropical sea-surface temperature anomalies) (e.g., Rowntree, 1972, 1976; Opsteegh and Van Den Dool, 1980; Webster, 1981). The time scale for changes in such forcing, however, is presumably much longer than the time scales that we typically find for the development and decay of persistent anomalies. This suggests that these persistent patterns often, and perhaps primarily, grow and decay while the external forcing remains nearly fixed. Further support for the view that the patterns evolve mainly by internal processes comes from the modelling study presented in Chapter 6. In addition to providing internal mechanisms for variability, a theory of persistent anomalies must account for the strong regionality and co-location of the positive anomaly and negative anomaly frequency maxima. Whether these characteristics can be explained in present models by changing the structure of the basic state or the form or location of the forcing, or whether more sophisticated approaches are required, remains to be seen.

Finally, we note that the relatively small shift in time scales between non-persistent anomalies and persistent anomalies appears somewhat discouraging from the point of view of long-range forecasting. Nevertheless, we find some encouragement in the ability of the above theoretical models to qualitatively replicate important features of structure and development. We are also heartened that the observed features of persistent anomalies are often rather simple; indeed, in many respects far simpler than we might originally have anticipated.

Acknowledgements

I particularly acknowledge the influence of Professor Jule G. Charney, who served as advisor for most of this research. His profound insights and infectious enthusiasm, so characteristic of all his endeavours, were both inspirational and enlightening. The thesis research was supported through National Science Foundation grant 76-20070 ATM. Ms. Isabelle Kole drafted the figures.

References

CHARNEY, J. G. and DeVORE, J. G. (1979). Multiple flow equilibria in the atmosphere and blocking. *J. atmos. Sci.*, **36**, 1205–1216.
CHARNEY, J. G. and STRAUSS, D. M. (1980). Form-drag instability, multiple equilibria and propagating planetary waves in baroclinic, orographically forced, planetary wave systems. *J. atmos. Sci.*, **37**, 1157–1176.

DAVIS, R. E. (1976). Predictability of sea surface temperature and sea level pressure anomalies over the North Pacific Ocean. *J. phys. Oceanogr.*, **6**, 249–266.

DICKSON, R. R. and NAMIAS, J. (1976). North American influences on the circulation and climate of the North Atlantic sector. *Mon. Weath. Rev.*, **104**, 1255–1265.

DOLE, R. M. (1982). Persistent anomalies of the extratropical Northern Hemisphere wintertime circulation. Ph.D. thesis. Mass. Inst. of Technology, Cambridge, Mass.

ELLIOTT, R. D. and SMITH, T. B. (1949). A study of the effects of large blocking highs on the general circulation in the Northern Hemisphere westerlies. *J. Meteor.*, **6**, 67–85.

GROSE, W. L. and HOSKINS, B. J. (1979). On the influence of orography on large-scale atmospheric flow. *J. atmos. Sci.*, **36**, 223–234.

HOREL, J. D. and WALLACE, J. M. (1981). Planetary-scale atmospheric phenomena associated with the Southern Oscillation. *Mon. Weath. Rev.*, **109**, 813–829.

HOSKINS, B. J. (1978). Horizontal wave propagation on a sphere. *The General Circulation: Theory, modeling and observations*. N.C.A.R. Summer Colloquium, 1978. Boulder, Co.

HOSKINS, B. J. and KAROLY, D. (1981). The steady, linear response of a spherical atmosphere to thermal and orographic forcing. *J. atmos. Sci.*, **38**, 1179–1196.

HOSKINS, B. J., SIMMONS, A. J. and ANDREWS, D. G. (1977). Energy dispersion in a barotropic atmosphere. *Qt. Jl R. met. Soc.*, **103**, 553–567.

LORENZ, E. N. (1956). Empirical orthogonal functions and statistical weather prediction. *Sci. Rep. No. 1, M.I.T. Statistical Forecasting Project*, contract no. AF19 (604), 49 pp.

MARTIN, D. E. (1953). Anomalies in the Northern Hemisphere 5-day mean circulation patterns. Air Weather Service Rep. No. 105-100, 39 pp.

NAMIAS, J. (1947). Extended range forecasting by mean circulation methods. Washington, D.C. US Weather Bureau, 55 pp.

NAMIAS, J. (1950). The index cycle and its role in the general circulation. *J. Meteor.*, **7**, 130–139.

O'CONNOR, J. T. (1969). Hemispheric teleconnections of mean circulation anomalies at 700 mb. *ESSA Tech. Rep. WB-10*, 103 pp.

OPSTEEGH, J. D. and VAN DEN DOOL, H. M. (1980). Seasonal differences in the stationary response of a linearized primitive equation model: Prospects for long-range forecasting? *J. atmos. Sci.*, **37**, 2169–2185.

REX, D. P. (1950). Blocking action in the middle troposphere and its effect on regional climate. II. The climatology of blocking action. *Tellus*, **2**, 275–301.

REX, D. P. (1951). The effect of Atlantic blocking action upon European climate. *Tellus*, **3**, 1–16.

ROWNTREE, P. R. (1972). The influence of tropical east Pacific Ocean temperature on the atmosphere. *Qt. Jl R. met. Soc.*, **98**, 290–321.

ROWNTREE, P. R. (1976). Tropical forcing of atmospheric motions in a numerical model. *Qt. Jl R. met. Soc.*, **102**, 583–605.

SUMNER, E. J. (1954). A study of blocking in the Atlantic–European sector of the Northern Hemisphere. *Qt. Jl R. met. Soc.*, **80**, 402–416.

WALLACE, J. M. and GUTZLER, D. S. (1981). Teleconnections in the geopotential height field during the Northern Hemisphere winter. *Mon. Weath. Rev.*, **109**, 784–812.

WEBSTER, P. J. (1981). Mechanisms determining the atmospheric response to sea surface temperature anomalies. *J. atmos. Sci.*, **38**, 554–571.

WHITE, W. B. and CLARK, N. E. (1975). On the development of blocking ridge activity over the central North Pacific. *J. atmos. Sci.*, **32**, 489–502.

— 5 —

Mid-latitude wintertime circulation anomalies appearing in a 15-year GCM experiment

NGAR-CHEUNG LAU

5.1 Introduction

The past decade has seen a rapid expansion of our empirical knowledge of the observed circulation anomalies. Various diagnostic studies using historical weather records have collectively demonstrated that the dominant modes of variability are characterized by well-defined and spatially coherent structures. Kutzbach (1970), Kidson (1975) and Trenberth and Paolino (1981) showed that the most recurrent fluctuation in the sea-level pressure field of the mid-latitude Northern Hemisphere takes the form of north–south seesaws centred over the North Atlantic and North Pacific. Van Loon and Rogers (1978) provided additional evidence linking this pressure oscillation with east–west temperature seesaws across the North Atlantic. As described in Section 3.2, the results presented by Wallace and Gutzler (1981) revealed that the surface phenomena described above are associated with a teleconnection pattern in the 500 mb geopotential height field. The latter pattern is characterized by geographically fixed centres of action where preferred development of mid-tropospheric pressure ridges and troughs occurs. The spatial and temporal characteristics of blocking ridges are also documented in other recent studies, notably that described in Chapter 4.

Efforts aimed at identifying the causes of observed circulation anomalies generally fall into two broad categories:

(a) *Modelling of, and establishing empirical evidence for, the influences of perturbed external conditions*

Those involved in extended-range forecasting have long regarded changes in the surface properties of continents and oceans as important predictors of

111

atmospheric conditions. Many synoptic cases of apparently strong association between sea-surface temperature fluctuations and the atmospheric circulation have been documented in the voluminous works of Namias (1975). Horel and Wallace (1981) provided new observational evidence linking warm-water episodes in the equatorial Pacific with mid-latitude circulation anomalies. The numerical experiments performed by Webster (1972, 1981), Opsteegh and van den Dool (1980) and Hoskins and Karoly (1981) using linear mechanistic models, and by Rowntree (1972, 1976a, 1976b), Kutzbach et al. (1977) and Chervin et al. (1980) using general circulation models (GCMs), have also demonstrated the sensitivity of the atmospheric flow pattern to the location and structure of the anomalous forcing.

(b) Associating the persistent anomalies with internal dynamics

The pioneering work of Charney and DeVore (1979) indicated the existence of a multiplicity of equilibrium states in a nonlinear, highly truncated flow system with a fixed set of external parameters. Subsequent studies, notably those presented by Charney and Straus (1980) and Källén (1981), also identified multiple equilibrium phenomena in more complicated systems. These investigations suggest that the intermittent changes in circulation types could result from transitions between different equilibrium states.

A natural question arises when an assessment of these two contending hypotheses is being attempted: is the presence of fluctuations in external forcing an essential prerequisite for anomalous atmospheric behaviour? Some light may be shed on this issue by determining whether the observed anomalous phenomena mentioned above can be generated internally in an atmosphere subjected to fixed boundary forcing. Attempts to undertake such a study using observational data of the natural atmosphere would be complicated by the necessity to identify extended periods with largely similar boundary conditions. On the other hand, such conditions can easily be achieved in the controlled setting of a GCM. By analysing the history tapes of a 15-year GCM simulation with fixed boundary constraints, an effort is made in the present study to diagnose the nature of fluctuating phenomena occurring in an 'unperturbed' model environment. Particular emphasis is placed on comparing the prominent modes of oscillation in the model atmosphere with those observed in nature, and on examining the synoptic characteristics in the outstanding anomalous episodes.

Only a few highlights pertaining to the mid-latitude wintertime circulation in the Northern Hemisphere are included in this chapter. More detailed results of the full study have been presented by Lau (1981).

5.2 Data sets

The GCM that generated the history tapes for this study was developed at the Geophysical Fluid Dynamics Laboratory. The meteorological fields in the model are defined at nine sigma levels and their horizontal distributions are represented by spherical harmonics with rhomboidal truncation at wave-number 15. Topography and ocean–continent contrast are incorporated in the lower boundary of the global model domain, and the essential physical and hydrological processes are included. In the course of the model integration, the prescribed insolation and sea-surface temperature evolved through successive annual cycles that are identical to each other, so that no nonseasonal variation of the boundary forcing was introduced. The data analysed in this study pertain to the last 15 years of a $17\frac{3}{4}$-year simulation. Further details of the model structure and climatology have been described by Gordon and Stern (1974), Manabe and Hahn (1981) and Lau (1981).

The model results are to be compared with their observed counterparts, which are based on synoptic analyses prepared by the US National Meteorological Center for the 15 winters from 1962–63 to 1976–77.

5.3 Eigenvector patterns

In this study the most recurrent anomalies appearing in the model simulation are identified using empirical eigenvectors. A concise description of this analysis technique and the physical interpretation of results have been given by Kutzbach (1967). This objective technique enables us to represent a given fraction of the variance of a meteorological field by the least number of patterns (or 'eigenvectors') and time series (or 'coefficients'). Here we shall focus our attention on the first eigenvector, which explains the largest percentage of the total variance. It is worth noting that this leading eigenvector is free from the somewhat artificial orthogonality constraints to which the remaining eigenvectors are subjected. However, in attempting to maximize the fraction of explained variance, the first eigenvector in some cases tends to emphasize overly those features with spatial scales comparable with that of the domain under investigation. Hence certain interesting relationships of a more local nature may not manifest themselves clearly in the first eigenvector. Such regional features can be made more discernible by other analysis tools, such as one-point teleconnection maps (Wallace and Gutzler, 1981) and rotated eigenvectors (Horel, 1981). This issue does not seem to have a strong bearing on the validity of the results presented here, since it has been demonstrated in Lau (1981) that the patterns of the first eigenvectors resemble those of teleconnection maps for appropriately chosen reference points.

The leading eigenvectors presented here were based on monthly averaged data for December, January and February at selected grid points in the extratropical Northern Hemisphere (indicated by solid dots in Fig. 5.1a). The seasonal cycle has been removed and all data were normalized by the local standard deviations. The model eigenvectors were obtained individually for each atmospheric variable, and the percentages of explained variance are given at the upper right corner of the model patterns. The corresponding observed patterns were obtained using a combined eigenvector representation of the three variables examined here (see Lau, 1981, Section 3c and Appendix). The leading observed eigenvector based on this combined representation, shown in Figs. 5.1(b), 5.2(b) and 5.3(b), explains 13.7% of the total variance.

In Figs. 5.1–5.3 are shown the leading eigenvectors for 500 mb height, 1000 mb height and 850 mb temperature, based on (a) model data and (b) observational data. It is seen that the simulated and observed patterns exhibit a considerable degree of resemblance in the sector extending eastward from eastern Asia to the western Atlantic. Some noticeable discrepancies between model and observations are found over Europe and central Asia. Moreover, there is much less longitudinal variation in the model pattern for the eigenvector of 1000 mb height (Fig. 5.2a) than in the corresponding observed pattern (Fig. 5.2b). The following features in these figures are of particular interest:

1. The polarity of the centres of extremum over the central Pacific and Atlantic in the eigenvector pattern for 500 mb height [Fig. 5.1, hereafter referred to as $e_1(z_{500})$] is opposite to that over Greenland and the Arctic region north of Alaska and eastern Siberia. A belt of positive values extends southward from Alaska along the west coast of North America. One may infer from this configuration that amplified ridges over western North American and Greenland tend to be accompanied by amplified troughs over the central Pacific and Atlantic. Wallace and Gutzler (1981) have associated the observed eigenvector (Fig. 5.1b) with the Pacific/North American pattern, which is discussed in Section 3.2.

2. Over the maritime areas, the eigenvector patterns for 1000 mb height [$e_1(z_{1000})$, shown in Figs. 5.2(a) and 5.2(b)] display a strong spatial correspondence with the respective patterns for 500 mb height (Figs. 5.1a and 5.1b). The model pattern (Fig. 5.2a) is dominated by north–south seesaws, with the surface pressure tendency over the high latitude oceans being the opposite of that over the mid-latitude Pacific and Atlantic.

3. The eigenvector patterns for 850 mb temperature [$e_1(T_{850})$, shown in Figs. 5.3(a) and 5.3(b)] indicate that warm anomalies over western Greenland and regions in the vicinity of the Gulf of Alaska and the Bering Strait tend

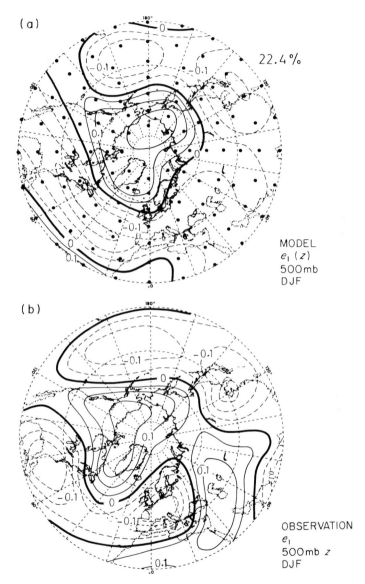

Fig. 5.1. Distribution of the first eigenvectors of normalized monthly mean 500 mb height for winter, based on (a) model data and (b) observational data. The dots in (a) indicate the grid points used in the eigenvector analysis. Lines of latitude and longitude are drawn every 20° with the outer latitudinal circle being 20°N.

116

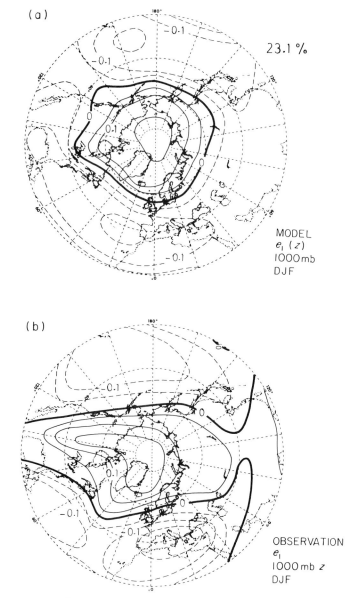

Fig. 5.2. As in Fig. 5.1, but for 1000 mb height.

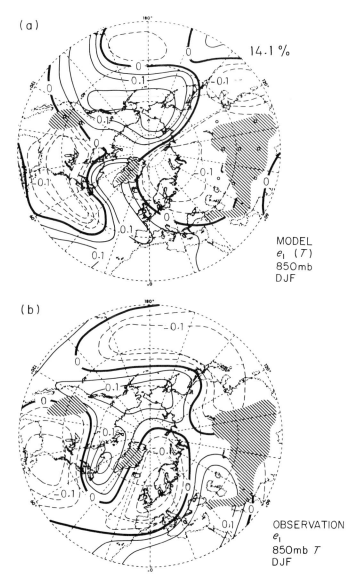

Fig. 5.3. As in Fig. 5.1, but for 850 mb temperature. The shaded areas indicate local topographic heights greater than 1500 m. The open circles in (a) indicate those grid points that are not used in the eigenvector analysis.

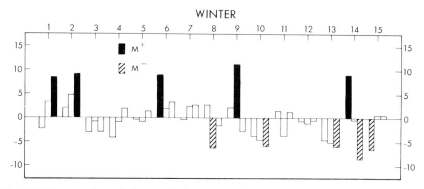

Fig. 5.4. Time series of the coefficients associated with the first eigenvector of wintertime 500 mb height, 850 mb temperature and 1000 mb height, as determined by a combined 3-variable representation of the monthly averaged model data. The three columns for each winter correspond (from left to right) to December, January and February. The solid (striped) columns in this time series indicate the five winter months with the highest positive (negative) coefficients.

to coexist with cold anomalies over northern Europe and the eastern seaboards of China and the United States, or *vice versa*.

5.4 Circulation features in anomalous months

The model eigenvectors $e_1(z_{500})$, $e_1(z_{1000})$ and $e_1(T_{850})$ discussed in the previous section are evidently related to each other. The similarity between $e_1(z_{500})$ and $e_1(z_{1000})$ suggests that the geopotential height fluctuations associated with this mode are essentially equivalent barotropic. Moreover, the sites of stronger than normal southerly and northerly flows, as inferred from $e_1(z_{500})$ and $e_1(z_{1000})$ using the geostrophic relationship, appear to be linked to the warm and cold anomalies in $e_1(T_{850})$, respectively. These impressions are further substantiated by the fact that the coefficients associated with $e_1(z_{500})$, $e_1(z_{1000})$ and $e_1(T_{850})$ exhibit strong temporal correlations with each other (see Lau, 1981, Table 2 and Fig. 12). The eigenvectors in Figs. 5.1 to 5.3 hence describe different aspects of the same anomaly complex having a coherent three-dimensional structure.

The temporal variation of this anomaly complex during the 15 simulated winters is described by the time series of coefficients based on a combined 3-variable eigenvector representation, shown in Fig. 5.4. The months with large positive coefficients (hereafter referred to as M^+, indicated by solid columns in Fig. 5.4) correspond to those periods when the anomaly patterns bear a strong resemblance to the model eigenvectors in Figs. 5.1 to 5.3. Conversely, the anomaly patterns in the months with large negative coefficients (hereafter referred to as M^-, indicated by striped columns in Fig. 5.4) exhibit a strong

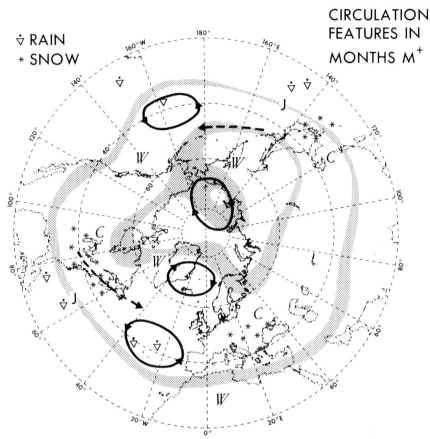

Fig. 5.5. Schematic diagram depicting the anomalous circulation features in months M^+. The shaded belts are bounded by selected isolines in the composite charts of 500 mb height based on months M^+. Locations of the jetstream cores are denoted by the letter J. The closed loops are inferred from the anomaly centres of the 1000 mb height field. The cold and warm anomalies are identified by the letters C and W, respectively. Locations of the 500 mb storm tracks are indicated by arrows with dashed shafts. Above normal snowfall and precipitation are indicated by * and $\dot{V}$, respectively.

negative spatial correlation with the eigenvectors.

The synoptic features associated with the leading eigenvectors may be delineated by compositing the model data separately for months M^+ and M^-. The anomaly patterns for months M^+ are largely similar to those for months M^-, except for a reversal in sign. The distinctive circulation features in months M^+ are depicted schematically in Fig. 5.5 and are summarized as follows:

1. The shape of the midtropospheric circumpolar vortex, as represented by the shaded belts, indicates a strongly perturbed wave pattern with

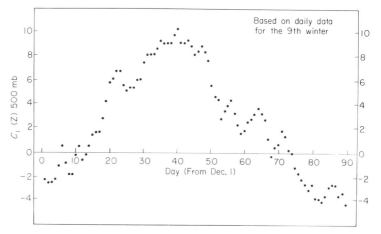

Fig. 5.6. Variation during the ninth model winter of the coefficients associated with the first eigenvector of 500 mb height, as determined using daily data.

intensified ridges over Greenland and the East Siberian Sea, and intensified troughs over eastern Asia, eastern North America and the central Pacific. The ridge over Greenland is paired with a trough further south, whereas the axis of the ridge over Alaska extends all the way down the west coast of North America. The jetstream cores (denoted by the letter J) are displaced south of their climatological positions.

2. The anomalous geostrophic flow at sea level, as depicted by the closed loops, bears a strong relationship to the upper-level circulation (shaded belts). The strengthening of anticyclonic and cyclonic circulations over Iceland and the mid-latitude Atlantic, respectively, correspond to one phase of the north–south sea-level pressure seesaw over this area. The sites of below normal sea-level pressure over the mid-latitude oceans are associated with enhanced precipitation (indicated by the symbol $\bar{V}$).

3. The passage of transient cyclones across the mid-latitude oceans is blocked by amplified quasi-stationary ridges over the eastern Atlantic and eastern Pacific. This confinement of transient activity to the western oceans results in noticeably truncated storm tracks, which are indicated by arrows with dashed shafts.

5.5 Synoptic description of an outstanding blocking episode

The time series of coefficients associated with the leading eigenvector (Fig. 5.4) indicates that the most outstanding anomalous episode occurred in January of the ninth simulated winter. In Fig. 5.6 is shown the day-to-day variation of the coefficients associated with $e_1(z_{500})$ for the entire length of the ninth winter (December 1 to February 28, hereafter referred to as Day 1 to Day 90). This time series suggests that the episode may be characterized by three distinct stages: the build-up stage (before Day 34), the quasi-steady stage (Days 34–48) and the break-down stage (beyond Day 48). In Fig. 5.7 is shown the instantaneous charts of the 500 mb height for Days 28–50 at 2-day intervals. The following synoptic events are highlighted by this sequence of maps:

1. Days 28–34. The North Atlantic ridge centred at 40°–60°W evolved into a flow pattern characterized by a closed high-pressure centre lying north of a closed low. The high-pressure ridge over the west coast of North America underwent considerable amplification. The circulation in the Pacific sector is dominated by the presence of two pressure troughs, with axes centred at 130°E and 150°W.

2. Days 36–48. The most prominent features during this period are the persistent high-pressure ridges over western North America and Europe. Both of these features occasionally extended far into the Arctic region. Such events were often followed by the appearance of retrograding high pressure centres in the high latitudes. A succession of cyclone waves made their passage from eastern North America to the North Atlantic.

3. Days 48–50. The intrusion of a deep cyclone trough into the Gulf of Alaska signalled the termination of the blocking action over western North America. A more zonal circulation over western North America and Europe was re-established by Day 52 and Day 56 (not shown), respectively.

5.6 Discussion

The pattern of the leading eigenvector for 500 mb height exhibits a strong sensitivity to the season and hemispheric domain under consideration. The dominant modes of variability for Northern Hemisphere summer, as well as for Southern Hemisphere winter and summer, are characterized by a much more zonally symmetric pattern than that shown in Fig. 5.1(a). There exists no evidence in the model data on the east–west sea-level pressure seesaw associated with the observed Southern Oscillation over the equatorial Pacific.

The essential features of the model eigenvectors discussed in Section 5.3 are just as evident in the analysis of monthly averaged, 5-day averaged and daily data. This apparent lack of frequency dependence is checked by spectral and lagged autocorrelation analyses of the coefficients associated with these

122

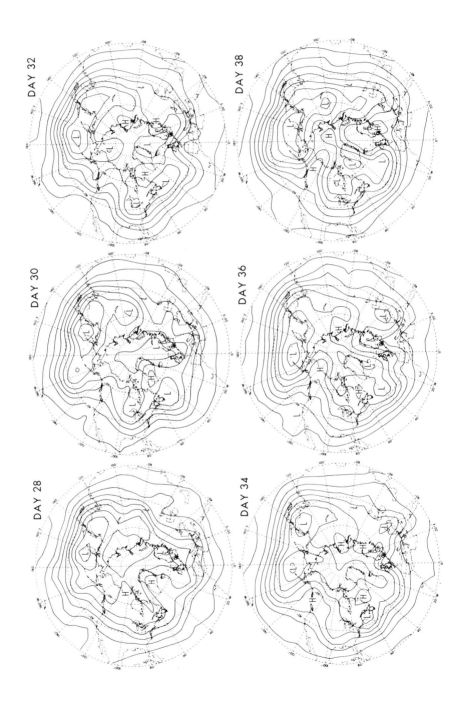

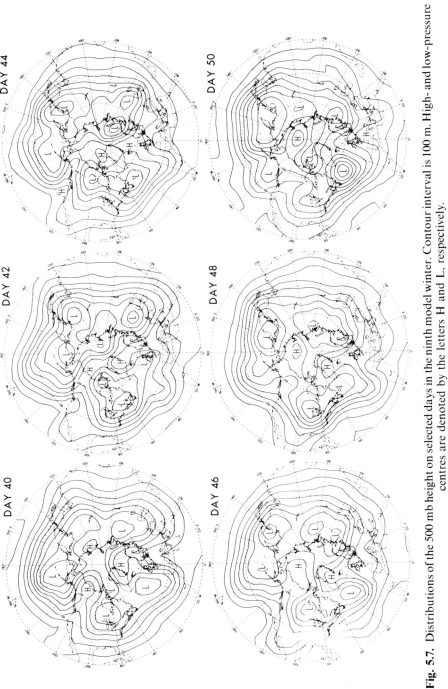

Fig. 5.7. Distributions of the 500 mb height on selected days in the ninth model winter. Contour interval is 100 m. High- and low-pressure centres are denoted by the letters H and L, respectively.

eigenvectors. The autocorrelation method indicates that the characteristic time scale between independent samples is approximately 15 days. The results of the present study demonstrate that the internal dynamics and physics of the model atmosphere are by themselves able to generate quite realistic circulation anomalies. Whereas these results suggest that certain mid-latitude anomalies do not derive their existence solely from nonseasonal perturbations in the boundary forcing, one can certainly not dismiss the potential role of external influences such as sea surface temperature changes in modifying the character of these anomalies at various stages of their life history. An assessment of the impact of fluctuating external conditions on atmospheric variability will have to await the diagnostics of other model simulations which allow for full interactions between the atmosphere and its environment.

Acknowledgements

I wish to thank Professors John M. Wallace and Brian J. Hoskins for their comments on the manuscript, and the staff members of GFDL for their encouragement to undertake this study. My appointment at the Geophysical Fluid Dynamics Program, Princeton University, is supported by NOAA grant 04-7-022-44017.

References

CHARNEY, J. G. and DEVORE, J. G. (1979). Multiple flow equilibria in the atmosphere and blocking. *J. atmos. Sci.*, **36**, 1205–1216.

CHARNEY, J. G. and STRAUS, D. M. (1980). Form-drag instability, multiple equilibria and propagating planetary waves in baroclinic, orographically forced, planetary wave systems. *J. atmos. Sci.*, **37**, 1157–1176.

CHERVIN, R. M., KUTZBACH, J. E., HOUGHTON, D. D. and GALLIMORE, R. G. (1980). Response of the NCAR general circulation model to prescribed changes in ocean surface temperature. Part II: midlatitude and subtropical changes. *J. atmos. Sci.*, **37**, 308–332.

GORDON, C. T. and STERN, W. F. (1982). A description of the GFDL global spectral model. *Mon. Weath. Rev.*, **110**, 625–644.

HOREL, J. D. (1981). A rotated principal component analysis of the interannual variability of the Northern Hemisphere 500 mb height field. *Mon. Weath. Rev.*, **109**, 2080–2092.

HOREL, J. D. and WALLACE, J. M. (1981). Planetary-scale atmospheric phenomena associated with the southern oscillation. *Mon. Weath. Rev.*, **109**, 813–829.

HOSKINS, B. J. and KAROLY, D. J. (1981). The steady linear response of a spherical atmosphere to thermal and orographic forcing. *J. atmos. Sci.*, **38**, 1179–1196.

KÄLLÉN, E. (1981). The nonlinear effects of orographic and momentum forcing in a low-order, barotropic model. *J. atmos. Sci.*, **38**, 2150–2163.

KIDSON, J. W. (1975). Eigenvector analysis of monthly mean surface data. *Mon. Weath. Rev.*, **103**, 177–186.

KUTZBACH, J. E. (1967). Empirical eigenvectors of sea-level pressure, surface temperature and precipitation complexes over North America. *J. appl. Meteor.*, **6**, 791–802.

KUTZBACH, J. E. (1970). Large-scale features of monthly mean Northern Hemisphere anomaly maps of sea-level pressure. *Mon. Weath. Rev.*, **98**, 708–716.

KUTZBACH, J. E., CHERVIN, R. M. and HOUGHTON, D. D. (1977). Response of the NCAR general circulation model to prescribed changes in ocean surface temperature. Part I: Mid-latitude changes. *J. atmos. Sci.*, **34**, 1200–1213.

LAU, N.-C. (1981). A diagnostic study of recurrent meteorological anomalies appearing in a 15-year simulation with a GFDL general circulation model. *Mon. Weath. Rev.*, **109**, 2287–2311.

MANABE, S. and HAHN, D. G. (1981). Simulation of atmospheric variability. *Mon. Weath. Rev.*, **109**, 2260–2286.

NAMIAS, J. (1975). *Short period climatic variations. Collected Works of J. Namias.* University of California, San Diego, 905 pp.

OPSTEEGH, J. D. and VAN DEN DOOL, H. M. (1980). Seasonal differences in the stationary response of a linearised primitive equation model: Prospects for long-range weather forecasting? *J. atmos. Sci.*, **37**, 2169–2185.

ROWNTREE, P. R. (1972). The influence of tropical East Pacific Ocean temperatures on the atmosphere. *Qt. Jl R. met. Soc.*, **98**, 290–321.

ROWNTREE, P. R. (1976a). Tropical forcing of atmospheric motions in a numerical model. *Qt. Jl R. met. Soc.*, **102**, 583–606.

ROWNTREE, P. R. (1976b). Response of the atmosphere to a tropical Atlantic Ocean temperature anomaly. *Qt. Jl R. met. Soc.*, **102**, 607–626.

TRENBERTH, K. E. and PAOLINO, Jr., D. A. (1981). Characteristic patterns of variability of sea level pressure in the Northern Hemisphere. *Mon. Weath. Rev.*, **109**, 1169–1189.

VAN LOON, H. and ROGERS, J. C. (1978). The seesaw in winter temperatures between Greenland and Northern Europe. Part I: General description. *Mon. Weath. Rev.*, **106**, 296–310.

WALLACE, J. M. and GUTZLER, D. S. (1981). Teleconnections in the geopotential height field during the Northern Hemisphere winter. *Mon. Weath. Rev.*, **109**, 784–812.

WEBSTER, P. J. (1972). Response of the tropical atmosphere to local, steady forcing. *Mon. Weath. Rev.*, **100**, 518–541.

WEBSTER, P. J. (1981). Mechanisms determining the atmospheric response to sea surface temperature anomalies. *J. atmos. Sci.*, **38**, 554–571.

— 6 —

Stationary and quasi-stationary eddies in the extratropical troposphere: theory

ISAAC M. HELD

6.1. Introduction

The climatological zonal asymmetries in the atmosphere result from asymmetries in the Earth's surface. The problem of explaining the atmospheric asymmetries is complicated by the fact that some of the relevant surface properties, such as ocean temperature, land-surface albedo and ground wetness, are at least partially conteolled by the atmosphere. Yet, even if one ignores these important feedback processes and simply prescribed boundary conditions, the construction of a theory explaining the atmospheric asymmetries remains a difficult challenge. This is particularly so for the asymmetries within the troposphere, where latent heat release and the mixing of momentum and heat by large-scale transients are both dependent on the boundary conditions and the forced wave structure in ways that are little understood.

There exist several recent reviews of problems associated with stationary eddies on various scales, most notably Smith (1979) and Dickinson (1980) on the effects of orography. The intention in this survey is not to present a systematic review of theory but simply to touch upon some of the theoretical issues that have arisen time and again since the pioneering work of Charney and Eliassen (1949), Bolin (1950) and Smagorinsky (1953) on planetary-scale stationary waves. Among these issues are the following:

1. What is the relative importance of topographic forcing and zonally asymmetric diabatic heating in determining the planetary stationary wave structure?

2. To what extent can the observed low-frequency variability in the atmosphere be associated with anomalous forced waves?

3. Why do barotropic models of topographically forced waves work as well as they do?

4. Are forced stationary waves potentially resonant?

Two additional questions concerning Rossby wave propagation arise naturally in discussions of 3 and 4:

5. Are stationary Rossby waves absorbed or reflected when they encounter low latitude easterlies?

6. In what sense, if any, does the tropopause act as if it were a rigid lid?

Without at least partial answers to basic questions such as these, one is not in a very good position to address the problems mentioned in the opening paragraph, concerning the various and subtle ways in which the stationary eddies interact with latent heat release, large-scale transients and the underlying surface.

We begin by describing the response to orography in three linear models: a barotropic model on a beta-plane channel; a barotropic model on a sphere; and a baroclinic, quasi-geostrophic model on a beta-plane channel. These models illustrate the zonal, meridional and vertical propagation of Rossby waves and raise questions 3–6 in relatively simple settings. The emphasis in these sections is on orographic forcing, not because thermal forcing is unimportant (it is not), but partly because the actual diabatic heating distribution in the troposphere is poorly known, and partly because the problem of thermal forcing is complicated by the dependence of the response on the vertical as well as horizontal structure of the forcing. The linear response to diabatic heating is then briefly discussed in Section 6.5.

Following this review of linear forced wave theory, results from atmospheric general circulation models (GCMs) of relevance for questions 1 and 2 are discussed. A comparison of GCMs with realistic orography and with a flat lower boundary is used to address 1, whereas statistics from a GCM with zonally symmetric lower boundary conditions and, therefore, a zonally symmetric climate are used to attempt to shed some light on 2.

The discussion throughout is restricted somewhat arbitrarily to the problem of explaining the stationary and quasi-stationary eddies in the extratropical Northern Hemisphere troposphere during winter. However, many of the ideas are relevant to the summer season and to the Southern Hemisphere.

6.2 Barotropic model in a beta-plane channel:
zonal propagation of Rossby waves

The classic paper by Charney and Eliassen (1949) contains the first attempt at quantitative modelling of planetary-scale stationary waves. Using a barotropic model on a beta-plane, linearized about a uniform westerly zonal wind flowing over surface topography, these authors obtained a striking fit to the wintertime 500 mb stationary geopotential field along 45°N. It is worthwhile discussing this calculation in detail both because of its historical importance and because it raises the issue of the potential for resonant behaviour of stationary waves in a very simple context.

Postponing any attempt at justification, consider the equation for conservation of potential vorticity for the hydrostatic, depth-independent flow of a homogeneous layer of fluid with a free surface:

$$\left(\frac{\partial}{\partial t} + \mathbf{v}\cdot\nabla\right)\left(\frac{f+\zeta}{H}\right) = 0. \tag{6.1}$$

$H = h - h_T$, where H is the layer thickness, h the height of the free surface and h_T the height of the rigid lower boundary. The notation is otherwise conventional. In the quasi-geostrophic approximation on a beta-plane, (6.1) reduces to:

$$\left(\frac{\partial}{\partial t} + \mathbf{v}_g\cdot\nabla\right)q = 0, \tag{6.2}$$

where $q = f_0 + \beta y + \zeta - f_0(\eta - h_T)/h_0$, h_0 is the mean depth, $\eta = h - h_0$, $\mathbf{v}_g = g/f_0\mathbf{k}\times\nabla\eta$ and $\zeta = (g/f_0)\nabla^2\eta$ (compare with the general formulation in the Appendix—Eqns. A10 and A11). Linearizing about a zonal flow $[u] = -(g/f_0)\,\partial[\eta]/\partial y$ independent of x, y and t and retaining only the lowest-order term in the topographic forcing, one obtains for the perturbation (denoted by a *)

$$\left(\frac{\partial}{\partial t} + [u]\frac{\partial}{\partial x}\right)q^* + v^*\frac{\partial[q]}{\partial y} = 0, \tag{6.3a}$$

where $q^* = \zeta^* - f_0(\eta^* - h_T)/h_0$ and $\partial[q]/\partial y = \beta + [u]/\lambda^2$ with $\lambda^2 = gh_0/f_0^2$. Substituting for q^* and $\partial[q]/\partial y$ an alternative form of this equation is:

$$\frac{\partial}{\partial t}\left(\zeta^* - \frac{f_0}{h_0}\eta^*\right) + [u]\frac{\partial}{\partial x}\zeta^* + \beta v^* = -[u]\frac{f_0}{h_0}\frac{\partial h_T}{\partial x}. \tag{6.3b}$$

From (6.3a) and (6.3b), the dispersion relation for free plane waves, $\eta^* = \text{Re }\tilde{\eta} \exp i(kx + ly - \omega t)$, is

$$\omega = k\left\{[u] - \frac{\partial[q]}{\partial y}\bigg/(k^2 + l^2 + \lambda^{-2})\right\} \qquad (6.4a)$$

$$= k\{[u](k^2 + l^2) - \beta\}/(k^2 + l^2 + \lambda^{-2}). \qquad (6.4b)$$

(Note that ω is not equal to the frequency of a Rossby wave in the absence of mean flow, Doppler shifted by $k\bar{u}$. As is clear from (6.4a), a 'non-Doppler' effect results from the mean flow being balanced by a height gradient that alters $\partial[q]/\partial y$. As discussed by White (1977), precisely the same effect occurs in a continuously stratified flow, where it enters through the lower boundary condition and has caused some confusion in the literature.) If $[u] < 0$ the system cannot support a stationary Rossby wave, whereas if $[u] > 0$ the total wavenumber of the stationary wave is $k^2 + l^2 \equiv K^2 = \beta/[u] \equiv K_s^2$. This stationary wavenumber is independent of λ. If one is only interested in the stationary flow, and if the flow is quasi-geostrophic, there is no advantage in the use of the divergent rather than non-divergent vorticity equation.

Consider the steady version of Eqn. (6.3) in a channel geometry, assuming that η^* and h_T are both proportional to $\sin(ly)$. Fourier analysing in x and denoting Fourier amplitudes by tildes, one finds that:

$$\tilde{\eta} = \tilde{h}_T/[\lambda^2(K^2 - K_s^2)]. \qquad (6.5)$$

For $K > K_s$ ($K < K_s$), the response is exactly in (out of) phase with the topography, with the topographic source of vorticity balanced primarily by the zonal (meridional) advection of relative (planetary) vorticity. Following Charney and Eliassen, the resonance singularity is removed by adding Ekman pumping, which amounts to adding a linear damping on the vorticity in this barotropic model. For steady flow, Eqn. (6.3b) is modified to:

$$[u]\frac{\partial\zeta^*}{\partial x} + \beta v^* = -r\zeta^* - \frac{[u]f_0}{h_0}\frac{\partial h_T}{\partial x}, \qquad (6.6)$$

resulting in the modified Fourier amplitudes:

$$\tilde{\eta} = \tilde{h}_T/[\lambda^2(K^2 - K_s^2 - i\varepsilon)] \qquad (6.7)$$

with $\varepsilon \equiv rK^2/k[u]$.

Charney and Eliassen justify this barotropic model by starting with the quasi-geostrophic vorticity equation and assuming that the flow is separable, $\eta = A(x, y)B(z)$; the choice of vertical structure $B(z)$ is then based on observations. As described in the Appendix, $[u]$ on the lhs of Eqn. (6.6) is reinterpreted as a mid-tropospheric wind, whereas $[u]$ on the rhs is reinterpreted as a surface wind—and then replaced by $0.4[u]$, so that the implied vertical velocities are of reasonable magnitude. A more satisfying procedure for justifying barotropic calculations is discussed in Section 6.4.

The variance of the solution to (6.6), $[\eta^{*2}]$, is plotted in Fig. 6.1 as a function

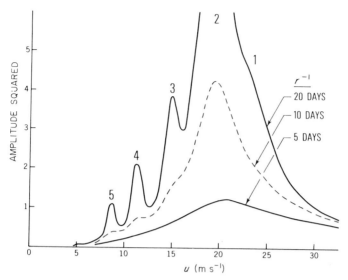

Fig. 6.1. The mean square height response, $[\eta^{*2}]$, in the Charney–Eliassen model as a function of $[u]$ for different strengths of dissipation, in units of 10^4 m^2. The integers mark the values of $[u]$ at which particular zonal wavenumbers resonate.

of $[u]$ for different values of r. The x-dependence of h_T used is obtained from the smoothed topography at 45°N in the GCM discussed in Section 6.6, and is plotted in Fig. 6.2. The meridional wavenumber is chosen so that one half-wavelength equals 35° latitude, while $f_0 = 2\Omega \sin (45°)$ and $h_0 = 8$km. The results show a clear resonance structure when $r^{-1} = 20$ days, with the wavenumber 2 resonance dominant but with 3, 4 and 5 also apparent. Little of this structure remains at $r^{-1} = 5$ days.

Figure 6.2 compares the solution $\eta^*(x)$ for $r^{-1} = 5$ days and $[u] = 17$ m s^{-1} with the climatological 500 mb height at 45° for January. (As in Charney and Eliassen, $[u]$ on the rhs of Eqn. (6.6) has been multiplied by 0.4.) Since thermal forcing contributes a substantial part of this mid-tropospheric eddy field (see Section 6.6), one should consider the excellence of this fit as accidental. It is still of interest, however, that responses of the correct order of magnitude and structure are obtained for values of r sufficiently large that the system does not resonate.

To explore this point further, the total response has been split into two parts in Fig. 6.3, that due to the topography of the Eastern Hemisphere and that due to the topography of the Western Hemisphere. The total response is clearly the sum of two well-defined wavetrains, one emanating from the Rockies and the other, somewhat stronger, from the Tibetan Plateau. Each of the wavetrains decays eastward sufficiently rapidly that there is only modest interference between the two. For this large a dissipation, it is evidently more fruitful to

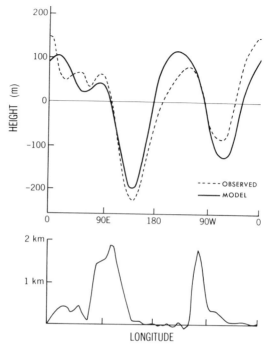

Fig. 6.2. Upper figure: the height response as a function of longitude in the Charney–Eliassen model for the parameters listed in the text (solid line), and the observed climatological 500 mb eddy heights at 45°N in January, from Oort (1982) (dashed line). Lower figure: the topography $h_T(x)$ used in the calculation.

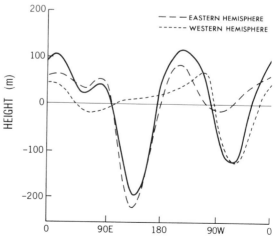

Fig. 6.3. The response shown in Fig. 6.2 split into the parts due to the topography of the western and eastern hemispheres.

think in terms of wavetrains emanating from localized features, that is, Green's functions, rather than Fourier components.

The solution to Eqn. (6.6) on a channel of infinite length can be written, for sufficiently small r, in the form:

$$\eta^*(x, y) = \sin (ly) \int_{-\infty}^{\infty} G(x - x')h_T(x') \, dx', \qquad (6.8)$$

where:

$$G(x - x') = 0 \quad \text{if } x \leqslant x',$$

$$= -\frac{1}{k_s \lambda^2} \sin (k_s(x - x')) \exp \left[-r(x - x')/c_x \right]$$

$$\text{if } x \geqslant x'.$$

$k_s = (K_s^2 - l^2)^{1/2}$ is assumed real, and c_x is the zonal group velocity for non-divergent Rossby waves evaluated at the stationary wavenumber:

$$c_x = \partial \omega / \partial k = 2[u]k_s^2/(k_s^2 + l^2). \qquad (6.9)$$

c_x is always positive and less than $2[u]$. It is because c_x is positive that the response is confined to the east of the source. As long as the width of the mountain is much less than the stationary wavelength, and as long as the wave is damped out before progressing around the latitude circle and returning to the source region, the response will resemble this Green's function. There will be anticyclonic vorticity over the mountain and a trough immediately downstream, with the stationary wavenumber clearly apparent in the response.

In this simple model, the parameter r determines whether or not the picture of the wave field as composed of potentially resonant Fourier components or of damped wavetrains is the most appropriate. The choice $r^{-1} = 5$ days, for which the latter picture seems more appropriate, is rather small for a spin-down time due to surface friction, but there are other potential sources of dissipation in the troposphere, most notably mixing due to large-scale transients. More importantly, waves can also disperse into low latitudes and into the stratosphere, from which they may be unable to return to the region of forcing. We return to this problem in the following sections.

The character of the long-period variability of the forced waves should be markedly different depending on whether or not the waves are potentially resonant. If they are potentially resonant, one expects these low-frequency anomalies to be of global scale. One also expects dramatic variations in amplitude and phase, with $180°$ phase reversals in individual zonal wavenumbers as the flow passes through resonance. In such circumstances, zonal Fourier analysis is clearly illuminating. In contrast, in the non-resonant dissipative or dispersive case one expects anomalies to be local, unless the

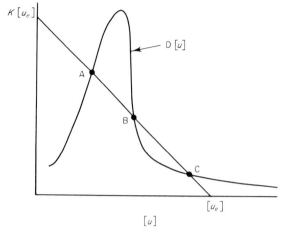

Fig. 6.4. A schematic of the graphic solution for the equilibrium states in the Charney–DeVore model.

mean flow variations happen to be coherent around a latitude circle, and one does not expect Fourier decomposition to provide the simplest picture of the wave field. A simple shift in the wavelength downstream of the Rockies, say, could be disguised as a complex redistribution among several Fourier components.

Resonant behaviour can have more subtle consequences for variability in the atmosphere, as is evident in the intriguing recent work of Charney and Devore (1979) (see also Charney *et al.*, 1981) on the possible importance of wave-mean flow interaction for the maintenance of anomalous forced wave patterns. In the presence of dissipation the steady forced waves generated by topography exert a drag on the zonal flow, $\mathscr{D}([u])$. Following Charney and Devore, suppose that in the absence of this drag the mean flow relaxes to some $[u_e] > 0$, i.e.:

$$\frac{\partial [u]}{\partial t} = -\kappa([u] - [u_e]) - \mathscr{D}([u]). \tag{6.10}$$

$\mathscr{D}([u])$ is the sum of the eddy momentum flux divergence and the form drag, or 'mountain torque',

$$\mathscr{D}([u]) = \frac{\partial}{\partial y}[u^*v^*] + \frac{1}{h_0}[p^* \, \partial h_T/\partial x]$$

$$= -[v^*\zeta^*] - \frac{f_0}{h_0}[v^*h_T]$$

$$= -[v^*q^*]$$

where p^* is the perturbation pressure and q^* is the perturbation potential vorticity. If the wave consists of only one meridional mode, sin (ly), one can show that:

$$\mathscr{D}([u]) = \frac{rf_0^2 K^2 |\tilde{h}_T|^2 \sin^2 (ly)}{2[u]h_0^2((K^2 - K_s^2)^2 + \varepsilon^2)}. \qquad (6.11)$$

Thus, $\mathscr{D}([u])$ has basically the same shape as $[\eta^{*2}]$ in Fig. 6.1.

A graphical solution to the steady-state equation, $\mathscr{D}([u]) = \kappa([u_e] - [u])$ is shown in Fig. 6.4. (For the purposes of this qualitative argument, we ignore the y-variation in $\mathscr{D}([u])$.) If the parameters are such as to produce the graph shown, three possible equilibria exist, labelled A, B and C in the figure. It is intuitively clear from Eqn. (6.10) that if $\kappa + \mathrm{d}\mathscr{D}/\mathrm{d}[u]$ is negative, as it is for B, the state is unstable. The high-index, small-wave amplitude state C is associated by these authors with the 'normal' flow, and the low-index, amplified-wave state A with 'blocking'. Without the resonant structure in $\mathscr{D}([u])$, multiple equilibrium states of this sort do not occur.

6.3 Barotropic model on a sphere: meridional propagation of Rossby waves

6.3.1 Numerical results

These results for barotropic flow in a beta-plane channel are substantially modified when account is taken of the spherical geometry and mean flow variation with latitude, as recently described by Grose and Hoskins (1979). Although these authors study a divergent shallow water model, the free surface has no effect on quasi-geostrophic stationary waves, as described in Section 6.2, and we therefore restrict the following discussion to the spherical version of Eqn. (6.6):

$$\frac{[u]}{a \cos \phi} \frac{\partial \zeta}{\partial \lambda} + v^* \left(\frac{1}{a} \frac{\partial [\zeta]}{\partial \phi} + \beta \right) = \frac{-[u_s]f}{h_0 a \cos \phi} \frac{\partial h_T}{\partial \lambda} - r\zeta^* \qquad (6.12)$$

with

$$(u, v) \equiv \left(-\frac{1}{a} \frac{\partial \psi}{\partial \phi}, \frac{1}{a \cos \phi} \frac{\partial \psi}{\partial \lambda} \right).$$

The coefficient r is taken to be independent of latitude.

We describe responses to this equation with h_T equal to the Northern Hemisphere topography used in the GCM described in Section 6.6 and with $h_T = 0$ in the Southern Hemisphere. The topography is shown in Fig. 6.5. $[u]$ and $[u_s]$ are set equal to the zonally averaged 300 mb and surface winds produced by the 'no-mountain' model in Section 6.6, and are shown in Fig. 6.6. The choice of 300 mb winds is guided by the experience of Grose and Hoskins, who

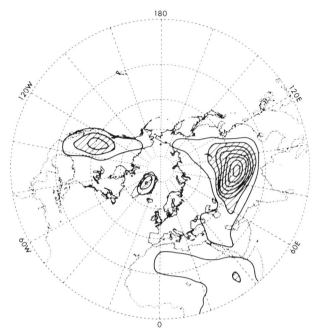

Fig. 6.5. The topography used in the spherical barotropic calculations, taken from the low resolution spectral model described in Section 6.6. The contour interval is 500 m and heights over 1 km are shaded. The zero contour is not shown. The concentric dotted circles are 20° latitude apart, the outer circle being the equator.

find that this choice produces realistic wave patterns. The model's $[u]$ is everywhere positive, so problems associated with 'critical latitudes' (latitudes at which $[u] = 0$) do not arise. The streamfunctions obtained by solving Eqn. (6.12) for two values of r^{-1}, 5 days and 20 days, are shown in Fig. 6.7. The corresponding height fields, computed from the approximation $\eta = f\psi/g$, are shown in Fig. 6.8. Several aspects of these responses are worth noting:

1. The height field response is everywhere less than 60 m. The amplitude at 45°N is more than a factor of three smaller than that obtained with the Charney–Eliassen model, using the identical topography along 45°. Roughly half of this difference in amplitude can be explained by setting h_T in the spherical model equal to $h_T(45°) \sin (ly)$ in the latitude band corresponding to the domain of the Charney–Eliassen model, with $h_T \equiv 0$ outside of this band and with l and $[u_s]$ as in that model. The rest of the difference must be due to the two-dimensional rather than the one-dimensional dispersion of Rossby waves away from localized sources. This particular barotropic model on a sphere, at least, seriously underestimates the amplitude of the orographically forced waves in the mid-troposphere (see Section 6.6).

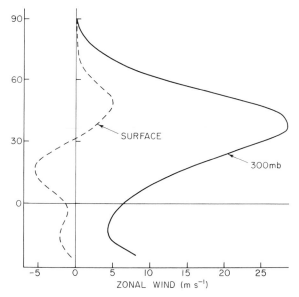

Fig. 6.6. The upper tropospheric and surface winds in the Northern Hemisphere used for $[u]$ and $[u]_s$ in Eqn. (6.12), taken from the 'no-mountain' model described in Section 6.6.

STREAM FUNCTION

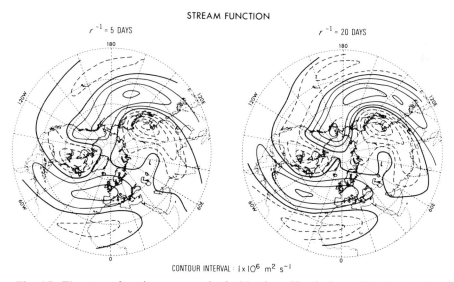

Fig. 6.7. The streamfunction response in the Northern Hemisphere of the barotropic model, using the topography and mean winds in Figs. 6.5 and 6.6, for two different levels of dissipation. Solid contours are positive or zero; dotted contours are negative.

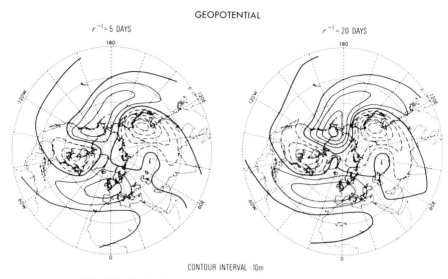

GEOPOTENTIAL

$r^{-1}=5$ DAYS $r^{-1}=20$ DAYS

CONTOUR INTERVAL : 10m

Fig. 6.8. As in Fig. 6.7, but for the height response, $f\psi/g$.

However, comparison with Fig. 2.2 shows the very realistic nature of the pattern obtained.

2. The amplitudes of the dominant troughs downstream of the Tibetan Plateau and the Rockies are nearly identical for the two values of r shown. The perturbations further downstream are naturally more sensitive to the level of dissipation, as is evident in the streamfunction maps in which one can follow the equatorward propagating wavetrains into the tropics. This contrasts sharply with the channel model, in which the amplitude of the entire response is sensitive to the value of r in this range, as is clear from Fig. 6.1.

3. As pointed out by Grose and Hoskins, the geopotential response gives the misleading impression of predominantly zonal propagation because of the factor $f \propto \sin \phi$ multiplying the streamfunction. In contrast, the stream-function responses seem to be dominated by two wavetrains propagating southeastwards, one produced by the Rockies and the other by the Tibetan plateau. Besides these two dominant groups, there are other highs and lows that are with more difficulty associated into wavetrains. The separate responses to the topography of the Eastern and Western Hemispheres are plotted in Fig. 6.9 using $r^{-1} = 20$ days. The interference between these two parts of the response is modest: the Tibetan plateau response decreases the strength of the low in eastern North America by 15%, whereas the response to the Rockies has almost no effect on the low on the east Asian coast. (This result should be sensitive to the structure of the zonal flow, however.) In a

EASTERN   WESTERN

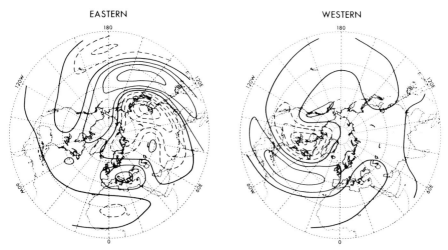

Fig. 6.9. The streamfunction response of Fig. 6.7 with $r^{-1} = 20$ days, split into the responses to the topography of the Eastern and Western Hemispheres.

channel model with the same value of r, interference between the Rocky and Tibetan responses is much more substantial, due to the meridional confinement of the waves.

6.3.2 Meridional dispersion

As noted by Hoskins and Karoly (1981), Rossby wave ray tracing theory explains these structures rather well. For this purpose it is advantageous to transform the linearized unforced inviscid vorticity equation into Mercator coordinates:

$$x = a\lambda; \quad \frac{dy}{a} = \frac{d\phi}{\cos \phi},$$

so that:

$$\hat{u} \frac{\partial}{\partial x}\left(\frac{\partial^2}{\partial x^2} + \frac{\partial^2}{\partial y^2}\right)\psi^* = -\hat{\beta} \frac{\partial \psi^*}{\partial x} \tag{6.13}$$

where $\hat{u} \equiv [u]/\cos \phi$ and $\hat{\beta} \equiv \cos \phi(\beta + 1/a \, \partial[\zeta]/\partial\phi)$.

For a wave of the form $\psi' = \mathrm{Re}\, \tilde{\psi} \exp(ikx)$, one has

$$\frac{\partial^2 \tilde{\psi}}{\partial y^2} = (k^2 - \hat{\beta}/\hat{u})\tilde{\psi}. \tag{6.14}$$

Given a source localized in latitude, zonal wavenumbers $k > k_s \equiv (\hat{\beta}/\hat{u})^{1/2}$ are meridionally trapped in the vicinity of the source, while wavenumbers $k < k_s$

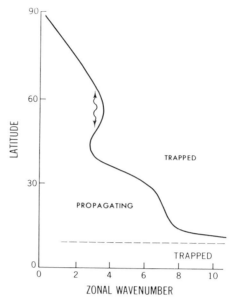

Fig. 6.10. The wavenumber, $n_s = ak_s$, dividing propagating from evanescent solutions for a mean wind resembling the wintertime flow at 300 mb, from Karoly (1978). The arrows indicate the potential for meridional confinement of wavenumber 3 near 55°.

propagate away from the source. Fig. 6.10 taken from Karoly (1978), shows $n_s = ak_s$ as a function of latitude for a particular wintertime 300 mb zonally averaged wind. $n_s \rightarrow \infty$ near 10°N, where $[u] = 0$. In the easterlies south of this latitude, all waves are trapped.

A source localized in longitude as well as latitude can be thought of as producing two rays for each $n < n_s$, corresponding to the two possible meridional wavenumbers, $al = \pm(n_s^2 - n^2)^{1/2}$. Since the meridional group velocity of non-divergent Rossby waves, $c_y = \partial\omega/\partial l = 2\hat{\beta}kl(k^2 + l^2)^{-2}$, has the same sign as l, the positive (negative) sign corresponds to poleward (equatorward) propagation. The rays can be traced downstream in the usual way: the zonal wavenumber characterizing the wave group remains unchanged because the mean flow is independent of x, whereas the meridional wavenumber adjusts to satisfy the local dispersion relation; the ray path is tangent to the vector group velocity (c_x, c_y) at this k and l. The amplitude of the streamfunction envelope is proportional to $l^{-1/2}$. One can show further that ray paths are always refracted toward larger 'refractive index' $(n_s^2 - n^2)^{1/2}$, i.e., towards larger n_s, this result being equivalent to Snell's law in optics. At a 'turning latitude', where $n = n_s$, one finds that $l \rightarrow 0$, $c_y \rightarrow 0$ and an incident poleward propagating wavetrain is simply reflected and continues propagating eastward. At a critical latitude, where $[u] = 0$, $l \rightarrow \infty$ but $c_y \rightarrow 0$ once again. Rays are refracted into the large region near the critical latitude. In fact,

$c_x/c_y \to 0$ as $l \to \infty$, so the rays approaching the critical latitude are oriented north–south. Examples of ray path calculations can be found in Hoskins and Karoly (1981). For zonal flows with uniform angular rotation the rays are great circles. In other cases this is qualitatively true away from critical latitudes, in good agreement with the observations described in Chapter 3.

It is possible for rays to be trapped away from the critical latitude; for example, the ray corresponding to $n = 3$ in Fig. 6.10 is just barely trapped in the high n_s region near 55°. However, waves can be expected to tunnel through the small region of low n_s near 45° and continue propagating into the tropics, so this flow will not provide a very efficient cavity for capturing an $n = 3$ resonance. The form of $n_s(y)$ is sensitive to the curvature of $[u]$, and using climatological wintertime 300 mb winds in Oort (1982), one does not find a well-defined dip in n_s in mid-latitudes. On the other hand, one suspects that substantial changes in curvature occur at particular times, creating more imposing barriers to equatorward propagation, with modest changes in $[u]$ itself. Meridional trapping of this kind is a distinct possibility and may be of importance for the variability of forced waves in the atmosphere. In its absence, however, all rays are eventually attracted into the zero-wind line. The possibility of resonance behaviour then depends on whether or not the complex dynamics in the vicinity of the zero-wind line results in significant reflection of the incident waves.

6.3.3 The critical latitude

If y_c is this critical latitude and if $\partial \hat{u}/\partial y \neq 0$ at $y = y_c$, then $l \propto (y - y_c)^{-1/2}$ and $c_y \propto (y - y_c)^{3/2}$ as $y \to y_c$. One is tempted to conclude that the packet will never reach y_c, since $\int^y c_y^{-1} \to \infty$ as $y \to y_c$. However, this is not the case, since this 'WKB' theory breaks down: WKB predicts that streamfunction amplitude is proportional to $l^{-1/2} \propto (y - y_c)^{1/4}$, so that $(\partial^2 A/\partial y^2)/A \propto (y - y_c)^{-2}$, which eventually grows larger than $l^2 \propto (y - y_c)^{-1}$ as $y \to y_c$, implying substantial fractional changes in amplitude over one wavelength. With the very small meridional scales generated as y_c is approached, one expects nonlinearity or dissipation of some sort to come into play.

By adding a linear damping $-r\zeta$, to the vorticity equation and then taking the limit of vanishing r, one obtains the familiar result that such a linear dissipative critical layer absorbs incident Rossby waves, at least partially. An indication of this behaviour can be gained without obtaining the explicit solution. Consider the simplest problem of a mean flow with $\hat{u} = u_1 > 0$ for $y > y_1$ and $\hat{u} = u_2 < 0$ for $y < y_2$, where u_1 and u_2 are constants, with a wave incident on the zero-wind line between y_1 and y_2 from the north. From the vorticity equation:

$$\tilde{\zeta} = i\tilde{\beta}\tilde{v}/(k\hat{u} - ir),$$

one has:

$$[v^*\zeta^*] = \frac{-\hat{\beta}|\tilde{v}|^2}{2k} \, Im\left(\frac{1}{\hat{u} - irk^{-1}}\right).$$ (6.15)

But:

$$\lim_{\varepsilon \to 0} Im \frac{1}{\hat{u} - i\varepsilon} = \frac{\pi}{|\partial\hat{u}/\partial y|} \delta(y - y_c),$$

where δ is the Dirac delta-function. Since $[v^*\zeta^*] = -(\partial/\partial y)[u^*v^*]$, $[u^*v^*]$ must be constant with latitude in the limit $r \to 0$, except for a jump at the critical latitude:

$$[u^*v^*]|_{y < y_c} + \frac{\pi\hat{\beta}|\tilde{v}|^2}{2k|\partial\hat{u}/\partial y|}\bigg|_{y = y_c}$$ (6.16)

In the constant $\hat{u}$ easterly region, $\tilde{\psi} \propto \exp(l_2 y)$, so that $[u^*v^*] = 0$, whereas in the constant $\hat{u}$ westerly region the solution has the form of an incident wave (with amplitude normalized to unity) plus a reflected wave, $\tilde{\psi} \propto \exp(-il_1 y)$ $+ R \exp(il_1 y)$, so that

$$[u^*v^*]|_{y > y_c} = \frac{kl_1}{2}(1 - |R|^2) = \frac{\pi\hat{\beta}|\tilde{v}|^2}{2k|\partial\hat{u}/\partial y|}\bigg|_{y = y_c}.$$ (6.17)

Assuming that $\hat{\beta} > 0$ and $|\tilde{v}|^2 \neq 0$ at y_c, we must have $|R|^2 < 1$. Part of the incident wave has been absorbed, decelerating the mean flow in the process. The amount of absorption is dependent on $|\tilde{v}|^2$, and its determination generally requires an explicit solution. Dickinson (1968) shows that the absorption is complete if the region in which WKB is valid overlaps with the region around the critical latitude within which $\hat{\beta}/\hat{u}$ can be replaced by $\gamma(y - y_c)^{-1}$, with γa constant.

The physical relevance of this absorbing critical layer has been questioned in light of the work of Benney and Bergeron (1969) on the possibility of a nonlinear balance in the critical layer. For a discussion of this theory in an atmospheric context, see Tung (1979). Benney and Bergeron's solution matches asymptotically to a linear solution far from the critical layer, consisting of incident and reflected waves of equal magnitudes. Simplified wave-mean flow calculations (Geisler and Dickinson, 1974) suggest one way of understanding these reflections: the waves modify the mean flow in the vicinity of y_c in such as way as to drive $\hat{\beta}$ close to zero; Eqn. (6.17) then implies that $|R|$ = 1. The realizability of a strongly reflecting critical layer remains a problem of fundamental importance for stationary wave theory.

If the equatorward propagating waves are organized into localized wavetrains, several additional problems arise concerning the critical latitude. The nonlinear critical layer solution matches onto incident plus reflected waves of a single zonal wavenumber. It is not clear that the behaviour of an

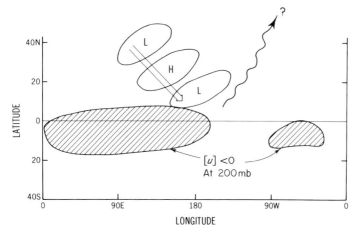

Fig. 6.11. Schematic of the Rossby wavetrain generated by the Tibetan plateau propagating into the tropics.

incident packet localized in longitude can be inferred from this solution or if nonlinearity results in significant coupling of different zonal wavenumbers. Also, the two dominant wavetrains in Fig. 6.9 enter the tropics in the central Pacific and in the Atlantic and should, therefore, be more sensitive to the structure of the zonal wind in these regions than to the zonal mean of the zonal wind. The areas of easterly zonal flow at 200 mb in the time mean January flow taken from Newell *et al.* (1974) are indicated by the shaded areas in Fig. 6.11. The regions of westerlies in the equatorial western Pacific and the Atlantic suggest the possibility that these wavetrains partially avoid zero-wind lines. This idea is returned to in Section 9.5.1.

Some of the low-frequency variability in the extratropical height field is known from the work of Horel and Wallace (1981) and others to be correlated with equatorial Pacific sea-surface temperatures and the phase of the southern oscillation (Fig. 3.10). The direct forcing by a tropical thermal source of a poleward propagating Rossby wave seems to be the simplest explanation for this teleconnection, as suggested by Hoskins and Karoly (1981). But the southern oscillation also involves large changes in the upper tropospheric zonal winds over the Pacific. If one accepts the idea of significant reflection from the easterlies, one can conceive of the amount of reflection of the incident wavetrain as being dependent on the phase of the southern oscillation—an alternative way of generating variability in the poleward propagating waves over North America, as illustrated schematically in Fig. 6.11.

The solutions described by Figs. 6.7–6.9 are obtained for a flow with $\bar{u} > 0$ everywhere, so the waves simply propagate into the Southern Hemisphere. The values of r chosen are sufficiently large that the waves do not return to the Northern Hemisphere with significant amplitude. Thus, these solutions

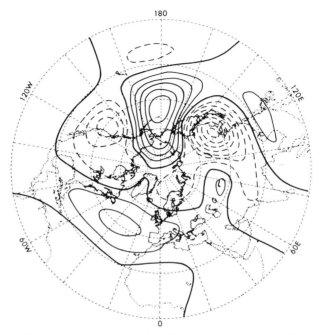

Fig. 6.12. The height response with $r^{-1} = 20$ days in a calculation identical to that in Fig. 6.8, except for the placement of a perfectly reflecting wall at the equator.

should be very similar to those obtained for a mean flow with an absorbing zero-wind line in low latitudes. In order to examine the implications of significant reflections, we modify these calculations by placing a perfectly reflecting wall at the equator. [As Tung (1979) points out, a reflecting wall does not produce the correct phase of the reflected wave, so this should not be viewed as an attempt at computing the detailed effects of a nonlinear critical layer.] With $r^{-1} = 5$ days, this modification produces little change in the midlatitude response, there being insufficient time for propagation to the tropics and back. The height response for $r^{-1} = 20$ days is shown in Fig. 6.12, to be compared with Fig. 6.8. With this small dissipation, considerable modification of amplitude and phase results in midlatitudes, along with a reduction in the southwest–northeast tilt of the highs and lows. If one repeats these calculations with $[u]$ multiplied everywhere by a constant μ, one finds very little sensitivity to these mean flow changes in the absence of reflections, but with a reflecting wall one finds a modest resonant-like response at $\mu \approx 1.1$, in particular. Not surprisingly, the introduction of the reflecting wall reintroduces the sensitivity to mean winds and to the level of dissipation seen in the beta-channel model, although not in as extreme a form.

Is there evidence in observations for the reflection of Rossby waves? One clue is the observed correlation coefficient between the stationary eddy u^* and

v^* at the maximum in poleward momentum flux near the tropopause at 30°N. Pure equatorward propagation in an inviscid atmosphere in the WKB approximation produces a correlation of $+1$, whereas perfect reflection from the tropics results in standing oscillations with $[u^*v^*] = 0$. The observed correlation is ≈ 0.4 in Oort and Rasmusson (1971), which implies predominantly equatorward propagation but with significant admixture of a poleward component. [This value for the correlation coefficient could be a significant underestimate, however; Mak (1978) finds a much larger $[u^*v^*]$ than do Oort and Rasmussen, for instance.] The linear calculations without a reflecting wall, Figs. 6.7–6.9, yield correlations at 30° close to 0.9. In the calculations with a reflecting wall, the correlation coefficient decreases from ≈ 0.9 at $r^{-1} = 5$ days to ≈ 0.5 at $r^{-1} = 20$ days. It seems that reflection sufficient to produce a correlation similar to that observed is also sufficient to produce weak resonant-like behaviour. However, an alternative explanation for the observed low correlation exists: the poleward propagating component could simply be thermally forced from the tropics (see Section 9.5.2). Indeed, Simmons (1982) has recently argued that tropical diabatic heating may force a significant part of the extratropical stationary eddy field.

6.4 Baroclinic model in a beta-plane channel: vertical propagation of Rossby waves

Just as meridional dispersion of Rossby waves is capable of fundamentally altering the response to topography, so is vertical dispersion. A quasi-geostrophic model on a beta-plane channel, linearized about a mean flow dependent only on height, is the simplest framework in which one can discuss vertical propagation and structure. The concept of vertically propagating Rossby waves was introduced by Charney and Drazin (1961). The use of a baroclinic quasi-geostrophic model for the stationary wave problem was initiated by Smagorinsky (1953) in his study of the response to thermal forcing.

In the absence of diabatic heating, the quasi-geostrophic thermodynamic equation with $z \equiv H \ln (p_*/p)$ as vertical coordinate,

$$\frac{\partial}{\partial t}\left(\frac{\partial \psi}{\partial z}\right) + J\left(\psi, \frac{\partial \psi}{\partial z}\right) = -\frac{N^2}{f_0} W \qquad (6.18)$$

when combined with the vorticity equation,

$$\frac{\partial}{\partial t} \nabla^2 \psi + J(\psi, \nabla^2 \psi + \beta y) = \frac{f_0}{\rho_0}\frac{\partial}{\partial z}(\rho_0 W), \qquad (6.19)$$

yields the pseudo-potential vorticity equation:

$$\frac{\partial q}{\partial t} + J(\psi, q) = 0,$$

where

$$q \equiv \nabla^2 \psi + \beta y + \frac{f_0^2}{\rho_0} \frac{\partial}{\partial z} \left(\frac{\rho_0}{N^2} \frac{\partial \psi}{\partial z} \right),$$

$$W \equiv dz/dt,$$

and

$$\rho_0 \equiv \exp(-z/H).$$

The lower boundary condition is obtained from Eqn. (6.18), after recognizing the relation between W and the actual vertical velocity w,

$$w = \frac{f_0}{g} \frac{\partial \psi}{\partial t} + \frac{RT}{gH} W.$$

Choosing $H = RT(0)/g$, then

$$W = -\frac{f_0}{g} \frac{\partial \psi}{\partial t} + \mathbf{v} \cdot \nabla h_T + \alpha \zeta \qquad (6.20)$$

at the ground, in the presence of topography and Ekman pumping represented by the last term. Linearizing about a mean wind $[u]$ dependent only on z, and once again retaining only the lowest order term in the topographic forcing, one finds:

$$\frac{\partial q^*}{\partial t} + [u] \frac{\partial q^*}{\partial x} + v^* \frac{\partial [q]}{\partial y} = 0 \qquad (6.21)$$

with the boundary condition:

$$\frac{\partial}{\partial t} \left(\frac{\partial \psi^*}{\partial z} - \frac{N^2}{g} \psi^* \right) + [u] \frac{\partial}{\partial x} \frac{\partial \psi^*}{\partial z} - v^* \frac{\partial [u]}{\partial z}$$

$$= -\frac{N^2}{f_0} \left([u] \frac{\partial h_T}{\partial x} + \alpha \nabla^2 \psi^* \right) \quad \text{at } z = 0, \qquad (6.22)$$

where:

$$\frac{\partial [q]}{\partial y} \equiv \beta - \frac{f_0^2}{\rho_0} \frac{\partial}{\partial z} \left(\frac{\rho_0}{N^2} \frac{\partial [u]}{\partial z} \right).$$

Stationary solutions to Eqn. (6.21)–(6.22) exist of the form:

$$\psi^* = \sin(ly) \, \text{Re} \left[\tilde{\psi}(z) \exp(ikx) \right]$$

provided that:

$$\frac{f_0^2}{\rho_0} \frac{\partial}{\partial z} \left(\frac{\rho_0}{N^2} \frac{\partial \tilde{\psi}}{\partial z} \right) = \tilde{\psi} \left(K^2 - \frac{\partial [q]/\partial y}{[u]} \right) \qquad (6.23)$$

and

$$[u] \frac{\partial \tilde{\psi}}{\partial z} - \frac{\partial [u]}{\partial z} \tilde{\psi} = -\frac{N^2[u]}{f_0} \tilde{h}_T - \frac{i\alpha N^2 K^2}{k f_0} \quad \text{at } z = 0.$$

In addition, a radiation condition must be satisfied at infinity.

In the simplest case of constant $[u]$ and N^2, and with $\alpha = 0$, the change of variables $\tilde{\xi} \equiv \tilde{\psi} \exp(-z/2H)$ yields:

$$\frac{\partial^2 \tilde{\xi}}{\partial z^2} = \tilde{\xi} \frac{N^2}{f_0^2} (K^2 + \gamma^2 - \beta/[u]) \tag{6.24a}$$

and

$$\frac{\partial \tilde{\xi}}{\partial z} + \frac{1}{2H} \tilde{\xi} = -\frac{N^2}{f_0} \tilde{h}_T \quad \text{at } z = 0, \tag{6.24b}$$

where $\gamma \equiv f_0/(2NH)$. If $K^2 + \gamma^2 - \beta/[u] < 0$, that is, if $0 < [u] < \beta/(K^2 + \gamma^2)$, the interior equation has the solutions $\tilde{\xi} = \tilde{\xi}_0 \exp(imz)$ with:

$$m = \pm \frac{N}{f_0} \left(\frac{\beta}{[u]} - K^2 - \gamma^2 \right)^{1/2}. \tag{6.25}$$

From the time-dependent problem [Eqn. (6.21)] one finds the dispersion relation for vertically propagating waves:

$$\omega = [u]k - \frac{\beta k}{K^2 + \dfrac{m^2 f^2}{N^2} + \gamma^2}, \tag{6.26}$$

from which it follows that the vertical group velocity $(\partial \omega / \partial m)$ has the same sign as m. Therefore, the plus sign in Eqn. (6.25) corresponds to a wave propagating upwards from the source at the ground and is the acceptable solution. If $K^2 + \gamma^2 - \beta/[u] > 0$, that is, if either $[u] < 0$ or $[u] > \beta/(K^2 + \gamma^2)$, then the solution if $\tilde{\xi} = \tilde{\xi}_0 \exp(-\mu z)$ with:

$$\mu \equiv \frac{N}{f_0} (K^2 + \gamma^2 - \beta/[u])^{1/2}. \tag{6.27}$$

Employing the boundary condition [Eqn. (6.24b)] one finds:

$$\tilde{\xi}_0 = -\frac{N^2 h_T}{f_0} \left[\left(\begin{matrix} im \\ -\mu \end{matrix} \right) + \frac{1}{2H} \right]^{-1} \tag{6.28}$$

for the $\begin{pmatrix} \text{propagating} \\ \text{trapped} \end{pmatrix}$ case.

From Eqn. (6.28) one sees that a resonance occurs if $\mu = (2H)^{-1}$, corresponding to $K^2 = K_s^2 \equiv \beta/[u]$. This external Rossby wave is a solution of the unforced inviscid equations with $\tilde{\psi} \equiv 1$ and therefore has no vertical

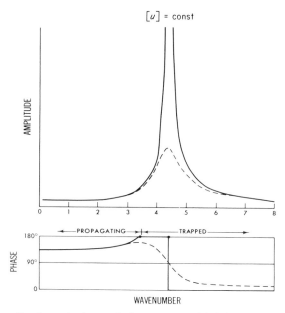

Fig. 6.13. Amplitude and phase of the geopotential height response at $z = 0$ for constant mean wind, as a function of zonal wavenumber. Solid line is for inviscid flow and dotted line for $H/(f_0\alpha) = 5$ days. A phase of $0°$ implies a response in phase with the topography. A phase between $0°$ and $180°$ implies the ridge in the response upstream from the ridge in the topography.

variation in amplitude or phase. For $K > K_s$, the forced wave solutions decay away from the surface with no phase variation with z, and are exactly in phase with the topography. For $(K_s^2 - \gamma^2)^{1/2} < K < K_s$, $\tilde{\psi}$ increases with height, although not so fast as to prevent $\rho_0|\tilde{\psi}|^2$ from decreasing with height. The solution is equivalent barotropic once again but is now $180°$ out of phase with the topography. For $K < (K_s^2 - \gamma^2)^{1/2}$, $\rho_0|\tilde{\psi}|^2$ is independent of z, the phase shifts westward with height and $\overline{[v^* \, \partial\psi^*/\partial z]} > 0$, and the surface response lags the topography by an amount between $90°$ and $180°$, depending on wavenumber.

Figure 6.13 shows the amplitude and phase of the height response at the ground as a function of total wavenumber for $\alpha = 0$ and for $r^{-1} = H/(\alpha f_0) = 5$ days. The other parameter values are $f_0 = 2\Omega \sin(45°)$, $H = R(275K)/g$, $[u] = 15$ m s^{-1}, $N^2 = 1.0 \times 10^{-4}$ s^{-2}, $h_T = $ constant independent of k, and $l = 0$. The structure is similar to that obtained in the barotropic channel model in that there is one and only one resonance for a given mean flow and meridional modal structure, with the response shifting in phase by $180°$ as the system passes through resonance. The important difference, of course, is the transition to upward propagating waves at small K. (It is because of the choice of zero meridional wavenumber that an equivalent barotropic spin down time

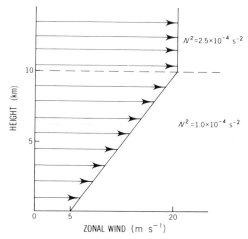

Fig. 6.14. Zonal wind profile and static stabilities used for calculations described in Figs. 6.15–6.18.

of 5 days has less of an effect in Fig. 6.13 than in the barotropic channel model. For non-zero l, the effective strength of the damping is increased by the factor $(k^2 + l^2)^{1/2}/k$.)

Consider now the more realistic model atmosphere shown in Fig. 6.14, with $[u]$ increasing linearly from 5 m s^{-1} at $z = 0$ to 20 m s^{-1} at $z = 10 \text{ km}$, and with $N^2 = 1.0 \times 10^{-4}$ and $2.5 \times 10^{-4} \text{ s}^{-2}$ for $z < 10 \text{ km}$ and $z > 10 \text{ km}$, respectively. The critical wavenumber:

$$K_c(z) \equiv \left(\frac{\partial[q]/\partial y}{[u]} - \frac{f_0^2}{4N^2H^2} \right)^{1/2} \tag{6.29}$$

dividing the locally wavelike from the locally evanescent solutions is plotted in Fig. 6.15. (All other parameters are identical to those used in the constant wind case.) Note that $\partial[q]/\partial y = \beta$ for $z > 10 \text{ km}$, while $\partial[q]/\partial y = \beta(1 + h/H)$ in the troposphere, where $h \equiv f_0^2 (\partial[u]/\partial z)/(\beta N^2)$. h/H is of order unity, so there is a substantial discontinuity in $\partial[q]/\partial y$ and, therefore, in K_c at the tropopause. (The discontinuity in $f_0^2/(4N^2H^2)$ in Eqn. (6.29) is of less consequence.) For $K < K_1 \simeq 3.7$, the forced wave propagates on both sides of the tropopause, for $K_1 < K < K_2 \simeq 5.9$, the wave has a turning point precisely at the tropopause, while for $K > K_2$, the wave has a turning point within the troposphere.

All of the waves that are evanescent above some height Z_T (that is, all $K > K_1$) are equivalent barotropic. The proof follows from the fact that $[v^*q^*] = 0$ (easily obtained from the equation of motion, $[u] \, \partial q^*/\partial x + v^* \, \partial[q]/\partial y = 0$) and the kinematic identity:

$$[v^*q^*] = -\frac{\partial}{\partial y}[u^*v^*] + \frac{f_0^2}{\rho_0} \frac{\partial}{\partial z}\left(\frac{\rho_0}{N^2}[v^* \, \partial\psi^*/\partial z] \right). \tag{6.30}$$

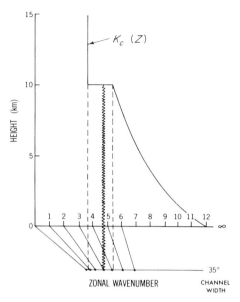

Fig. 6.15. The critical total wavenumber, $K_c(z)$, dividing the locally propagating ($K < K_c$) from the locally evanescent solutions ($K > K_c$). The zonal wavenumber which yields this total wavenumber if the meridional structure is a half wavelength sine wave in a channel of width 35° is also indicated. The wavy line marks the wavenumber of the stationary external Rossby wave.

$[u^*v^*] = 0$ for waves of the form $\sin(ly) \operatorname{Re} \{\tilde{\psi} \exp(ikx)\}$, so $\rho_0 N^{-2}[v^* \, \partial\psi^*/\partial z]$ must be independent of height. There is no heat flux at infinity if the waves are evanescent above Z_T; therefore, the heat flux and the related vertical phase variation must be identically zero at all heights, except for possible 180° phase shifts. In effect, these trapped waves are composed of upward and downward propagating components of equal magnitude.

Figure 6.16 shows the amplitude and phase of the height response at the ground for this profile. There is considerable similarity with the analogous plot for $[u]$ = constant. There is once again a single resonance, corresponding to a stationary external Rossby wave, located at $K = K_R = 4.7$ for the parameters chosen. The resonance happens to lie within the range $K_1 < K < K_2$ for which the waves have a turning point precisely at the tropopause. Figure 6.17 is a plot of the Rossby wave's vertical structure. The maximum at the tropopause can be understood from the WKB wavefunction noted in the figure: $\tilde{\psi}$ increases with increasing z within the troposphere partly because of the $\exp(z/2H)$ factor and partly because $m = Nf_0^{-1}(K_c^2 - K^2)$ decreases with height; the decay above the tropopause results from μ being larger than $(2H)^{-1}$. If $\partial[u]/\partial z$ were negative rather than zero in the model stratosphere, this decay would be more rapid.

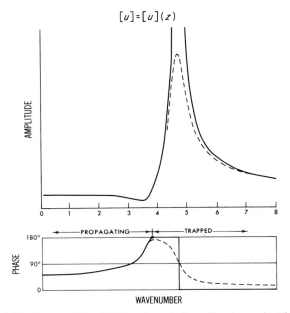

$[u] = [u](z)$

Fig. 6.16. Same as Fig. 6.13 for the wind profile shown in Fig. 6.14.

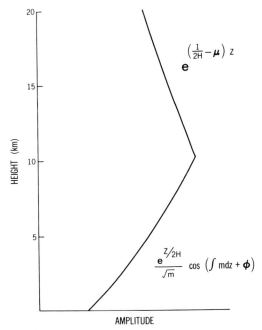

Fig. 6.17. The vertical structure of the external Rossby wave as computed numerically. The expressions alongside the curve are the WKB approximations to this structure.

The importance of the span of wavenumbers $K_1 < K < K_2$ for which the tropopause acts like a perfect reflector in this model atmosphere is enhanced by the fact that it contains the external Rossby wave. Furthermore, for a given meridional scale, $l \neq 0$, this range of total wavenumbers translates into a larger range of zonal wavenumbers. In Fig. 6.15, the lower horizontal axis is labelled with the appropriate zonal wavenumbers for the fundamental meridional mode in a channel of width 35°. For this mode, close to that used by Charney and Eliassen, zonal wavenumbers 1 through 4 all have turning points at the tropopause.

If one defines z_R such that $(\beta/[u(z_R)])^{1/2} = K_R$, then $z_R \simeq 6.7$ km ($p_R \sim 430$ mb) for these parameters. Repeating these calculations for various values of the tropospheric shear (holding all other parameters fixed, including the surface wind), one finds that larger shears result in larger z_R, but the changes are modest. z_R increases from $\simeq 6.2$ km at a shear of 1×10^{-3} s^{-1} to $\simeq 7.2$ km at 3×10^{-3} s^{-1}. Holding the shear fixed at 1.5×10^{-3} and varying $[u(0)]$, z_R varies monotonically from $\simeq 6.85$ km at $[u(0)] = 2.5$ m s^{-1} to 6.4 km at $[u(0)] = 10$ m s^{-1}. The application of Rossby's stationary wave formula in all cases requires the choice of an upper tropospheric wind in order to produce the correct wavelength of the stationary external wave.

Just as in the barotropic model, at times the response may be more easily understood in terms of Green's functions rather than Fourier components. The Green's functions for Eqns. (6.21)–(6.22) [more precisely, the solution when $h_T = \sin (ly) \delta(x)$] is shown in Fig. 6.18. l is chosen to correspond to the fundamental mode in a channel of width 42.5°, so that zonal wavenumbers 1 and 2 propagate into the stratosphere whereas 3 does not, and $H/(\alpha f_0)$ is set equal to 5 days. (This solution is obtained by Fourier analysing in x and retaining only the first 15 zonal harmonics, so the structure in the immediate vicinity of the source is somewhat distorted.) Within the troposphere one sees an equivalent barotropic wavetrain propagating eastward from the source with maximum amplitude near the tropopause, of wavelength close to that of the external mode. If one modifies l, the zonal wavelength of this tropospheric response changes so as to maintain constant K. Amplitudes increasing with height in the stratosphere are primarily confined to longitudes slightly east of the source. If one thinks of the source in Fig. 6.18 as the Tibetan plateau, then the dominant high in the stratosphere can be associated with the Aleutian anticyclone.

Some insight into the structure of this Green's function can be gained by using ray tracing theory in the zonal-vertical plane. The ratio of vertical to zonal group velocities $(\partial\omega/\partial m)/(\partial\omega/\partial k)$ is $(m/k)(f/N)^2$, where m is obtained from the local dispersion relation for stationary waves. Evaluating this ratio, one finds that the waves propagate through the troposphere while moving only 10°–30° longitude downstream, consistent with Fig. 6.18. A similar calculation can be found in Hayashi (1981), where one also finds a discussion of why this

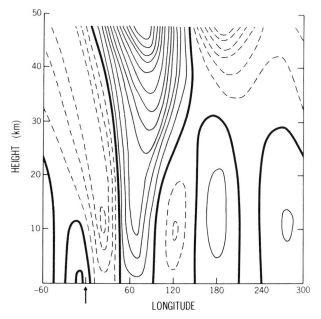

Fig. 6.18. The solution to Eqns. (6.21)–(6.22) with $h_T(x) = \delta(x) \sin(ly)$. The arrow marks the location of the source. Solid lines are positive contours and dotted lines negative.

group velocity argument works even though the 'group' consists only of wavenumbers 1 and 2. Because of this rapid vertical propagation, a short distance downstream from the source one is left with only those waves reflected from turning points at or below the tropopause. But of these, all are destroyed as they travel downstream by destructive interference between the upward and downward propagating components, except for the external Rossby wave, which then dominates the far field. This picture is confirmed by standard asymptotic analysis.

If $[u]$ is allowed to increase with height indefinitely, or if it is given a realistic structure with a mesospheric as well as tropospheric jet, then there can exist additional stationary normal modes (e.g., Geisler and Dickinson, 1975; Tung and Lindzen, 1979). These internal modes are still equivalent barotropic but differ from the external, or fundamental, vertical mode in having more structure in the vertical, with maximum amplitudes at turning points in the middle atmosphere. If significant reflections from upper level winds occur, whether or not the system is capable of exciting the vertical normal modes created by these reflections, stratospheric and mesospheric winds will have some impact on the tropospheric forced wave structure. Unless a particular internal mode is selectively amplified, however, it seems unlikely that waves reflected from the middle atmosphere can dominate over the more easily excited external Rossby waves.

The analysis of forced wave responses in numerical general circulation models is further complicated by the possibility of artificial internal modes created by reflection from the model's upper boundary. For example, in a two-layer model without vertical shear in $[u]$, the two vertical modes have phase speeds $[u] - \beta/(k^2 + l^2)$ and $[u] - \beta/(k^2 + l^2 + \lambda^{-2})$, where λ is here the internal radius of deformation. Comparison with the results for a semi-infinite atmosphere with constant $[u]$ shows that the second of these modes is artificial. Fortunately, this mode has minimum eastward phase speed $[u] - \beta\lambda^2$, so it can only be stationary if $[u] < \beta\lambda^2$. In the presence of vertical shear, one can demonstrate that only one of the two modes can be stationary if the *vertically averaged* zonal mean flow is larger than $\beta\lambda^2$. This condition is typically met in mid-latitudes: $\frac{1}{2}([u]_1 - [u]_2) > \beta\lambda^2$ is the condition for baroclinic instability, where subscripts 1 and 2 refer to the upper and lower layers, respectively. Therefore, an unstable or marginally stable state with $[u]_2 > 0$ also has $\frac{1}{2}([u]_1 + [u]_2) > \beta\lambda^2$. It is generally the case in multi-level models as well that the artificial internal modes have intrinsic phase speeds that are too small for them to be stationary in a midlatitude wintertime flow. Some distortion of the tropospheric stationary waves resulting from the artificial upper boundary condition imposed in GCM's is undoubtedly present, particularly in the longest waves, but it has yet to be evaluated in a realistic three-dimensional context.

As is clear from Fig. 6.5, topographic sources on the Earth are quite localized. If the external Rossby wave does indeed dominate the far field, as in the simple model atmosphere of Fig. 6.14, then the response in most regions should have the horizontal and vertical structure of this external mode. This picture suggests that ray tracing in the horizontal should be performed using the local dispersion relation of the stationary external Rossby wave as determined by the quasi-geostrophic model with local values of $[u]$, $\partial[q]/\partial y$, and N^2, thus eliminating any ambiguity in the choice of equivalent barotropic level, Z_R. In fact, Z_R will then be a slowly varying function of position. It is the rapid adjustment to this particular modal structure in the vertical, much more rapid than the adjustment in the horizontal, which is responsible for the relevance of barotropic calculations and two-dimensional ray tracing theory.

6.5 Thermal forcing

In the presence of diabatic heating, the quasi-geostrophic thermodynamic equation for linear, stationary eddies is modified to read:

$$[u]\frac{\partial}{\partial x}\frac{\partial \psi^*}{\partial z} - \frac{\partial \psi^*}{\partial x}\frac{\partial [u]}{\partial z} + \frac{N^2}{f_0}W^* = \frac{\kappa Q^*}{f_0 H} \equiv R^*, \qquad (6.31)$$

where Q is the heating rate per unit mass and $\kappa = R/c_p$. As in Section 6.4, we assume a channel geometry, with $[u]$ and N^2 functions of z only, and with $[u]$ > 0. The structure of the forced response depends crucially on whether the diabatic heating is balanced primarily by horizontal advection, either zonal or meridional, or primarily by adiabatic cooling. Which of these limiting balances holds, if either, is determined by the vertical structure of the heating as well as by the structure of the mean flow. The limit in which horizontal advection dominates is the more relevant for extratropical forcing, as argued below. Our discussion is based on that of Dickinson (1980) and Hoskins and Karoly (1981).

If horizontal advection balances the diabatic heating, one can obtain a particular solution by ignoring W^* in Eqn. (6.31) and solving for ψ^*. Noting that $[u] \, \partial\tilde{\psi}/\partial z - \tilde{\psi} \, \partial[u]/\partial z = [u]^2(\partial/\partial z)(\tilde{\psi}/[u])$, one finds for the Fourier amplitudes $\tilde{\psi}$:

$$\tilde{\psi}_p = \frac{i[u]}{k} \int_z^\infty \frac{\tilde{R}}{[u]^2} \, d\mu. \tag{6.32}$$

From the linearized vorticity equation:

$$ik[u](K_s^2 - K^2)\tilde{\psi}_p = \frac{f_0}{\rho_0} \frac{\partial}{\partial z}(\rho_0\tilde{W}_p) \tag{6.33}$$

where $K_s^2(z) \equiv \beta/[u]$, one then obtains for the vertical velocity:

$$\tilde{W}_p = \frac{-ik}{f_0\rho_0} \int_z^\infty \rho_0[u](K_s^2 - K^2)\tilde{\psi}_p \, d\mu. \tag{6.34}$$

The full linear response consists of this particular solution plus a homogeneous solution, denoted by a subscript h, chosen so that $W_h(0) + W_p(0) = 0$ (in the absence of topography and Ekman pumping). It is convenient to think of this homogeneous solution as forced by an 'equivalent topography':

$$\tilde{h}_T = -i\tilde{W}_h(0)/(k[u(0)]$$

$$= \frac{1}{f_0\rho_0(0)[u(0)]} \int_0^\infty \rho_0[u](K_s^2 - K^2)\tilde{\psi}_p \, d\mu. \tag{6.35}$$

The consistency of this solution can be checked by noting whether or not $N^2\tilde{W}_p/f_0$, with $\tilde{W}_p$ defined by Eqn. (6.34), is indeed negligible in Eqn. (6.31). This check is particularly simple if $[u]$ can be taken as uniform within the source region and Q can be assumed to have a simple vertical scale, $H_Q \propto |Q/(\partial Q/\partial z)|$ that is smaller than or comparable to the scale height of the atmosphere. For then:

$$\tilde{\psi}_p \propto RH_Q/(k[u]) \quad \text{and} \quad \tilde{W}_p \propto Rf_0^{-1}H_Q^2(K_s^2 - K^2),$$

and we have consistency if and only if:

$$\varepsilon \equiv N^2 f_0^{-2} H_Q^2 (K_s^2 - K^2) \ll 1. \tag{6.36}$$

In the extratropical lower troposphere, $K_s^2 \ll K^2$ for the planetary scale waves of interest. Making use of this approximation, one finds for the parameters in Fig. 6.14 that Eqn. (6.36) reduces approximately to $(H_Q/6 \text{ km})^2 \ll 1$.

If $[u]$ cannot be taken as constant with height this consistency condition will be altered. In particular, if the second rather than the first term in Eqn. (6.31) dominates, that is, if $H_u \propto |[u]/(\partial[u]/\partial z)| \ll H_Q$ within the region of diabatic heating, then $\tilde{\psi}_p \propto RH_u/(k[u])$, and H_u replaces H_Q in Eqn. (6.36). For the not untypical parameters in Fig. 6.14, $H_u \simeq 3$ km in the lower troposphere, so we still expect the consistency condition to be reasonably well satisfied for extratropical lower tropospheric heating.

The particular solution [Eqn. (6.32)] always possesses a low pressure centre a quarter wavelength downstream from the heating maximum. (We assume for simplicity that $\tilde{Q}$ has no phase variation with height.) Whether this is a cold or warm low depends on the vertical structure of the heating. If $H_Q \ll H_u$, so that zonal temperature advection is dominant in Eqn. (6.31), then the downstream low is warm. If meridional advection is dominant, $H_u \ll H_Q$, then the low is warm (cold) if Q decreases (increases) with height.

According to Eqn. (6.35), the equivalent topography is in phase with $\tilde{\psi}_p$ if $K < K_s$. Using Fig. 6.16 as a guide, one sees that for relatively short waves (but not so short that $K > K_s$ at low levels) the particular and homogeneous solutions are in phase, whereas for the longer waves that propagate into the stratosphere the homogeneous solution is shifted less than half a wavelength upstream. The scaling arguments of the preceding paragraphs suggest that within the source region,

$$|\psi_h/\psi_p| \propto \varepsilon/(m \min(H_Q, H_u)),$$

where m is the local vertical wavenumber of $\tilde{\psi}_h$. Even though ε may be considerably less than unity, m^{-1} is generally larger than H_Q or H_u, so ψ_h need not be entirely negligible within the heated region. If ψ_p does dominate, then one should see a gradual transition from ψ_p to ψ_h as z increases. For the long propagating waves, this implies a rapid westward phase shift with height that is distinct from the westward phase shift in the homogeneous solution itself.

Whatever the relative importance of ψ_p and ψ_h within the heated region, the magnitude of ψ_p will decrease away from the source and ψ_h will eventually dominate, having precisely the same characteristics as the orographically forced waves discussed in Sections 6.3 and 6.4. For a localized source, the far field will once again be dominated by equivalent barotropic external Rossby waves with the appropriate local stationary wavelength and with maximum amplitude near the tropopause.

In the opposite extreme, in which adiabatic cooling is primarily responsible for balancing the diabatic heating, a particular solution can be obtained by setting $\tilde{W}_p = f_0 N^{-2} \tilde{R}$ and substituting into the vorticity equation to obtain:

$$\tilde{\psi}_p = \frac{-if_0^2}{[u]k\rho_0(K_s^2 - K^2)} \frac{\partial}{\partial z}\left(\frac{\rho_0 \tilde{R}}{N^2}\right). \tag{6.37}$$

If the heating is non-zero at the ground, a homogeneous solution with $\tilde{W}_h(0) = -f_0 N^{-2} \tilde{R}(0)$ must be added, corresponding to flow over the equivalent topography

$$\tilde{h}_T = \frac{if_0}{N^2 k[u]} \tilde{R}(0).$$

A similar argument to that leading to Eqn. (6.36) yields $\varepsilon \gg 1$ for the consistency condition. Typically, this condition will be met for deep sources in subtropical latitudes. Assuming once again that $K \ll K_s$, one finds from Eqn. (6.37) that low pressure is found a quarter wavelength downstream (upstream) from the heating maximum if $\rho_0 R N^{-2}$ decreases (increases) with height. A solution for a tropical heat source is shown in Fig. 9.22. The feature which is most apparent in the height field perturbation is the 'great circle' poleward propagation of the equivalent barotropic external Rossby waves.

6.6 The relative importance of thermal and orographic forcing

Whereas with linear models one often starts with the simplest dynamical framework and then adds complications one by one, in research with general circulation models one often takes the opposite approach, starting with a very detailed model with which one attempts to simulate the observed climate as best one can. One then uses this model as a surrogate atmosphere to be experimented upon, removing some physical process or simplifying some boundary condition to determine its importance for maintaining the climate. The comparison of a GCM with realistic orography and one with a flat lower boundary is an excellent example of this approach.

Several 'mountain'–'no-mountain' comparisons of this kind have been published, the most detailed being that of Manabe and Terpstra (1974). Rather than attempting to review these studies, a few preliminary results are presented from recent calculations performed at the Geophysical Fluid Dynamics Laboratory/NOAA, conducted with a relatively low-resolution spectral model (15 wave rhomboidal truncation). The distinctive feature of these calculations is the length of the integrations: the two experiments were each integrated for 15 model years after having reached statistically steady states. The reduction in sampling errors hopefully puts the dynamical interpretation of the difference between the two time-averaged flows on a surer footing. The long integrations also allow one to analyse the model's low-frequency variability, as described by Manabe and Hahn (1982) and in Chapter 5. Both models have the identical sea-surface temperatures prescribed as a function of time of year and both have the same realistic continental distribution. The calculations were performed by S. Manabe and D. Hahn.

It has often been pointed out that the difference between the mountain and no-mountain climates cannot be interpreted as simply due to the mechanical diversion of flow by the mountains, since the insertion of the mountains has some effect on the distribution of the diabatic heating as well. Figure 6.19 shows the diabatic heating at 850 mb in both experiments, averaged over the 15 winter seasons. Although there is more small-scale structure in the heating when mountains are present, the large-scale patterns are quite similar. On the bases of such comparisons, we suspect that the interaction between the orographic and thermal sources may be of less importance than the nonlinear interaction between the forced wave fields themselves. With these provisos in mind, we shall persist in referring to the eddy field in the no-mountain model as the 'thermally forced' component and the difference between the mountain and no-mountain eddy fields as the 'orographically forced' component, the total eddy field being the sum of these two parts.

Figure 6.20 displays these two components, as well as the total field, for the geopotential at 45°N averaged over the 15 winter seasons. The total field can be compared with the observations in Fig. 2.3. The model provides a reasonably good approximation to the amplitudes of the major highs and lows in this section, the most noticeable discrepancy being the underestimation of the strength of the high centered at 20W and 250 mb at this latitude in the observations. The tendency for these geopotential fields to reach their maximum amplitudes near the tropopause is apparent both in the model and the observations, implying that much of this wave field can be thought of as vertically propagating in the troposphere but evanescent in the stratosphere (see Section 6.4). It is precisely the disturbances trapped within the troposphere that can be expected to be handled well by a GCM with most of its resolution within the troposphere (6 of 9 vertical levels in this case) and with an artificially reflecting upper boundary condition. Propagation into the stratosphere is more evident at higher latitudes, as is clear in the observed cross-section at 60°N.

One sees from the figure that the two components have rather different structures. The thermally forced eddies are of larger scale than the orographically forced eddies, particularly in the upper troposphere. They also have a larger and more systematic westward phase shift with height and show somewhat less of a tendency toward maximum amplitude at the tropopause—all plausible consequences of the larger zonal scale.

That the heating has larger scale than the orography might seem surprising, at least if one thinks of the very localized sensible heating of the atmosphere off

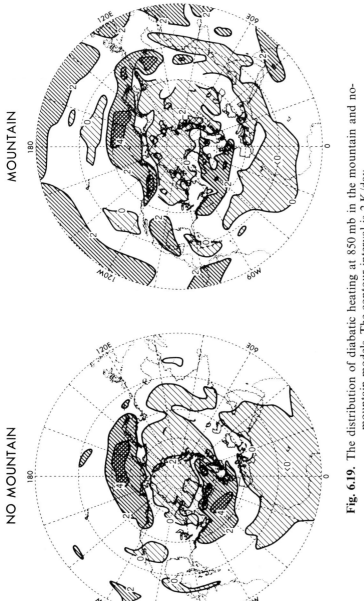

Fig. 6.19. The distribution of diabatic heating at 850 mb in the mountain and no-mountain models. The contour interval is 2 K/day.

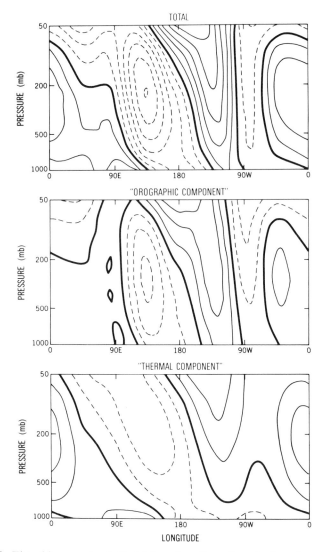

Fig. 6.20. The eddy geopotential height as a function of longitude and pressure at 45°N for the mountain model ('total'), the mountain minus no-mountain flow ('orographic component') and the no-mountain model ('thermal component'). The contour interval is 50 m with negative contours dashed. Model fields in Figs. 6.19–6.23 are averaged over the three winter months in each of the 15 model years.

the east coasts of the continents in the winter as being dominant. In fact, this sensible heating is dominant in the model only below 850 mb. At 850 mb (Fig. 6.19), the very large heating rates over the oceans at 45°, exceeding 5 K d^{-1}, are almost entirely due to latent heat release, whereas radiative cooling

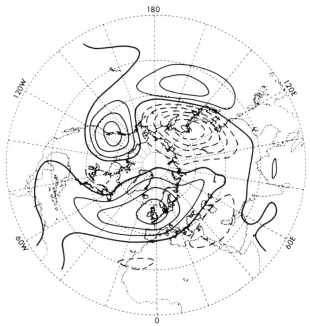

Fig. 6.21. The stationary eddy geopotential height at 300 mb in the mountain model. The contour interval is 50 m. Solid contours are positive or zero; dashed contours are negative.

dominates in the centres of the large continents. The latent heating maxima correspond roughly to the two major oceanic storm tracks. A theory for the thermally forced stationary eddies thus requires a theory for the location of the storm tracks.

Comparison with the topography used by the model (Fig. 6.5) shows that the scale of the topography is indeed somewhat smaller than that of the time-averaged heating. Even if the heating and orography had similar horizontal scales, the response to the orography would have the smaller scale, since it is $\partial h_T/\partial x$ rather than h_T itself that enters the boundary condition (cf. the k^{-1} factor in the 'equivalent topography' for thermal sources defined in Section 6.5). We have argued in Section 6.3 that the topography is sufficiently localized that the scale of the response is the scale of the stationary external Rossby wave rather than that of the topography itself. The scale of the time-averaged heating, on the other hand, seems to be sufficiently large compared with this stationary wavelength that the scale of the thermally forced component does reflect the scale of the heating. (Anomalies in the heating can have much smaller scales than the time-averaged heating, of course.)

The 300 mb geopotential height in the mountain experiment is shown in Fig. 6.21 and the separation into thermal and orographic components in Fig.

THERMAL COMPONENT   OROGRAPHIC COMPONENT

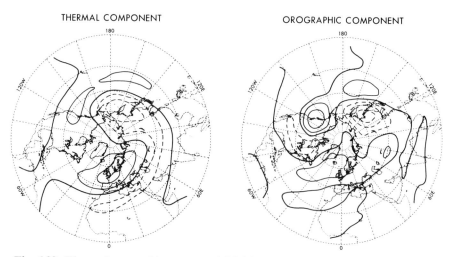

Fig. 6.22. The stationary eddy geopotential field of Fig. 6.21 split into its 'thermal' and 'orographic' components.

6.22. The two parts of the eddy field are of the same order of magnitude, but the smaller scale of the orographic component is again apparent. Because of this smaller scale, the orographic component should be more dominant in the meridional velocity and vorticity fields than in the geopotential. Constructive interference between the two components is evident in the Western Pacific and in the high at the Greenwich meridian and 50°N. There is destructive interference, if anything, over North America.

The streamfunction for the difference in the flows at 300 mb between the mountain and no-mountain models is shown in Fig. 6.23. The wavetrain structure is now more evident. There is a fairly close correspondence between the structure of the wavetrain between 120W and 60W and the linear waves forced by the Rockies in Fig. 6.9, and similarly for the high–low couplet in the Eastern Pacific and the linear response to the Tibetan plateau. The wavetrain oriented north–south near 0° longitude also seems to resemble a wavetrain at this longitude in the linear Tibetan response, although not in detail.

A comparison of the sea-level pressures in the two experiments in Fig. 6.24 shows that the Icelandic and Aleutian lows are present in full force in the no-mountain model. In fact, the Icelandic low is weakened substantially when mountains are inserted (while the Aleutian low is strengthened slightly). These experiments are consistent with the long-held belief that thermal effects are dominant near the surface, a belief based on the seasonal reversal of the surface pressure pattern. As is clear from Fig. 6.20, however, the topographic response is not negligible at the surface.

An interesting feature of the models' stationary waves is the correlation

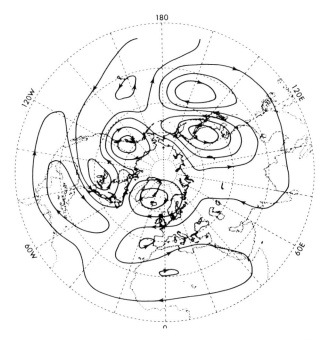

Fig. 6.23. The streamfunction for the mountain minus no-mountain flow at 300 mb. Contour interval is $4 \times 10^6 \text{ m}^2 \text{ s}^{-1}$.

between u^* and v^* near the maximum in the poleward momentum flux. The mountain and no-mountain models produce similar correlation coefficients of $\simeq 0.8$, as does the difference flow field. This is much higher than the observed value of $\simeq 0.4$ in Oort and Rasmussen (1971) discussed in Section 6.3. Primarily as a result of this difference, the model overestimates the standing eddy momentum flux by nearly a factor of three (the eddy velocities also being somewhat too large). This error in the stationary eddy flux is nearly compensated by an error of opposite sign in the transient eddy flux. Why compensation of this sort should occur is not clear.

Following the discussion in Section 6.3, we interpret this anomalously high correlation as resulting from the absence of Rossby waves of sufficient amplitude propagating polewards from the tropics into midlatitudes. Two alternative explanations exist for this deficiency: either tropical thermal sources of poleward propagating Rossby waves are not sufficiently active; or reflection of the equatorward propagating wavetrains is not occurring because of the absence of upper tropospheric easterlies (see Fig. 6.6). In any case, one can conclude from the absence of significant poleward propagation in the sub-tropics that the asymmetries in the extratropics of the no-mountain GCM are forced by thermal asymmetries in the extratropics rather than the tropics.

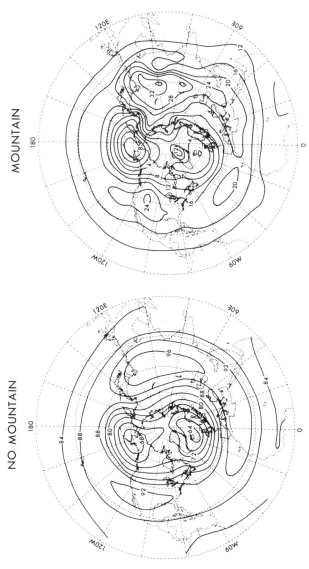

Fig. 6.24. The sea-level pressure for the mountain and no-mountain calculations. xx mb should be read 10xx mb for the mountain model and 9xx mb for the no-mountain model.

6.7   Quasi-stationary eddies in a zonally symmetric climate

It is tempting to assume that the quasi-stationary eddies in the atmosphere are closely related to the forced waves discussed in previous sections, the absence of strict stationarity being due either to variations in thermal sources and in the flow over the mountains, or to variations in the environmental flow through which the waves propagate. The teleconnection patterns described, for example, by Wallace and Gutzler (1981) and Horel and Wallace (1981) do resemble the Rossby wavetrains that, according to the arguments in Section 6.4, dominate the response to a localized source. The structure of the quasi-stationary eddies in a model that has no strictly stationary eddies, that is, in a model with a zonally symmetric lower boundary and zonally symmetric climate, clearly has some bearing on this hypothesis.

The model analysed is identical to the spectral GCM discussed in Section 6.6 and in Chapter 5, except that the realistic lower boundary condition is replaced by a flat land surface, and the seasonally varying solar flux is replaced by its annual mean values. The model statistics are thus stationary in time as well as homogeneous in longitude. In addition, surface albedos are fixed and the land is assumed to be always saturated, removing the potential sources of low-frequency variability in snow albedo-feedback and ground hydrology. Cloudiness is fixed in all of these models. The magnitude of the low-frequency variability in this idealized model, as measured by the standard deviation of monthly-mean geopotential height, is comparable to that in the model discussed in Chapter 5 that produces blocking episodes that are at least superficially realistic.

One aspect of the structure of the low-frequency variability in the idealized model is illustrated in Fig. 6.25, a plot of the correlation coefficient between the 300 mb monthly-mean meridional velocity at a reference point ($\simeq 45°$ latitude, $0°$ longitude) with the 300 mb monthly-mean meridional velocity at all other points on the sphere. (This field is computed using as a reference point each of the model's grid points along the latitude circle at $\simeq 45°$S; making use of the homogeneity of the statistics, these fields are all shifted to the same reference point and then averaged to produce the figure shown.) A similar plot using instantaneous rather than monthly-mean fields looks very different, the correlations dropping off more rapidly, with small negative correlations on either side of the peak at the reference point and only a hint of a wavelike structure.

This correlation field suggests that the low-frequency variability in the model is dominated by external Rossby wavetrains that happen to be nearly stationary because of their horizontal scale. The horizontal wavenumber in midlatitudes in Fig. 6.25 is 4–5, which is roughly what is needed for stationarity, the shape of the wavetrain resembles some of the ray paths

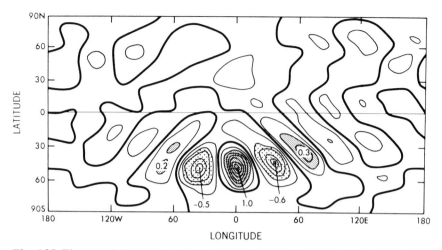

Fig. 6.25. The correlation coefficient between the monthly-mean 300 mb meridional velocity $v(\lambda, \phi)$ and $v(0, 45°S)$ for the model described in the text. The values of the local extrema of the correlation field are indicated.

computed by Hoskins and Karoly, and the phase tilts are just those expected for a wave propagating from west to east and reaching its turning latitude near the reference point. Examination of the vertical structure of the low-frequency variability shows these wavetrains to be equivalent barotropic with maximum amplitude near the tropopause, as expected for external Rossby waves.

It is rather surprising that the monthly-mean variability is dominated by such a well defined wavetrain, rather than some complex superposition of wavetrains passing through the reference point in different directions. One possible explanation is that it is that quasi-stationary wavetrain which has its turning latitude at the reference point that dominates the low-frequency variability at this point because meridionally propagating wavetrains generally attain their largest amplitudes near the turning latitude (see Hoskins and Karoly for a discussion of this point). In addition, Rossby waves are transverse; for a given streamfunction amplitude, meridional velocities are maximized if the wave is propagating east–west. Thus, one expects the variance of the meridional velocity to be dominated by that wave propagating east–west at the reference point, this being once again that wavetrain with a turning latitude at the reference point. The relevance of this last consideration is indicated by the fact that similar correlation maps computed using geopotential height or zonal velocity show a less wavelike pattern.

Discussion of the free external Rossby waves in the atmosphere is generally limited to the global normal modes which propagate westward sufficiently rapidly that they are not strongly affected by the structure of the mean winds, the wavenumber 1, 5-day wave being the best documented example (e.g.,

Madden, 1979). But Fig. 6.25 reminds us that a greater part of the energy in the external Rossby wave field is likely to be in somewhat smaller scales, in waves with small phase speeds with respect to the ground that are strongly refracted and perhaps absorbed by the mean flow. It is difficult, if not impossible, for these waves to organize themselves into global or hemispheric normal modes in the absence of significant reflection from low-latitude easterlies. They evidently exist primarily as wavetrains of the sort depicted in Fig. 6.25.

This analysis suggests an alternative to the point of view that low frequency circulation anomalies such as blocking highs are directly related to anomalous forced waves. Just as high-frequency cyclone waves would certainly exist in the absence of asymmetries in the lower boundary, but are organized into storm tracks by these asymmetries, so might the pre-existing external Rossby wave field be organized by these asymmetries to create preferred regions for low-frequency variability. How this might occur remains to be elucidated, although the large differences obtained by Simmons (1982) between small amplitude waves on zonally symmetric and asymmetric basic states is certainly suggestive.

Acknowledgements

I thank N.-C. Lau for assistance with some of the calculations and for helpful suggestions on a number of aspects of this work. Conversations with S. Manabe, S. Fels, B. Hoskins, G. Branstator, M. McIntyre, R. Dole, J. Roads and K. Trenberth were also extremely helpful. D. Linder assisted with the analysis of the GCM discussed in Section 6.7. The figures were drafted by the Scientific Illustrations Group at the Geophysical Fluid Dynamics Laboratory/NOAA.

References

BENNEY, D. J. and BERGERON, R. F. (1969). A new class of nonlinear waves in parallel flows. *Studies in appl. Math.*, **48**, 181–204.

BOLIN, B. (1950). On the influence of the Earth's orography on the general character of the westerlies. *Tellus*, **2**, 184–195.

CHARNEY, J. G. and DEVORE, J. G. (1979). Multiple flow equilibria in the atmosphere and blocking. *J. atmos. Sci.*, **36**, 1205–1216.

CHARNEY, J. G. and ELIASSEN, A. (1949). A numerical method for predicting the perturbations of the middle latitude westerlies. *Tellus*, **1**, 38–54.

CHARNEY, J. G., SHUKLA, J. and MO, K. C. (1981). Comparison of barotropic blocking theory with observation. *J. atmos. Sci.*, **38**, 762–779.

DICKINSON, R. E. (1968). Planetary Rossby waves propagating vertically through weak westerly wind wave guides. *J. atmos. Sci.*, **25**, 984–1002.

DICKINSON, R. E. (1980). *Orographic Effects on Planetary Flows*. GARP Publication Series, No. 23. W.M.O., Geneva.

GEISLER, J. E. and DICKINSON, R. E. (1974). Numerical study of an interacting Rossby wave and barotropic zonal flow near a critical level. *J. atmos. Sci.*, **31**, 946–955.

GEISLER, J. E. and DICKINSON, R. E. (1975). External Rossby modes on a β-plane with realistic vertical wind shear. *J. atmos. Sci.*, **32**, 2082–2093.

GROSE, W. L. and HOSKINS, B. J. (1979). On the influence of orography on large-scale atmospheric flow. *J. atmos. Sci.*, **36**, 223–234.

HAYASHI, Y. (1981). Vertical-zonal propagation of a stationary planetary wave packet. *J. atmos. Sci.*, **38**, 1197–1205.

HOREL, J. D. and WALLACE, J. M. (1981). Planetary-scale atmospheric phenomena associated with the southern oscillation. *Mon. Weath. Rev.*, **109**, 813–829.

HOSKINS, B. J. and KAROLY, D. J. (1981). The steady linear response of a spherical atmosphere to thermal and orographic forcing. *J. atmos. Sci.*, **38**, 1179–1196.

KAROLY, D. J. (1978). Unpublished lecture notes, NCAR Boulder, Colorado.

MADDEN, R. A. (1979). Observations of large-scale travelling Rossby waves. *Rev. Geophys. and Space Phys.*, **17**, 1935–1949.

MAK, M. (1978). On the observed momentum flux by standing eddies. *J. atmos. Sci.*, **35**, 340–346.

MANABE, S. and HAHN, D. G. (1981). Simulation of atmospheric variability. *Mon. Weath. Rev.*, **109**, 2260–2286.

MANABE, S. and TERPSTRA, T. B. (1974). The effects of mountains on the general circulation of the atmosphere as identified by numerical experiments. *J. atmos. Sci.*, **31**, 3–42.

NEWELL, R. E., KIDSON, J. W., VINCENT, D. G. and BOER, G. J. (1974). *The General Circulation of the Tropical Atmosphere and Interactions with Extra-tropical Latitudes*, Vol. 2. M.I.T. Press, Cambridge.

OORT, A. H. (1982). *Global Atmospheric Circulation Statistics, 1958–1973*. NOAA Professional Paper, U.S. Govt. Printing Office, Washington, D.C. (in press).

OORT, A. H. and RASMUSSON, E. M. (1971). *Atmospheric Circulation Statistics*. NOAA Professional Paper No. 5, U.S. Govt. Printing Office, Washington, D.C.

SIMMONS, A. J. (1982). The forcing of stationary wave motion by tropical diabatic heating. *Qt. Jl R. Met. Soc.*, **108**, 503–534.

SMAGORINSKY, J. (1953). The dynamical influence of large-scale heat sources and sinks on the quasi-stationary mean motions of the atmosphere. *Qt. Jl R. met. Soc.*, **79**, 342–366.

SMITH, R. B. (1979). In *Advances in Geophysics* (B. Saltzman, Ed.), Vol. 21, 87–230. Academic Press, New York and London.

TUNG, K. K. (1979). A theory of stationary long waves. Part III: Quasi-normal modes in a singular wave-guide. *Mon. Weath. Rev.*, **107**, 751–774.

TUNG, K. K. and LINDZEN, R. S. (1979). A theory of stationary long waves. Part II: Resonant Rossby waves in the presence of realistic vertical shears. *Mon. Weath. Rev.*, **107**, 735–750.

WALLACE, J. M. and GUTZLER, D. S. (1981). Teleconnections in the geopotential height field during the Northern Hemisphere winter. *Mon. Weath. Rev.*, **109**, 784–812.

WHITE, A. A. (1977). Modified quasi-geostrophic equations using geometric height as vertical coordinate. *Qt. Jl R. Met. Soc.*, **103**, 383–396.

— 7 —

Modelling of the transient
eddies and their
feedback on the mean flow

B. J. HOSKINS

7.1 Introduction

The monograph by Lorenz (1967) on the general circulation of the atmosphere gives an interesting account of the history of the ideas on the maintenance of the zonally averaged atmospheric flow. For most of the last 50 years it has been agreed that the transports of heat and momentum due to deviations from the zonal mean are crucial in the balance. These eddies have been mostly associated with the transient synoptic waves. In recent years, as reflected in this book, the emphasis has moved away from averaging together the very different climates along a line of latitude towards a consideration of the local (in latitude and longitude) weather averaged over a certain period. The role of the transient eddies in this view of the atmospheric circulation is less clear.

In this Chapter, an attempt is made to draw together the theories and ideas that may help in understanding the role of the transient eddies and, perhaps, in parameterizing their feedback on the mean flow. In Section 7.2 some general arguments that are useful either in zonal or time-averaged problems are put forward. The zonal and time-averaging operators are considered in Sections 7.3 and 7.4, respectively, concentrating on synoptic-scale eddies. The latter section contains less hard results, which is a reflection on the present state of the theory. The chapter finishes with a short discussion in Section 7.5.

7.2 Some general arguments

Before dealing with specific averaging operators and the net transports by the deviations from the chosen average, it is helpful to consider general arguments about these transports. Such arguments are made easier if the quantity under consideration is conservative under certain conditions. Accordingly, we shall use as our starting point the quasi-geostrophic, p-coordinate equations for flow on a rotating β-plane (see the Appendix):

$$(\partial/\partial t + \mathbf{v}\cdot\nabla)q = \mathscr{D}, \tag{7.1}$$

where $\mathscr{D}$ represents thermodynamic and mechanical forcing, including dissipative processes. The geostrophic motion $\mathbf{v}$ is horizontal and non-divergent, i.e., $\nabla_h\cdot\mathbf{v} = 0$, and thus we may write $\mathbf{v} = (-\psi_y, \psi_x, 0)$. The quasi-geostrophic potential vorticity, q, is related to this streamfunction, ψ, by:

$$q = f_0 + \beta y + \psi_{xx} + \psi_{yy} + f_0^2[\psi_p/(-\hat{R}\Theta_p)]_p. \tag{7.2}$$

Here $\hat{R}$ is the gas constant times $(p_0/p)^{1/\gamma}p_0^{-1}$, γ being the ratio of specific heats, $\Theta(p)$ is a standard potential temperature distribution, and the deviation of potential temperature from this distribution is:

$$\theta = -f_0\hat{R}^{-1}\psi_p. \tag{7.3}$$

Equation (7.2) is an elliptic equation for ψ if q is known and solution is possible given the provision of suitable boundary conditions. Approximating the condition $w = 0$ on a horizontal boundary by $\omega = 0$ at $p = p_0$, the necessary information is provided by the thermodynamic equation:

$$(\partial/\partial t + \mathbf{v}\cdot\nabla)\theta = \mathscr{Q} \quad \text{on } p = p_0, \tag{7.4}$$

where $\mathscr{Q}$ represents diabatic effects.

Let $(\bar{\ })$ define, for the moment, any simple averaging process, say zonal or time average, and $(\ )'$ the deviation from this average. From Eqns. (7.1) and (7.4) and the nondivergence of $\mathbf{v}$, the mean flow potential vorticity equation and boundary conditions may be written:

$$(\partial/\partial t + \bar{\mathbf{v}}\cdot\nabla)\bar{q} = \overline{\mathscr{D}} - \nabla\cdot(\overline{\mathbf{v}'q'}), \tag{7.5a}$$

with:
$$(\partial/\partial t + \bar{\mathbf{v}}\cdot\nabla)\bar{\theta} = \overline{\mathscr{Q}} - \nabla\cdot(\overline{\mathbf{v}'\theta'}) \quad \text{on } p = p_0. \tag{7.5b}$$

The presence of the eddies changes the mean flow through the eddy flux of potential vorticity, $\overline{\mathbf{v}'q'}$, in the interior and of potential temperature, $\overline{\mathbf{v}'\theta'}$, on the boundaries. They may also change the mean flow by modifying the forcing terms $\overline{\mathscr{D}}$ and $\overline{\mathscr{Q}}$.

Some general indications of the sense of these eddy fluxes can be obtained from the linearized eddy equations. Subtracting Eqn. (7.5a) from Eqn. (7.1) and Eqn. (7.5b) from Eqn. (7.4) and neglecting quadratic perturbation terms gives:

$$(\partial/\partial t + \bar{\mathbf{v}}\cdot\nabla)q' + \mathbf{v}'\cdot\nabla\bar{q} = \mathscr{D}', \tag{7.6a}$$

with:
$$(\partial/\partial t + \bar{\mathbf{v}}\cdot\nabla)\theta' + \mathbf{v}'\cdot\nabla\bar{\theta} = \mathscr{Q}' \quad \text{on } p = p_0. \tag{7.6b}$$

Multiplication of Eqns. (7.6a,b) by q' and θ', respectively, and averaging gives equations for the potential enstrophy $\overline{q'^2}$ and boundary potential temperature variance $\overline{\theta'^2}$:

$$(\partial/\partial t + \bar{\mathbf{v}}\cdot\nabla)\tfrac{1}{2}\,\overline{q'^2} + \overline{\mathbf{v}'q'}\cdot\nabla\bar{q} = \overline{\mathcal{D}'q'}, \tag{7.7a}$$

with:
$$(\partial/\partial t + \bar{\mathbf{v}}\cdot\nabla)\tfrac{1}{2}\,\overline{\theta'^2} + \overline{\mathbf{v}'\theta'}\cdot\nabla\bar{\theta} = \overline{\mathcal{Q}'\theta'} \quad \text{on } p = p_0. \tag{7.7b}$$

Before discussing these equations we note that the operator $(\partial/\partial t + \bar{\mathbf{v}}\cdot\nabla)$ indicates a derivative 'following the mean flow'. In the case in which the averaging process is a time average, then the $\partial/\partial t$ term may be dropped. When zonal averaging is used, $\bar{v} = \bar{\psi}_x = 0$ and the advective term is omitted so that the operator is a simple time derivative.

Where or when $\overline{q'^2}$ is increasing following the mean flow and dissipation is negligible, then Eqn. (7.7a) shows that the component of $\overline{\mathbf{v}'q'}$ parallel to the gradient in $\bar{q}$ is negative. The flux is said to be downgradient and acts in the same sense as a simple mixing process, acting to decrease $\nabla\bar{q}$. If the dissipation is negligible but there is a cascade to small scales with potential enstrophy conserved, then $\overline{\mathbf{v}'q'}$ is neither down- nor upgradient. However, there is a decrease in eddy activity as measured by, for instance, the eddy energy because of the smaller number of spatial derivatives involved. If the potential enstrophy on the small scales is dissipated then the original downgradient flux is irreversible. Similar considerations apply to the boundary temperature flux: if eddy growth is followed by a cascade to small scales and dissipation then irreversible downgradient boundary temperature fluxes occur.

However, an alternative decay mechanism is possible. In the absence of dissipation, fluid particles may return towards their original $\bar{q}(\bar{\theta})$ lines and thereby reduce $\overline{q'^2}$ $(\overline{\theta'^2})$. From Eqn. (7.7), this process is accompanied by upgradient transfers. Thus if there is reversible eddy growth with particles deviating from their mean positions and returning to them, then downgradient transfers are followed by upgradient fluxes.

These ideas will be returned to in later sections.

7.3 Zonal averaging

7.3.1 Basic equations

In this section we shall start from the primitive equations of motion and employ a zonal averaging operator $[\ ]$ and deviation $(\ )^*$. Using mass conservation, the x-momentum and thermodynamic equations may be written:

$$[u]_t = (f - [u]_y)[v] - [\omega][u]_p - [u^*v^*]_y - [u^*\omega^*]_p + [\mathscr{F}], \quad (7.8)$$

$$[\theta]_t = -[v][\theta]_y - [\omega][\theta]_p - [v^*\theta^*]_y - [\omega^*\theta^*]_p + [\mathscr{Q}], \quad (7.9)$$

where $\mathscr{F}$ denotes the contribution to $[u]_t$ from small-scale processes. Application of standard scaling arguments to the y-momentum equation suggests that $[u]$ is geostrophic to a very good approximation. Combining with the hydrostatic relation gives the 'thermal wind' equation in the form:

$$f_0[u]_p = \hat{R}[\theta]_y. \quad (7.10)$$

Mass conservation implies

$$[v]_y + [\omega]_p = 0. \quad (7.11)$$

As shown in Eqns. (7.8) and (7.9), eddies interact with the zonally averaged flow through the meridional and vertical fluxes of u and θ.

Before concentrating on eddy effects in the extratropics, it is worth noting that Eqns. (7.8)–(7.11) with Eqn. (7.10) modified to:

$$\{(f + [u] \tan \phi/a)[u]\}_p = \hat{R}[\theta]_y \quad (7.10')$$

and large-scale eddy fluxes omitted have been used by Schneider (1977), Schneider and Lindzen (1977), and Held and Hou (1980) to discuss the dynamics of the Hadley Cell. As well as exhibiting numerical solutions for particular parameter values, they produce a simple argument for the existence and approximate width of the cell. According to their analysis, $[u]$ is small everywhere at low levels and at the equator at upper levels. Approximate angular momentum conservation, Eqn. (7.8), at upper levels as air moves away from the equator is then consistent, from Eqn. (7.10'), with temperature gradients that are much smaller than those implied by the diabatic sources. Thus the ascent in the equatorial region and descent in the sub-tropical region must be of a magnitude to reduce the temperature gradients sufficiently. Further, the extent of the Hadley Cell is likely to be approximately up to that latitude at which momentum conservation and the thermal wind implied by the diabatic sources are compatible. It is of interest that simple calculations omitting the eddy terms show zonal winds at the polar extremity of the Hadley Cell much larger than those observed. Schneider (personal communication) has attempted to incorporate transient eddy effects into such calculations and Pfeffer (1981) has used observational data to compute the net effect of the eddies in Eqns. (7.8) and (7.9). They are found to produce a weakening and slight poleward movement of the jet.

Turning to the extratropical region, we first assume that quasi-geostrophic scaling and dynamics apply to both the eddies and the zonally averaged flow. In this case, all the underlined terms are neglected, and $[\theta]_p$ is replaced by a standard value $\Theta_p(p)$. Equations (7.8) and (7.9) then become:

$$[u]_t = f_0[v] - [u^*v^*]_y + [\mathscr{F}], \qquad (7.12a)$$

$$[\theta]_t = -[\omega]\Theta_p - [v^*\theta^*]_y + [\mathscr{Q}]. \qquad (7.12b)$$

The relevant eddy fluxes are now the poleward fluxes of u and θ. We may define a meridional circulation streamfunction ψ such that $[v] = \psi_p$, $[\omega] = -\psi_y$. From Eqn. (7.12), maintenance of the thermal wind balance, Eqn. (7.10), implies that:

$$f_0^2\psi_{pp} + \hat{R}(-\Theta_p\psi_y)_y = f_0[u^*v^*]_{yp} - \hat{R}[v^*\theta^*]_{yy} - f_0[\mathscr{F}]_p + \hat{R}[\mathscr{Q}]_y. \qquad (7.13)$$

On a lower boundary, on which $[\omega] = 0$, we have $\psi = 0$. This equation is a simpler form of the one used by Pfeffer (1981) and others to study the forcing of the mean meridional circulation.

Assuming approximate horizontal non-divergence in the eddies, integration by parts shows that $-[u^*v^*]_y$ may be replaced in Eqns. (7.12a) and (7.13) by a poleward flux of relative or absolute vorticity $[v^*\zeta^*]$. In this formulation, it is the poleward fluxes of vorticity and potential temperature that modify the zonally averaged flow.

An alternative method of approach is to introduce the quasi-geostrophic potential vorticity q, defined in Eqn. (7.2) and the Appendix. It may be shown, following Bretherton (1966) and Green (1970) that the poleward flux of q by the eddies is given by:

$$[v^*q^*] = -[u^*v^*]_y + f_0([v^*\theta^*]/\Theta_p)_p. \qquad (7.14)$$

Following Andrews and McIntyre (1976) and Edmon et al. (1980), we may define a non-divergent 'residual circulation':

$$[\tilde{v}] = [v] - ([v^*\theta^*]/\Theta_p)_p, \qquad (7.15a)$$

$$[\tilde{\omega}] = [\omega] + ([v^*\theta^*]/\Theta_p)_y. \qquad (7.15b)$$

The reason for this definition is that it enables Eqn. (7.12) to be written in the concise form:

$$[u]_t = [v^*q^*] + f_0[\tilde{v}] + [\mathscr{F}] \qquad (7.16a)$$

$$[\theta]_t = -[\tilde{\omega}]\Theta_p + [\mathscr{Q}]. \qquad (7.16b)$$

Elimination of the residual circulation from Eqn. (7.16) using its non-divergence is easily shown to give the zonal mean version of the quasi-geostrophic potential vorticity equation (7.5a) and Eqn. (7.16b) applied on $p = p_0$ gives the boundary condition (7.5b). Defining a residual circulation streamfunction $\tilde{\psi}$ such that $[\tilde{v}] = \tilde{\psi}_p$ and $[\tilde{\omega}] = -\tilde{\psi}_y$, maintenance of the thermal wind balance (Eqn. 7.10) from Eqn. (7.16) implies that $\tilde{\psi}$ must satisfy:

$$f_0^2\tilde{\psi}_{pp} + \hat{R}(-\Theta_p\tilde{\psi}_y)_y = -f_0[v^*q^*]_p - f_0[\mathscr{F}]_p + \hat{R}[\mathscr{Q}]_y. \qquad (7.17a)$$

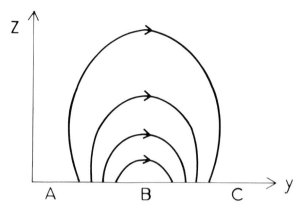

Fig. 7.1. A sketch of the residual circulation associated with a boundary poleward heat flux which is a positive maximum at B and zero at A and C. The positive values of $\bar{v}$ lead, from Eqn. (7.15), to a tendency to increase $[u]$ there.

On a lower boundary on which $[\omega] = 0$, from Eqn. (7.15b) we may take

$$\tilde{\psi} = -[v^*\theta^*]/\Theta_p. \tag{7.17b}$$

Thus, apart from the frictional and diabatic terms, the determination of the behaviour of the zonal flow from Eqns. (7.14)–(7.17) requires specification of the poleward eddy flux of q as a function of y and p, and the boundary poleward flux of θ. The effect of a potential vorticity flux is distributed in the vertical through the residual circulation. In the absence of boundary and forcing terms, a localized positive flux results in an increase in $[u]$ in the region of the maximum flux. In this formalism a positive boundary flux of heat by the eddies, from Eqns. (7.17b) and (7.17a) with zero on the right-hand side, is associated with a residual streamfunction which is sketched in Fig. 7.1, and thus tends to increase $[u]$ in the interior. Alternatively, we may note that the vertical average (denoted by $\{\ \}$) of Eqn. (7.12a) shows that, since $\{[v]\} = 0$ from Eqn. (7.11), the net acceleration of $\{[u]\}$ by the eddies is $-\{[u^*v^*]\}_y$. The second, temperature flux term on the right-hand side of Eqn. (7.14) produces no net acceleration of $\{[u]\}$. It merely produces a change in the distribution of $[u]$ with height as is intuitive from the thermal wind equation.

The foregoing analysis links directly with the concept of the Eliassen–Palm flux (Eliassen and Palm, 1961; Andrews and McIntyre, 1976; Edmon et al., 1980). At the level of quasi-geostrophic theory, we define a vector in the y, p plane:

$$\mathbf{E} = (-[u^*v^*], f_0[v^*\theta^*]/[\Theta_p]). \tag{7.18}$$

Then $\nabla \cdot \mathbf{E} = [v^*q^*]$ from Eqn. (7.14) so that 'EP' vector arrows $\mathbf{E}$ and their convergence indicate the relative importance of momentum and heat flux in their contribution to the potential vorticity flux. The vertical portion of the EP arrows near the boundary shows the boundary heat flux.

As shown by Eqn. (7.7a) with zonal averaging as the averaging operator, in the absence of wave transience or dissipation $\nabla \cdot \mathbf{E} = [v^*q^*]$ is zero, i.e., $\mathbf{E}$ is non-divergent. If the zonal flow is steady, Eqn. (7.6) gives:

$$A_t + \nabla \cdot \mathbf{E} = D \tag{7.19}$$

where the wave activity $A = \frac{1}{2}[q^{*2}]/[q]_y$ for $[q_y] \neq 0$. In cases in which the concept of group velocity $\mathbf{c}$ is applicable it can be shown, as discussed in Edmon et al. (1980), that:

$$\mathbf{E} = \mathbf{c}A. \tag{7.20}$$

Thus $\mathbf{E}$ quite generally gives a measure of net wave propagation.

There are several warnings and reservations concerning the above quasi-geostrophic analysis that need to be stressed. For the troposphere, in this formulation, the boundary heat flux is a crucial part of the total effect of the eddies. In the stratosphere this is not so, and Chapter 10 includes a more straightforward version of this analysis. Secondly, in a steady state in which friction in the free atmosphere is not dominant, the tendency for cancellation between the eddy flux and mean circulation terms in the momentum equation in the form (7.16a) is just as large as in the form (7.12a). Thirdly, it is the vertical integral of $-[u^*v^*]_y$ which balances the surface stress in the long-term average. In the potential vorticity formulation this would reappear as

$$\int_0^{p_0} [v^*q^*]\, dp - f_0([v^*\theta^*]/[\Theta_p])_{p=p_0}.$$

The final problem is the applicability of the quasi-geostrophic analysis to the eddies and to the mean flow. For small amplitude eddies Eqn. (7.6a) with $\mathscr{D} = 0$ gives:

$$(\partial/\partial t + [u]\, \partial/\partial x)q^* + v^*[q]_y = 0, \tag{7.21}$$

which is probably a good approximation provided that $\Theta_p(p)$ in $[q]_y$ is replaced by the zonally averaged $[\theta]_p$ which is a function of y as well as p, giving:

$$[q]_y = \beta - [u]_{yy} - f_0^2\{[u]_p/(-\hat{R}[\theta]_p)\}_p. \tag{7.22}$$

By returning to the linearized vorticity and thermodynamic equations and eliminating ω, it can be shown that this form is entirely consistent. For large-amplitude eddies it is less clear whether quasi-geostrophic theory is applicable. As shown by Williams (1972) and Hoskins and Bretherton (1972), quasi-geostrophic theory does not give the asymmetry between high- and low-pressure systems and the cascade to small scales in frontal regions. Given the importance of such cascades in irreversible fluxes as discussed in Section 7.2, the latter deficiency may be crucial. However, it is possible that a good indication of the fluxes may be obtained from quasi-geostrophic theory with dissipation included on a larger length scale.

Straightforward application of quasi-geostrophic analysis to the zonally averaged flow is almost certainly not valid for several reasons. Firstly, the convergence of the vertical eddy flux of θ is usually not negligible in Eqn. (7.9). In any quasi-geostrophic discussion of the eddy and $[q]_y$ behaviour, changes in the latter are conveniently discussed with the diabatic term in the thermodynamic equation taken as $[\mathcal{Q}] - [\omega^*\theta^*]_p$. Secondly, allowing the stability to vary with y and t does not lead to a consistent mean flow quasi-geostrophic potential vorticity equation. Further, if the zonal mean flow equations are intended to be applied over a region larger than a mid-latitude belt, then the underlined mean flow terms in the momentum equation (7.8), which are crucial in the tropical region, must also be included.

Finally, we should note the simple energy-balance climate models (e.g., North, 1975) include only a thermodynamic variable at one level. To discuss the behaviour of this variable, only the poleward eddy flux of heat is required. The details of the vertical structure are specified in the formulation of the model.

7.3.2 Some heat flux parameterization ideas

The equations for the zonally averaged flow contain fluxes due to the time-averaged and low-frequency eddies, but here, as mentioned above, the emphasis is on the transient synoptic scale eddies whose existence is due to baroclinic instability (for more details see, for example, Holton, 1979, and Pedlosky, 1979b).

The simplest example of baroclinic instability is, perhaps, that shown by Eady (1949). Using a Boussinesq approximation, an atmosphere with zonal flow having uniform shear in z, no shear in y, uniform static stability, and $\beta = 0$ has a most unstable mode with growth rate:

$$\sigma \simeq 0.31 f[u_z]/N = 0.31(g/\theta_0)|[\theta_y]|/N, \tag{7.23}$$

where θ_0 is a standard temperature and N the buoyancy frequency $\{(g/\theta_0)[\theta_z]\}^{1/2}$. Green (1960) and Lindzen and Farrell (1980a) have shown that Eqn. (7.23) is a good approximation to the largest growth rate in many simple situations. Rapid growth is associated with large baroclinity and small static stability. The solution for the most unstable Eady mode and the energy arguments in Green (1960) also suggest that trajectories in the interior of the fluid will tend to be at an angle approximately half-way between the horizontal and the basic isentropes. This implies that well away from the boundaries:

$$[w^*\theta^*]/[v^*\theta^*] \simeq \tfrac{1}{2}[\theta_y]/[\theta_z].$$ (7.24)

Green (1970) and Stone (1972) have given arguments for the parametric form of the horizontal heat flux which are summarized here in the manner of Held (1978). Assume that the latitudinal displacements of particles are similar to the longitudinal scale of the most unstable mode which is the Rossby radius of deformation $L_R = Nd/f$, d being the depth of the mode. Then $\theta^* \sim L_R[\theta_y]$. An assumption that the kinetic energy of eddies is of the order of their potential energy gives a magnitude for $v^* \sim g\theta^*/(\theta_0 N)$. Then:

$$g/\theta_0[v^*\theta^*] \sim (Nd^2/f^2)\{g/\theta_0[\theta_y]\}^2 = Nd^2[u_z]^2.$$ (7.25)

The sense of the flux is assumed to be down the gradient in $[\theta_y]$. The parameterizations of the heat flux by Green (1970) and Stone (1972) were of the form (7.25) with d taken as H, the scale height of the atmosphere. This gives a poleward flux proportional to the square of the baroclinity. Green (*loc. cit.*) produced some verification of this squared dependence on a global scale by comparing values of the heat flux at 50°N and the average baroclinity between 20°N and 70°N for various seasons.

Held (1978), generalizing results of Green (1960), pointed out that for simple basic states with small non-dimensional shear,

$$\lambda = [u_z]/u_{-_c},$$ (7.26)

where:

$$u_{-_c} = \beta N^2 H/f^2,$$

the horizontal and vertical scales of the most unstable mode are proportional to λ. From Eqn. (7.25) this implies that for small λ the heat flux would vary with the fourth power of the baroclinity. The depth-integrated heat flux would have another factor d that would imply a fifth power dependence. The two-layer model (Phillips, 1954) is unable to resolve these shallow disturbances and gives stability for λ less than a constant which depends on the particular finite differences used.

The picture that emerges is one in which large baroclinity leads to vigorous baroclinic eddies that transport significant amounts of heat and reduce the baroclinity. Small temperature gradients result in weak shallow eddies that transport little heat and allow a build-up in baroclinity due to diabatic processes. The point of equilibration of the eddy fluxes with the diabatic processes might be expected to depend crucially on the time-scales of these latter processes in restoring baroclinity.

However, Stone (1978) produced a simpler baroclinic adjustment hypothesis that the presence of the eddies demands that $\lambda \simeq 1$. We note that a little manipulation leads to the form:

$$\lambda = (\tan\phi a/H)[-\theta_y]/[\theta_z],$$ (7.27)

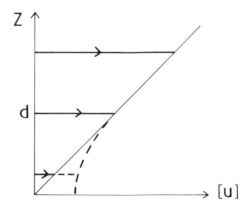

Fig. 7.2. A zonal flow $[u] = \Lambda z$ which is baroclinically unstable, and the minimum modification which will render it stable (dashed curve).

where a is the planetary radius and ϕ the latitude. At 45°, it is a measure of the slope, relative to the Earth, of the isentropes compared with H/a, which is approximately the slope of a line from the ground in the sub-tropics to the polar tropopause. The requirement $\lambda = 1$ necessitates a larger isentropic slope equatorward of 45° and a smaller slope poleward of that. Stone (1978) found that λ was approximately unity throughout the year from 40°N to 65°N. However, inspection of the Southern Hemisphere data shown in Fig. 1.4 suggests that agreement with Stone's postulate is less good there.

A baroclinic adjustment hypothesis that λ assumes a particular value only gives the isentropic slope. If the static stability can be determined by other means then the hypothesis gives the baroclinity. Conversely, as in Stone and Carlson (1979), if the horizontal temperature gradient is specified then the hypothesis allows determination of the static stability.

Another theory that attempts to go straight to the implied temperature field is that due to Lindzen and Farrell (1980b). The initial aim was to find the smallest adjustment to a zonal flow profile $[u] = \Lambda z$ with baroclinity parameter λ which will render it neutral to baroclinic instability (see Fig. 7.2). Sufficient conditions for stability are zero or reversed thermal gradient on the lower boundary and non-negative potential vorticity gradient in the interior. The minimum adjustment is obtained by setting $[u]_z = 0$ at $z = 0$ and demanding that the flow curvature be small enough for $0 \leqslant z \leqslant d$ that $[q]_y = 0$ there. The flow is unmodified for $z \geqslant d$ and continuity of $[u]$ and $[u]_z$ at $z = d$ is required. Making a Boussinesq approcimation, for a flow independent of y and static stability independent of z, Eqn. (7.22) may be written:

$$[q]_y = \beta - f^2[u]_{zz}/N^2. \tag{7.28}$$

It is easily shown that the above conditions give for the modified profile (m):

$$d = f^2\Lambda/(\beta N^2) = \lambda H, \quad [u]_{z=0}^m = \tfrac{1}{2}[u]_{z=d}. \tag{7.29}$$

In order to apply these ideas to the modification of thermal gradients due to the presence of baroclinic waves, Lindzen and Farrell (1980b) assumed that the reduction in baroclinity implied by the modification is in reality distributed over the depth H_T of the troposphere. This gives an average reduction of the baroclinity by a factor $1 - \tfrac{1}{2}\lambda H/H_T$.

Difficulties with this appealing baroclinic adjustment process are (i) the neglect of y variation, (ii) depending on λ, the adjustment depth d can be greater than the tropospheric depth, (iii) the arbitrary vertical redistribution of the modification, and (iv) the need to specify the static stability. Of these, (iii) is the most disturbing.

7.3.3 Some potential vorticity and momentum flux parameterization ideas

As discussed in Section 7.3.1, in order to discuss the zonally averaged flow as a function of height as well as y and t we require knowledge of the meridional distribution of the momentum flux $[u^*v^*]$, the vorticity flux $[v^*\zeta^*]$ or the potential vorticity flux $[v^*q^*]$. The Eady model is now not relevant because it gives zero interior fluxes of these quantities. The option chosen by Green (1970) was to argue using the quasi-conservative quantity potential vorticity and simple models of its transfer by baroclinic waves. For a linear normal mode perturbation to a basic flow with zonal velocity $[u]$ and poleward potential vorticity gradient $[q]_y$, the eddy equation (7.21) gives:

$$ik([u] - c)q^* + ik[q]_y\psi^* = 0. \tag{7.30}$$

Then

$$[v^*q^*] = \tfrac{1}{2}\mathscr{R}\{(ik\psi^*)_{\text{conj.}} \times q^*\}$$

$$= -\frac{\tfrac{1}{2}\sigma|\psi^*|^2[q]_y}{([u] - c_r)^2 + \sigma^2/k^2}. \tag{7.31}$$

$\sigma = kc_i$ is the growth rate, c_r the phase speed and k the wavenumber. An alternative form of Eqn. (7.31), suggested by the work of Taylor (1915) and highlighted by Bretherton (1966), is obtained by introducing the meridional displacement η^* of a particle from its initial position. Then $q^* = -\eta^*[q]_y$ and $v^* = d\eta^*/dt$. Thus $[v^*q^*] = -\partial/\partial t[\tfrac{1}{2}\eta^{*2}][q]_y$. Large potential vorticity transfers are expected where the amplitude of the mode is large. Also, since σ/k is generally small compared with typical values of $[u] - c_r$, large transfers occur in this *linear* theory near the 'steering level' at which $[u] = c_r$. Physically this is

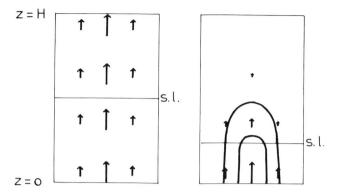

Fig. 7.3. Schematic EP fluxes (arrows) and their divergence, which is the same as the poleward flux of quasi-geostrophic potential vorticity (continuous contours at 0.4 and 0.8 of the maximum) for an Eady wave (left) and a Charney wave (right) in a channel. Steering levels are marked s.l. on these latitude-height sections. This picture is based on that in Edmon *et al.* (1980).

because air remains in the same phase of the wave for a long time and thus experiences large meridional excursions.

The EP flux and potential vorticity flux pictures for typical Eady and Charney waves are shown in Fig. 7.3. Neither mode has momentum transfers, but the Charney mode, on a flow with $[q]_y = \beta$, has significant potential vorticity fluxes between the surface where the amplitude is a maximum and the steering level in agreement with Eqn. (7.31).

Such considerations were used by Green (1970) to construct a transfer theory for potential vorticity and potential temperature and thus, from Eqn. (7.14), for momentum. The vertical integral of this provided Green (1970) and White (1977) with a check on the theory by comparing predicted with observed surface stresses and winds. The net result of using a formula (7.31) in the form $-K[q]_y$ resembles a classical mixing theory. However, the basic concept is one of organized transfer by growing baroclinic eddies followed by an assimilation of particles into their new environment. As implied by the work of Andrews and McIntyre (1976) and shown by Eqn. (7.7), in a statistically steady state the latter process is necessary for the transfers to be irreversible.

Clearly some caution is needed in application of this theory. The arguments given in Section 7.2 as applied to the zonally averaged situation show that in a region where the wave activity is decreasing with time, associated with particle displacements becoming smaller, an upgradient transfer is to be anticipated. Also, the theory as developed so far is based on transfers deduced from *linear* mathematical analysis applied to situations with trivial y dependence. However, for the real atmosphere the fluxes are required for nonlinear waves on flows with significant y variation.

A general linear theory deduction may be made from Eqn. (7.31) and the vertical integral of Eqn. (7.14) in regions where the fluid is stable to internal baroclinic instability ($[q]_y > 0$ for all z) and the vertical integral of the last term in Eqn. (7.14), i.e., the surface $[v^*\theta^*]$ term, is negligible. Then Eqn. (7.31) shows that $[v^*q^*]$ is negative for growing linear waves and Eqn. (7.14) that the depth averaged momentum flux convergence is negative. Held (1975) used this reasoning to show that in a two-layer model the momentum transport is into an unstable region from lateral stable regions. Hollingsworth *et al.* (1976) used the same argument on the sphere, where the boundary heat flux term becomes negligible sufficiently close to the equator, to show that the momentum flux in that region must be poleward.

7.3.4 Non-linear results

Most of the analyses of non-linear baroclinic waves that have appeared in the literature are for cases in which they are only marginally unstable. Such studies are useful for exhibiting some of the possible modes of behaviour and the results have been used to study the parameterization of heat flux (Pfeffer and Barcilon, 1978 ; Barcilon and Pfeffer, 1979) and the criticality of the baroclinity (Pedlosky, 1979a).

However, the very unstable waves that synoptic observations suggest probably dominate in producing fluxes in the atmosphere are really only amenable to controlled numerical simulations. Gall (1976) has shown that though for many realistic profiles the short waves are most unstable, in non-linear calculations they soon stabilize leaving the longer waves 5–7 whose eddy kinetic energy continues to grow particularly at upper levels. Similar results have been suggested also by Simmons and Hoskins (1978, 1980). Some new insights into one of their calculations (the basic case in Simmons and Hoskins, 1980) will be presented here.

The method used was to specify the zonal flow shown in Fig. 7.4(a), determine the most unstable mode in zonal wavenumber 6 and then perform a non-linear integration using as initial conditions the zonal flow plus a small amplitude mode in wavenumber 6. The model used is a spectral primitive equation model on the sphere with fourth-order horizontal internal diffusion as the only diabatic process. The wave grows by baroclinic instability, reaches a maximum eddy kinetic energy at days 7–8 and then decays as rapidly, converting kinetic energy back into the kinetic energy of the zonal flow. At day 15, there is very little eddy energy and the zonal flow has been modified by the baroclinic wave to that shown in Fig. 7.4(b). The surface temperature gradient between 35° and 55°, the region of main eddy activity, has been removed and away from the surface it has been reduced. On the other hand, very tight baroclinic regions have been created at low levels to the north and south.

182

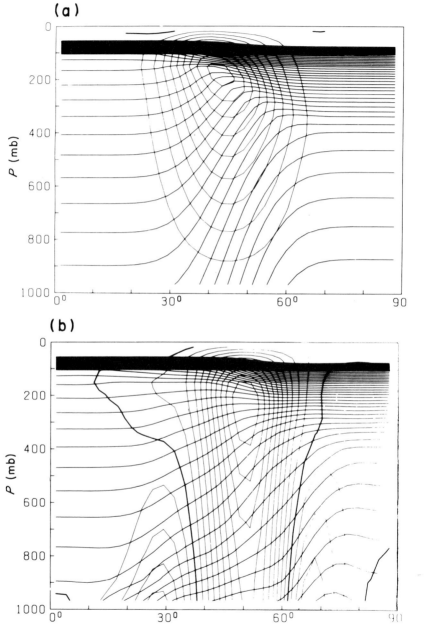

Fig. 7.4. Latitude–height sections showing potential temperature and zonal velocity in a baroclinic wave life-cycle experiment. Shown in (a) is the basic state and (b) the zonally averaged flow at day 15. Potential temperature contours, every 5 K, are drawn as medium lines. Zonal velocity contours, every 5 m s^{-1}, are drawn as lighter lines except that the zero contour is heavy.

Consistent with this cascade to small meridional scales, the zonal spectral coefficients of temperature at low levels have been generally decreased for total wavenumber $n \leqslant 8$ and increased for $n \geqslant 10$. In agreement with numerous observational studies, in particular Lorenz (1979), this suggests that a simple, uniform K, diffusive parameterization of baroclinic wave heat flux is incorrect. The other feature that has been markedly strengthened is the barotropic (depth-averaged) component of the flow.

To see to what extent the previous ideas are useful in interpreting the behaviour of the wave at various stages, Fig. 7.5 shows EP and potential vorticity fluxes and a Mercator coordinate version of the $[q]_y$ using the zonally averaged static stability as in Eqn. (7.22). The linear mode shows down-gradient fluxes similar to the Charney mode (Fig. 7.3b) except for a weak secondary maximum in q flux at the tropopause which, with its large $[q]_y$, acts almost as the lid in the Eady problem (Fig. 7.3a). The positive q fluxes in the polar stratosphere are downgradient, as $[q]_y$ is negative there. At day 5 the maximum heat flux is at a level away from the surface. The growth of the wave at upper levels may be viewed as the result of upward radiation into a region of stronger $[u]$, after cessation of linear growth (McIntyre and Weissman, 1978; Edmon et al., 1980), the 'wave packet' having a tendency to conserve the wave energy divided by the Doppler shifted frequency (Bretherton and Garrett, 1968). The negative potential vorticity flux is a strong maximum just below the tropopause and positive potential vorticity fluxes are apparent on the poleward side near the surface. Again, these are not upgradient because $[q]_y$ has become weak and mostly negative in this region. By day 8 growth of eddy kinetic energy has virtually ceased and wave activity is decreasing in middle latitudes as it is radiated towards the tropics. As expected from Eqn. (7.7) and the subsequent arguments, Rossby wave radiation leads to a positive (upgradient) flux of q in the middle latitudes and negative values at lower latitudes. Note from Eqns. (7.18) and (7.29) that this radiation towards the tropics is accompanied by large poleward momentum fluxes. The time-average flux picture shows the poleward potential vorticity flux at low levels, and the equatorward flux in the sub-tropical upper troposphere. There is cancellation in the middle latitude upper troposphere between the negative values typified by day 5 and the positive values typified by day 8, indicating the reversible, wave-like dynamics in that region. It might be expected that the residual of this cancellation might be sensitive to the actual flows used, as was indeed found by Wyatt (1981). It is interesting to compare the model EP fluxes in Fig. 7.5 with those for the Northern Hemisphere winter inferred from observations and given in Fig. 8.9.

We conclude that simple arguments are sufficient to qualitatively explain the behaviour of the fluxes at each stage in the wave development. However, the average picture, which is what might be required for parameterizations, is not simply explained in terms of the initial, final or average $[q]_y$. Nor is it

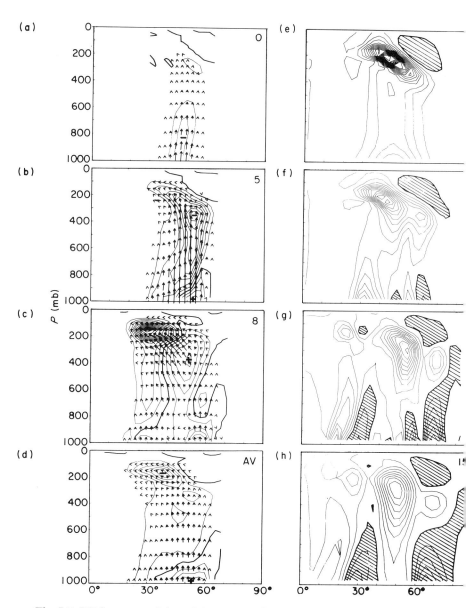

Fig. 7.5. EP flux, potential vorticity flux and $[q]_y$ during the life-cycle of a baroclinic wave. (a), (b) and (c) are the fluxes for days 0, 5 and 8 and (e), (f) and (g) the $[q]_y$ for the same days. (d) shows the time-averaged fluxes and (h) the day 15 $[q]_y$. The fluxes in (a) have been multiplied by 100 and the contour increment in (d) is reduced by a factor 0.375. The contour interval for $[q]_y$ is Ω/a. Zero contours are heavy and negative regions hatched.

explained solely in terms of either linear growth, or Rossby wave propagation, but rather (a) linear growth, then (b) non-linear cessation of growth leading to Rossby wave radiation upwards and then equatorwards and finally (c) an absorption process akin to critical layer absorption on the tropical flank of the jet. We also note that the changes in the $[q]_y$ picture are qualitatively explicable in terms of $[v^*q^*]$ plus a knowledge that the vertical flux of heat makes the atmosphere more statically stable near $45°$, leading to larger values of potential vorticity there and decreased $[q]_y$ on the poleward side. This is particularly important in causing the reversed $[q]_y$ at low levels by day 5.

Some major points from this experiment are:

1. the importance of the upward and equatorward propagation of wave activity after cessation of linear growth;

2. the heat flux peaks earlier than the momentum flux and linear theory will tend to underestimate the latter;

3. the vertical flux of heat is important and continues approximately to obey Eqn. (7.23) throughout the life cycle;

4. the pattern of the flux $[u^*w^*]$ is qualitatively similar to that of $[u^*v^*]$ and its sign is the same, during the whole life cycle. The maximum convergence of the vertical flux is a factor of 4 smaller than that of the horizontal flux;

5. parameterizations should not be local in y for typical jets with half widths, as here, comparable with the radius of deformation (NH/f). This contrasts with the very broad flow considered by Haidvogel and Held (1980). Equally, one would not expect that parameterizations can be global in character, depending only on differences between polar and equatorial values of basic parameters.

Life-cycle experiments have also been performed (Simmons and Hoskins, 1980) on zonal flows with the same baroclinity but added barotropic components with maximum velocities of 10 m s^{-1}. The life-cycle averaged potential vorticity and heat fluxes were found to vary by up to a factor of $2-3$ between these experiments though the pattern was the same except in one experiment. In this case the radiation towards the tropics appeared to be inhibited by a minimum in refractive index $\{[q]_y/([u] - c_r)\}^{1/2}$. The notion of a refractive index for Rossby wave propagation is discussed in Karoly and Hoskins (1982) and its successes and limitations (as well as many references to the stratospheric use of the idea) by McIntyre (1982).

Wyatt (1981) has used similar normal mode and life-cycle calculations to investigate Stone's baroclinic adjustment hypothesis and in particular whether the strength of the eddies depends strongly on λ, where λ is defined in Eqn. (7.26) or (7.27). The zonal flows used were 0.8, 1.0 and 1.2 times the Northern

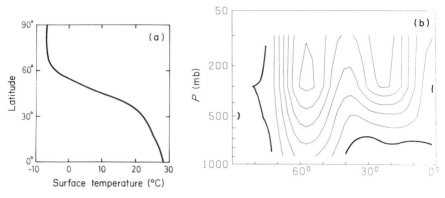

Fig. 7.6. The specified surface temperature distribution as a function of latitude and the day 4 to 67 time averaged, zonally averaged zonal wind as a function of latitude and pressure for the simple general circulation model described in the text. In (b) the zonal wind contours are drawn every 5 m s^{-1} with the zero contour heavy.

Hemisphere winter climatological zonal mean. The first two were also investigated using a planetary angular rotation 0.8 times that of the Earth. Linear calculations showed that shorter wavelength modes grew near 40°–45° and longer wavelengths near 60° with growth rates much as suggested by simple β-plane theory. However, the modes near 40°–45° were all shallow and those near 60° were all deep in response to the details of the atmospheric structure at that latitude. There was little indication of the variation with the baroclinity given by the simple theory. Non-linear life-cycle calculations always gave small heat fluxes for wavenumber 9. Longer wavelengths all exhibited essentially similar development though the time scale varied. Generally it did not appear that there was strong evidence to support the baroclinic adjustment hypothesis.

Finally in this section we shall very briefly describe the result of including in a simple manner various physical processes into a baroclinic wave life-cycle experiment. The calculation used a five-layer, spectral primitive equation model truncated at total wavenumber 42 on a domain symmetric about the equator and with sixfold symmetry in each hemisphere. The simple boundary layer scheme provided for vertical fluxes of momentum, sensible heat and water vapour into the interior. The surface temperature and humidity are fixed at their initial values, mimicking an ocean-covered surface. Water vapour is retained as one of the model variables and latent heat is released through convective and large-scale precipitation parameterizations. The radiation parameterization is a 1.25 K day^{-1} cooling everywhere except at the top level and near the pole. There is also an internal, fourth-order diffusion. In Fig. 7.6(b) is shown the day 4–67 average zonally averaged zonal wind from such a run starting from a realistic zonal flow and using the surface temperature shown in Fig. 7.6(a). There are two westerly jets. The existence of the subtropical jet depends on the imposed surface temperature distribution having a gradient in the tropical region. In experiments without this gradient, the convective maximum at the equator, the Hadley Cell and the subtropical jet are absent. This is entirely consistent with the arguments presented in Section 7.3.1. The polar jet owes its existence to the baroclinic eddies. The imposed surface temperature gradient is a maximum at 45°, the eddy kinetic energy is a maximum near 52° and the poleward westerly momentum flux convergence peaks near 61°.

The double jet structure obtained in this simple model is quite similar to that observed frequently in the Southern Hemisphere, particularly in the equinoctial seasons. This is consistent with the study of Williams (1981) which suggests that the parameters of the Earth's atmosphere are on the borderline of having two jets rather than one. In the Northern Hemisphere this is presumably masked to a large extent by the asymmetries introduced by the large continents and their topography.

7.4 Time-averaged three-dimensional flow

7.4.1 Basic equations

The longitudinal variations in weather and climate are so significant that the zonally averaged (y, z, t) problem could be considered to have little direct practical relevance. Because the zonal average covers such different regions, the parameterization of eddies in that case may be more subtle than in the three-dimensional space problem using time averaging. In this section we shall exhibit some of the equations analogous to those given in Section 7.3.1 but for the time-average operator ($^-$) and deviation ( )'.

Ignoring the vertical advection of momentum, the equations for the time averaged momentum and potential temperature may be written:

$$\bar{D}\bar{u} = f\bar{v} - \bar{\phi}_x - \overline{(u'u')}_x - \overline{(u'v')}_y + \bar{\mathscr{F}}, \qquad (7.33)$$

$$\bar{D}\bar{v} = -f\bar{u} - \bar{\phi}_y - \overline{(u'v')}_x - \overline{(v'v')}_y + \bar{\mathscr{G}}, \qquad (7.34)$$

$$\bar{D}\bar{\theta} = -\bar{\omega}\bar{\theta}_p - \overline{(u'\theta')}_x - \overline{(v'\theta')}_y - \overline{(\omega'\theta')}_p + \bar{\mathscr{Q}}, \qquad (7.35)$$

where $\bar{D} = \bar{u}\,\partial/\partial x + \bar{v}\,\partial/\partial y$. Again the momentum equations can be rewritten in terms of an eddy vorticity flux provided that the eddies are approximately horizontally non-divergent:

$$\bar{D}\bar{u} = f\bar{v} - (\bar{\phi} + K_e)_x + \overline{v'\zeta'} + \bar{\mathscr{F}}, \qquad (7.36)$$

$$\bar{D}\bar{v} = -f\bar{u} - (\bar{\phi} + K_e)_y - \overline{u'\zeta'} + \overline{\mathscr{G}} \tag{7.37}$$

where the transient eddy kinetic energy $K_e = \frac{1}{2}\overline{u'^2 + v'^2}$.

As in Section 7.3.1, it is of interest to obtain forms of the equations involving the quasi-geostrophic potential vorticity flux. We now use a quasi-geostrophic potential vorticity defined as in Eqn. (7.2). Taking the stability in Eqn. (7.35) to be a function of p only, Eqns. (7.33)–(7.35) become:

$$\bar{D}\bar{u} = \overline{v'q'} + f_0\tilde{v} + \beta y\bar{v} + \overline{\mathscr{F}}, \tag{7.38}$$

$$\bar{D}\bar{v} = -\overline{u'q'} - f_0\tilde{u} - \beta y\bar{u} + \overline{\mathscr{G}}, \tag{7.39}$$

and

$$\bar{D}\bar{\theta} = -\tilde{\omega}\Theta_p - (\overline{\omega'\theta'})_p + \overline{2}. \tag{7.40}$$

Here the 'residual ageostrophic motion' is:

$$\tilde{u} = u + \frac{1}{f_0}(\phi + K_e - P_e)_y - (\overline{u'\theta'}/\Theta_p)_p, \tag{7.41a}$$

$$\tilde{v} = v - \frac{1}{f_0}(\phi + K_e - P_e)_x - (\overline{v'\theta'}/\Theta_p)_p, \tag{7.41b}$$

$$\tilde{\omega} = \omega + \nabla_h \cdot (\overline{\mathbf{v}'\theta'}/\Theta_p), \tag{7.41c}$$

and

$$P_e = -\tfrac{1}{2}\hat{R}\,\overline{\theta'^2}/\Theta_p \quad \text{is the transient eddy potential energy.} \tag{7.42}$$

The 'omega equation' for $\tilde{\omega}$ with its boundary condition on $p = p_0$ is a simple extension of Eqns. (7.16) and (7.17), the equation being obtained by eliminating $\partial/\partial t$ between Eqns. (7.37)–(7.39). Eliminating the residual ageostrophic motion from Eqns. (7.38)–(7.40) using its non-divergence simply gives the mean flow potential vorticity equation (7.5a), provided that the vertical eddy heat flux convergence term is absorbed into the diabatic heating.

On the basis of the above equations, a parameterization of the transient eddy effects on the mean flow requires knowledge of either:

(a) $\overline{v'u'}, \overline{v'v'}, \overline{v'\theta'}, \overline{\omega'\theta'},$ (for (7.33)–(7.35))

or (b) $\overline{v'\zeta'}, \overline{v'\theta'}, \overline{\omega'\theta'},$ (for (7.35)–(7.37))

or (c) $\overline{v'q'}, \overline{\omega'\theta'}$ and boundary $\overline{v'\theta'}.$ (for (7.38)–(7.40))

Additionally, the presence of the eddies will lead to a modification of the frictional and diabatic terms $\overline{\mathscr{F}}, \overline{\mathscr{G}}$ and $\overline{2}$.

The simplest complete discussion is one based on quasi-geostrophic potential vorticity, but, even more than in the zonally averaged case, this suffers from the approximations that have to be made. In particular, the observed static stability varies significantly over synoptic length scales. Any discussion based on the more complicated Eqns. (7.33)–(7.35) or (7.35)–(7.37) suffers from not having used the observed fact that the time mean flow is approximately horizontally non-divergent. For this reason, it is worth considering the vorticity of the mean flow. Subtracting $\partial/\partial y$ of Eqn. (7.36) from $\partial/\partial x$ of Eqn. (7.37) gives:

$$\bar{D}\bar{\zeta} = -\beta\bar{v} - (f+\bar{\zeta})\overline{\mathscr{D}} - \nabla\cdot\overline{\mathbf{v}'\zeta'} + \overline{\mathscr{G}}_x - \overline{\mathscr{F}}_y, \tag{7.43}$$

where $\mathscr{D}$ is the divergence. Note that, from Eqns (7.38) and (7.39), the quasi-geostrophic form of Eqn. (7.43) may be written:

$$\bar{D}\bar{\zeta} = -\beta\bar{v} - \nabla\cdot\overline{v'q'} + f_0\,\partial\bar{\omega}/\partial p + \overline{\mathscr{G}}_x - \overline{\mathscr{F}}_y. \tag{7.43'}$$

As in the zonally averaged case, the vertical integral of the thermal portion of the eddy vorticity flux convergence is cancelled by $f_0\bar{\omega}|_{p=p_0}$. Again it is only the mechanical forcing by the eddies that occurs in the vertical average vorticity budget. It is clear from Eqns. (7.43) and (7.35) that it is only the divergent part of the ageostrophic motion that is really relevant. The rotational part may be determined by adding $\partial/\partial x$ of Eqn. (7.36) and $\partial/\partial y$ of Eqn. (7.37) and assuming approximate horizontal non-divergence. This gives the 'balance equation':

$$f\zeta_{ag} = 2(\bar{\psi}_{xy}^2 - \bar{\psi}_{xx}\bar{\psi}_{yy}) + \beta\bar{\psi}_y - \mathbf{k}\cdot\nabla \wedge \overline{\mathbf{v}'\zeta'} + \nabla^2 K_e - \overline{\mathscr{F}}_x - \overline{\mathscr{G}}_y, \tag{7.44}$$

where ζ_{ag} is the ageostrophic vorticity.

As pointed out by Holopainen (1978), for approximately horizontally non-divergent eddies, the horizontal eddy vorticity flux is given by:

$$\overline{\mathbf{v}'\zeta'} = (M_x + N_y, -M_y + N_x). \tag{7.45}$$

Here

$$M = \overline{u'v'}, \tag{7.46}$$

and

$$N = \tfrac{1}{2}(\overline{v'^2 - u'^2}). \tag{7.47}$$

The symmetry of Eqns. (7.46) and (7.47) may be demonstrated by considering the velocity components $\hat{u}$ and $\hat{v}$ along axes that are rotated from the basis set by $\pi/4$.
Then

$$M = -\tfrac{1}{2}(\overline{\hat{v}'^2 - \hat{u}'^2}), \tag{7.46'}$$

and

$$N = \overline{\hat{u}'\hat{v}'}. \tag{7.47'}$$

The divergence and curl of the vorticity flux that occur in Eqns. (7.43) and (7.44), respectively, may be written:

$$\nabla\cdot\overline{\mathbf{v}'\zeta'} = M_{xx} - M_{yy} + 2N_{xy}, \tag{7.48}$$

and:

$$\mathbf{k}\cdot\nabla \wedge \overline{\mathbf{v}'\zeta'} = -2M_{xy} + N_{xx} - N_{yy}. \tag{7.49}$$

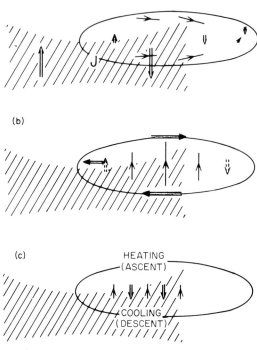

Fig. 7.7. An idealized upper tropospheric jetstream and storm-track in the Northern Hemisphere. The shaded region indicates the area between two streamlines, J the jet maximum, and the continuous contour the region of maximum synoptic time-scale eddy activity. Arrows with single shafts indicate vector eddy fluxes and those with double shafts represent the upper tropospheric ageostrophic vorticity that could balance specific terms in the equations. More details are given in the text. The budgets are for: (a) westerly momentum at upper tropospheric levels; (b) northerly momentum at upper tropospheric levels; and (c) heat at lower tropospheric levels.

7.4.2 Storm-track organization

As shown by Blackmon *et al.* (1977) and Lau (1979a) and supported by the study described in Chapter 1, the Northern Hemisphere synoptic eddy activity as evidenced by statistics based on time-filtered data is organized into storm tracks that are predominantly downstream and poleward of the upper tropospheric jet maxima. Such a situation is shown schematically in Fig. 7.7. In reality there is an angle of 5°–10° between the jet and storm track axes and the zonal direction, but this may be neglected in the simple budgets to be performed. We shall use typical values of quantities taken from these studies and others performed with the ECMWF data (as in Chapter 1) to illustrate the magnitudes of the terms in the mean flow equations derived above. We shall use length scales for both the jet and the storm track of $L_y \sim 750$ km in the cross-direction and $L_x \sim 3000$ km in the long-direction.

Concentrating on the x-momentum equation (7.33), the magnitude of each term in the upper troposphere will be expressed and illustrated in terms of an ageostrophic velocity necessary to balance the term, f being taken as 10^{-4} s^{-1}. For a zonal wind with characteristic value $\bar{u} \sim 40$ m s^{-1} and variation $\Delta\bar{u} \sim 20$ m s^{-1}, and a meridional wind $\bar{v} \sim 10$ m s^{-1} the mean x-acceleration is of the order $\bar{u}\,\Delta\bar{u}/L_x \sim 2.7 \times 10^{-4}$ m s^{-2} which would be balanced by $\bar{v}_a \sim 2.7$ m s^{-1} in the sense indicated in Fig. 7.7(a). For eddies of synoptic time scales, at upper tropospheric levels we may take a simple maximum for $\overline{u'^2} \sim 100$ m^2 s^{-2} in the storm-track. The poleward flux $\overline{u'v'}$ tends to have a dipole distribution centred slightly downstream of the middle of the storm-track and a predominance of positive values at the end of the track (as suggested by Fig. 1.13). This is reminiscent of the large poleward fluxes towards the end of the baroclinic wave life-cycle described in Section 7.2.4. Taking $\overline{u'v'} \sim 25$ m^2 s^{-2} the vector flux $\overline{\mathbf{v}'u'}$ looks as in Fig. 7.7(a). Its convergence would be balanced by $\bar{v}_a \sim 0.3$ m s^{-1} with a distribution as shown. Since this ageostrophic wind is an order of magnitude smaller than that associated with the mean flow acceleration, it appears that the eddies are of little direct importance in the long-track momentum equation.

The y-momentum equation [(7.34) and Fig. 7.7(b)] gives a mean acceleration at upper tropospheric levels $\sim \bar{u}\bar{v}/L_x$ which would be balanced by $\bar{u}_a \sim 1.3$ m s^{-1}. Taking $\overline{v'^2} \sim 200$ m^2 s^{-2}, this component dominates the transient eddy flux $\overline{\mathbf{v}'v'}$ and its convergence which would give $\bar{u}_a \sim 2.7$ m s^{-1}, as indicated in Fig. 7.7(b). The real importance of the eddies is probably rather less than this would suggest, because a horizontally non-divergent ageostrophic motion $f_0^{-1}(-(\overline{v'^2})_y, (\overline{v'^2})_x)$ could balance the large eddy term in Eqn. (7.34) and have a y-component ~ 0.7 m s^{-1} (dashed in Fig. 7.7b), which is small compared with the mean flow arrows shown in Fig. 7.7(a). Such an anticyclonic ageostrophic circulation around storm tracks is often found in studies of the ECMWF data like those in Chapter 1. An alternative view is gained by considering Eqn. (7.44). Using the magnitudes given above and Eqn. (7.49), the major eddy contribution is through the term $\nabla^2 K_e$. In a region of maximum K_e, we may expect this term to be negative and the implied ageostrophic vorticity to be negative also ($\sim -3 \times 10^{-6}$ s^{-1}).

We now consider the thermodynamic equation (7.35). In the lower troposphere we may take $\overline{v'\theta'} \sim (5, 15)$ m s^{-1} K so that the warming and cooling shown in Fig. 7.7(c) are of the order of 2×10^{-5} K s^{-1}. In the upper troposphere the heat flux convergence is probably an order of magnitude smaller. In order to compare the importance of the eddy heat flux with the eddy terms in the momentum budgets, we might assume that the eddy warming and cooling are balanced by ascent and descent, respectively, with equatorward ageostrophic motion at upper levels. A heating of 2 $\times 10^{-5}$ K s^{-1} could be balanced by ascent with $\omega \sim -0.05$ Pa s^{-1} corresponding to a divergence at upper levels $\mathscr{D} \sim 2 \times 10^{-6}$ s^{-1}. Taking $\partial \bar{v}_a / \partial y$ $\sim \mathscr{D}$ gives an upper bound (probably an overestimate by a factor of order 3) for the equatorward ageostrophic motion at upper levels of 1.5 m s^{-1}. This would be the largest eddy associated, meridional ageostrophic motion but still smaller than that implied by the changes in $\bar{u}$. Lau (1979b) produces convincing evidence that the transient eddies act as a damper not only on the zonally averaged baroclinity but also on the stationary long zonal wavelengths in the temperature field. The implied decay time is of the order of one week or less. This is also illustrated in Fig. 1.7.

The other direct eddy effect is to produce an upward flux of heat. Taking $\overline{\omega' \theta'} \sim -0.3$ K Pa s^{-1} gives a warming at upper levels and cooling at lower levels of the order of 1 $\times$ 10^{-5} K s^{-1}, which is very important in maintaining the static stability of the middle-latitude troposphere. The thermal advection by the mean flow, represented by the left-hand side of Eqn. (7.35), was found by Lau (1979b) at 700 mb to be generally of the same order as the eddy flux convergence. Higher in the atmosphere, the velocities are larger, but the mean temperature contours tend to be parallel to the mean streamlines, so that thermal advection by the mean flow does not increase.

Lau (1979b) also exhibits the 700 mb diabatic heating obtained as a residual. It is of similar magnitude to the other terms in Eqn. (7.35) and has a distribution very similar to that of the storm track. This result is true also for the GCM results illustrated in Fig. 6.19. It is most likely that this heating is due to the release of latent heat associated with precipitation in the synoptic systems moving along the storm-track. The mid-latitude mean diabatic heating thus owes much of its structure to the presence of the transient eddies.

The alternative to a consideration of the momentum balance is to evaluate the terms in the vorticity equation (7.43). The eddy vorticity flux (7.45) and its convergence (7.48) will be discussed by separating the contributions from M and N. Figure 7.8(a) shows a schematic upper tropospheric distribution of M and Fig. 7.8(b) the implied vorticity flux and convergence, neglecting the small x-component. Figure 7.8(c) shows an upper tropospheric distribution for N for which there is less observational basis but there is every indication that synoptic time-scale eddies are elongated in the y direction and thus have $\overline{v'^2}$ larger than $\overline{u'^2}$ (see Fig. 3.3b). The implied vorticity flux and convergence are given in Fig. 7.8(d). In Figs. 7.8(b) and (d) the convergences are all of the order 0.5×10^{-10} s^{-1} and could be balanced by a divergence $\mathscr{D} \sim 0.5 \times 10^{-6}$ s^{-1}, somewhat smaller than that which could balance the eddy heat flux convergence. The total vorticity flux convergence and the implied forcing of

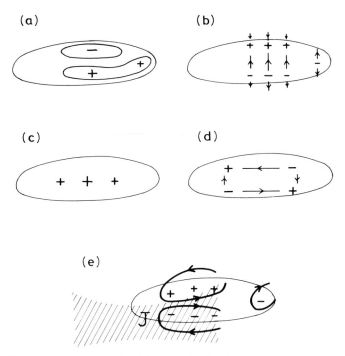

Fig. 7.8. The upper tropospheric eddy vorticity flux in a storm-track. Shown are the schematic distributions of: (a) $M = [u'v']$; (b) the eddy vorticity flux and convergence associated with M; (c) $N = \frac{1}{2}[v'^2 - u'^2]$; (d) the eddy vorticity flux and convergence associated with N; (e) the total eddy vorticity flux convergence and the implied forcing of the mean horizontal circulation.

the mean horizontal circulation are shown in Fig. 7.8(e). This illustrates the tendency for the eddies to move the jet polewards and to split the westerly flow at the diffluent end of the jet. Referring again to the mean vorticity equation (7.43), the mean vorticity advection and the β term are both very large but, as demonstrated by Lau (1979b), absolute vorticity contours tend to parallel streamlines so that they cancel to a large extent. The vertical integral of these terms, according to Holopainen and Oort (1981b), can also produce further cancellation. The vertical integral of the mean divergence term is negligible and that of the frictional terms reduces to the curl of the surface stress. As shown by Holopainen and Oort (1981a,b), the vertical integral of the eddy vorticity flux convergence appears to play a major role in balancing the surface stress. Figure 7.8(e) thus indicates how the transient eddies act to maintain surface low pressure to the north and high pressure to the south and east of the storm-track.

It is of interest to note, from Eqn. (7.48), that to a good approximation the

Fig. 7.9. An alternative view of the upper tropospheric time-mean mechanical forcing by eddies in a storm-track. The short vectors show schematic distributions of $[u'^2 - v^2]$, $[u'v']$. As indicated by Eqn. (7.50), their convergence implies a forcing of mean horizontal circulation consistent with increased westerlies. This circulation, identical with that in Fig. 7.8(e), is indicated.

eddy vorticity flux convergence may be written:

$$-M_{yy} + 2N_{xy} = -\partial/\partial y \nabla \cdot (\overline{u'^2 - v'^2}, \overline{u'v'}). \qquad (7.50)$$

The quantity in brackets may be thought of as the vector eddy flux of westerly momentum plus an extra term which comes as discussed above from the ageostrophic motion necessary to balance the y-momentum equation. The total mechanical forcing of the mean horizontal circulation by storm-track eddies given in Fig. 7.8(e) may be simply determined by plotting $(\overline{u'^2 - v'^2}, \overline{u'v'})$ vectors, and noting that their convergence implies a forcing of circulation consistent with increased westerlies. Further discussion of this is given in Hoskins *et al.* (1983).

Chapter 8 contains details of many observational studies of the mean effects of transients. It also introduces some new observational results based on a quasi-geostrophic potential vorticity analysis.

7.4.3 Some parameterization ideas

One plausible way to consider the parameterization of the transient fluxes in a storm-track is to hypothesize that the growth and decay of non-linear baroclinic waves as they move along the track is similar to that occurring in time in the life-cycle experiments, although an interesting complication is that the Rossby wave radiation effects which follow the cessation of linear growth must now take place in three dimensions rather than two. Then eddy fluxes at the beginning of the track should be similar to those during the initial stages of the life-cycle and those near the end might be similar to the decay stage. In particular, as was discussed by Lau (1979a), this would suggest a maximum in cross-track heat flux, followed downstream by a maximum in eddy kinetic energy and then a maximum in the poleward flux of westerly momentum in agreement with the observations. All the necessary flux quantities may be inferred from such a calculation assuming a slow variation in x. In reality there is a major problem associated with the fact that the eddy structure and transfers change markedly over the 3 days, during which it moves only one-half of a wavelength.

Potential vorticity transfer ideas may also be applied in the case of two horizontal dimensions. Equation (7.7) now tells us that in a storm-track entrance region we may expect downgradient transfers. In a storm-track exit, if eddies are decaying reversibly and not through enstrophy cascade and dissipation, upgradient transfer is anticipated. However, Eqn. (7.7) gives no indication of what transfers occur along $\bar{q}$ and boundary $\bar{\theta}$ contours. From the results quoted in Chapter 8, it is probable that convergences associated with long-$\bar{\theta}$ heat fluxes may be neglected, but it is less clear that this is true for long-$\bar{q}$ potential vorticity fluxes.

Another approach pioneered by Frekeriksen (1978, 1979a,b, 1980) is to study the linear stability of a basic state consisting of a long wave as well as a zonal flow. In general it was found that the normal mode eddy characteristics were linked with the baroclinity parameter λ. Maxima in eddy kinetic energy and in poleward fluxes of heat and westerly momentum occurred downstream of maxima in λ. Similar results have also been obtained by Niehaus (1980, 1981) for an extension of the Eady problem using numerical and two-scaling approaches.

Since non-linearity is so important in changing the pattern of the eddy fluxes on a zonal flow, it would seem that future efforts must be directed towards determining the behaviour of non-linear baroclinic waves on a flow with zonal variation.

As indicated by the above discussion, the quantity $\overline{u'^2 - v'^2}$ is important in determining the local vorticity flux convergence. It appears that a representation of this term for a long-period average of the synoptic eddies may be simpler than for the much discussed $\overline{u'v'}$ because these eddies are elongated predominantly in the meridional direction (see Fig. 3.3b).

7.4.4 Discussion

The idealized storm-track budgets discussed in Section 7.4.2 suggested that the transient eddy terms are secondary in the mean flow equations with the exception of the heat flux convergence which, as will be stressed in Chapter 8, tends to act as a damper on the long-wave flow. Consequently, one might speculate that a parameterization of the transient effects other than by increasing the damping coefficients is unnecessary. For various reasons this idea is probably incorrect. It is clear that the eddies are important in the budgets for zonal mean quantities (see for example Jeffreys, 1926; Pfeffer, 1981) and it would appear that the local climate could not be accurately simulated if the zonal mean was far from correct. Also, we have indicated how the eddy vorticity flux convergence is important in balancing the mean surface stress.

Opsteegh and Vernekar (1982) have used a steady-state, linearized model similar to those described in Chapter 6 to compute the response to observed vorticity and heat flux convergences by the eddies as well as to mountains and diabatic sources. They found the eddy vorticity forcing to be very important in obtaining a realistic January solution for the middle latitudes of the Northern Hemisphere.

Green (1977) and Illari (1981) have argued that momentum and vorticity fluxes were crucial in the maintenance of the persistent warm anticyclone near the British Isles during 1976. Unpublished diagnostics using the ECMWF data have also suggested that, in particular two-week periods, the transient eddies terms can be large, particularly in the jet-exit or end of storm-track regions.

An interesting new idea comes from the study by Reinhold and Pierrehumbert (1982). They have taken the highly truncated model of Lorenz (1963) which, as Charney and Strauss (1980) have shown, predicts multiple equilibria for given external forcing when the lower boundary is undulating. (Multiple equilibria in barotropic models are discussed in Chapter 6.) To it they have added a shorter wave which is baroclinically very unstable. The resulting model when integrated shows a distinct tendency for the long-wave flow to be in one of two patterns for a period of random duration. These 'weather regimes' are somewhat different from the multiple equilibria and exist in cases in which the equilibria are unstable. It is found that, when the model atmosphere is in one of the weather regimes, the mean effect of the baroclinic eddy term in the budget for the long-wave is small and may be damping or amplifying. However, the very existence of the weather regime depends on the presence of the eddy. Thus, although a budget study may show that the mean values of the term associated with synoptic scale transients are small, the presence of the transients may be crucial to the existence of the persistent long-wave flow. A further role played by the baroclinic eddy in this model is, by means of an extreme event, to move the flow out of or into a weather regime. The possibility of weather regimes in the atmosphere was discussed in Chapter 3 and tentative examples given in Figs. 3.14–3.17.

7.5 Conclusion

The modelling of the fluxes performed by the transient eddies is probably an important part of modelling the general circulation of the atmosphere. The same statement and the ideas discussed here are also relevant to the oceans. Our understanding of the role and behaviour of the transient eddies has progressed over the last decade, but much remains to be learnt about the 3-dimensional problem. One important aspect about which little has been written is the effect of the transient eddies on the mean forcing and dissipation terms, $\overline{\mathscr{F}}$, $\overline{\mathscr{G}}$ and $\overline{\mathscr{Q}}$. Because of the nonlinearity of surface stress and internal mixing, the presence of eddies will lead to different values of $\overline{\mathscr{F}}$ and $\overline{\mathscr{G}}$. Eddies also transport water vapour poleward and upward and greatly modify the mean heating $\overline{\mathscr{Q}}$, particularly through latent heat release but also through surface fluxes and radiational effects.

Finally, most of the specific parameterization ideas discussed here and in the literature have been oriented towards the fluxes due to synoptic eddies. However, as stressed by Chapters 3 and 8, fluxes due to eddies on longer time and space scales are very important in the general circulation.

Acknowledgements

The author is indebted to many people for discussions on the subject matter of this chapter. In particular, helpful comments on a draft of the chapter were made by Drs D. G. Andrews, J. S. A. Green, M. E. McIntyre, R. P. Pearce and J. M. Wallace.

References

ANDREWS, D. G. and McINTYRE, M. E. (1976). Planetary waves in horizontal and vertical shear: the generalized Eliassen–Palm relation and the mean zonal acceleration. *J. atmos. Sci.*, **33**, 2031–2048.

BARCILON, A. and PFEFFER, R. L. (1979). Further calculations of eddy heat fluxes and temperature variances in baroclinic waves using weakly nonlinear theory. *Geophys. Astrophys. Fluid Dyn.*, **12**, 45–60.

BLACKMON, M. L., WALLACE, J. M., LAU, N.-C. and MULLEN, S. L. (1977). An observational study of the Northern Hemisphere wintertime circulation. *J. atmos. Sci.*, **34**, 1040–1053.

BRETHERTON, F. P. (1966). Critical layer instability in baroclinic flows. *Qt. Jl R. met. Soc.*, **92**, 325–334.

BRETHERTON, F. P. and GARRETT, C. J. R. (1968). Wavetrains in inhomogeneous moving media. *Proc. R. Soc. A*, **302**, 529–554.

CHARNEY, J. G. and STRAUSS, D. M. (1980). Form-drag instability, multiple equilibria and propagating planetary waves in baroclinic, orographically forced, planetary wave systems. *J. atmos. Sci.*, **37**, 1157–1176.

EADY, E. T. (1949). Long waves and cyclone waves. *Tellus*, **1**, No. 3, 33–52.

EDMON, H. J., HOSKINS, B. J. and McINTYRE, M. E. (1980). Eliassen–Palm cross-sections for the troposphere. *J. atmos. Sci.*, **37**, 2600–2616 (see also Corrigendum, *J. atmos. Sci.*, **38**, 1115).

ELIASSEN, A. and PALM, E. (1961). On the transfer of energy in mountain waves. *Geophys. Publ.*, **22**, No. 3, 1–23.

FREDERIKSEN, J. S. (1978). Instability of planetary waves and zonal flows in two-layer models on a sphere. *Qt. Jl R. met. Soc.*, **104**, 841–872.

FREDERIKSEN, J. S. (1979a). The effects of long planetary waves on the regions of cyclogenesis: Linear theory. *J. atmos. Sci.*, **36**, 195–204.

FREDERIKSEN, J. S. (1979b). Baroclinic instability of zonal flows and planetary waves in multilevel models on a sphere. *J. atmos. Sci.*, **36**, 2320–2335.

FREDERIKSEN, J. S. (1980). Zonal and meridional variations of eddy fluxes induced by long planetary waves. *Qt. Jl R. met. Soc.*, **106**, 63–84.

GALL, R. (1976). A comparison of linear baroclinic instability theory with the eddy statistics of a general circulation model. *J. atmos. Sci.*, **33**, 349–373.

GREEN, J. S. A. (1960). A problem in baroclinic instability. *Qt. Jl R. met. Soc.*, **86**, 237–251.

GREEN, J. S. A. (1970). Transfer properties of the large-scale eddies, and the general circulation of the atmosphere. *Qt. Jl R. met. Soc.*, **96**, 157–184.

GREEN, J. S. A. (1977). The weather during July 1976: Some dynamical considerations of the drought. *Weather*, **32**, 120–126.

HAIDVOGEL, D. B. and HELD, I. M. (1980). Homogeneous quasi-geostrophic turbulence driven by a uniform temperature gradient. *J. atmos. Sci.*, **37**, 2644–2660.

HELD, I. M. (1975). Momentum transport by quasi-geostrophic eddies. *J. atmos. Sci.*, **32**, 1494–1497.

HELD, I. M. (1978). The vertical scale of an unstable baroclinic wave and its importance for eddy heat flux parameterisations. *J. atmos. Sci.*, **35**, 572–576.

HELD, I. M. and HOU, A. Y. (1980). Nonlinear axially symmetric circulations in a nearly inviscid atmosphere. *J. atmos. Sci.*, **37**, 515–533.

HOLLINGSWORTH, A., SIMMONS, A. J. and HOSKINS, B. J. (1976). The effect of spherical geometry on momentum transports in simple baroclinic flows. *Qt. Jl R. met. Soc.*, **102**, 901–911.

HOLOPAINEN, E. O. (1978). On the dynamic forcing of the long-term mean flow by the large-scale Reynolds stresses in the atmosphere. *J. atmos. Sci.*, **35**, 1596–1604.

HOLOPAINEN, E. O. and OORT, A. H. (1981a). Mean surface stress curl over the oceans as determined from the vorticity budget of the atmosphere. *J. atmos. Sci.*, **38**, 262–269.

HOLOPAINEN, E. O. and OORT, A. H. (1981b). On the role of large-scale transient eddies in the maintenance of the vorticity and enstrophy of the time-mean atmospheric flow. *J. atmos. Sci.*, **38**, 270–280.

HOLTON, J. R. (1979). *An Introduction to Dynamic Meteorology*, 2nd Edn. Academic Press. 391 pp.

HOSKINS, B. J. and BRETHERTON, F. P. (1972). Atmospheric frontogenesis models: Mathematical formulation and solution. *J. atmos. Sci.*, **29**, 11–37.

HOSKINS, B. J., JAMES, I. N. and WHITE, G. H. (1983). The shape, propagation and mean-flow interaction of large scale weather systems. *J. atmos. Sci.*, **40** (in press).

ILLARI, L. (1981). Warm blocking highs: A diagnostic study of the 1976 drought. Talk presented at the IAMAP Symposium on the Dynamics of the General Circulation of the Atmosphere, Reading.

JEFFREYS, H. (1926). On the dynamics of geostrophic winds. *Qt. Jl R. met. Soc.*, **52**, 85–104.

KAROLY, D. and HOSKINS, B. J. (1982). Three dimensional propagation of planetary waves. *J. met. Soc. Japan*, **60**, 109–123.

LAU, N.-C. (1979a). The structure and energetics of transient disturbances in the Northern Hemisphere wintertime circulation. *J. atmos. Sci.*, **36**, 982–995.

LAU, N.-C. (1979b). The observed structure of tropospheric stationary waves and the local balances of vorticity and heat. *J. atmos. Sci.*, **36**, 996–1016.

LINDZEN, R. S. and FARRELL, B. (1980a). A simple approximate result for the maximum growth rate of baroclinic instabilities. *J. atmos. Sci.*, **37**, 1648–1654.

LINDZEN, R. S. and FARRELL, B. (1980b). The role of polar regions in global climate, and a new parameterization of global heat transport. *Mon. Weath. Rev.*, **108**, 2064–2079.

LORENZ, E. N. (1963). The mechanics of vacillation. *J. atmos. Sci.*, **20**, 448–464.

LORENZ, E. N. (1967). *The Nature and Theory of the General Circulation of the Atmosphere*. World Meteorological Organization, Geneva, Switzerland. 161 pp.

LORENZ, E. N. (1979). Forced and free variations of weather and climate. *J. atmos. Sci.*, **36**, 1367–1376.

MCINTYRE, M. E. (1982). How well do we understand the dynamics of stratospheric warmings? *J. met. Soc. Japan*, **60**, 37–65.

MCINTYRE, M. E. and WEISSMAN, M. A. (1978). On radiating instabilities and resonant over-reflection. *J. atmos. Sci.*, **35**, 1190–1196.

NIEHAUS, M. C. W. (1980). Instability of non-zonal baroclinic flows. *J. atmos. Sci.*, **37**, 1447–1463.

NIEHAUS, M. C. W. (1981). Instability of non-zonal baroclinic flows: Multiple-scale analysis. *J. atmos. Sci.*, **38**, 974–987.

NORTH, G. R. (1975). Theory of energy-balance climate models. *J. atmos. Sci.*, **32**, 2033–2043.

OPSTEEGH, J. D. and VERNEKAR, A. D. (1982). A simulation of the January standing wave pattern including the effects of transient eddies. *J. atmos. Sci.*, **39**, 734–744.

PEDLOSKY, J. (1979a). Finite amplitude baroclinic waves in a continuous model of the atmosphere. *J. atmos. Sci.*, **36**, 1908–1917.

PEDLOSKY, J. (1979b). *Geophysical Fluid Dynamics*. Springer-Verlag. 624 pp.

PFEFFER, R. L. (1981). Wave-mean flow interactions in the atmosphere. *J. atmos. Sci.*, **38**, 1340–1359.

PFEFFER, R. L. and BARCILON, A. (1978). Determination of eddy fluxes of heat and eddy temperature variances using weakly nonlinear theory. *J. atmos. Sci.*, **35**, 2099–2110.

PHILLIPS, N. A. (1954). Energy transformations and meridional circulations associated with simple baroclinic waves in a two-level, quasi-geostrophic model. *Tellus*, **6**, 273–286.

REINHOLD, B. B. and PIERREHUMBERT, R. P. (1982). Dynamics of weather regimes: Quasi-stationary waves and blocking. *Mon. Weath. Rev.*, **110**, 1105–1145.

SCHNEIDER, E. K. (1977). Axially symmetric steady-state models of the basic state for instability and climate studies. Part II: Nonlinear calculations. *J. atmos. Sci.*, **34**, 280–296.

SCHNEIDER, E. K. and LINDZEN, R. S. (1977). Axially symmetric steady-state models of the basic state for instability and climate studies. Part I: Linearized calculations. *J. atmos. Sci.*, **34**, 263–279.

SIMMONS, A. J. and HOSKINS, B. J. (1978). The life cycles of some nonlinear baroclinic waves. *J. atmos. Sci.*, **35**, 414–432.

SIMMONS, A. J. and HOSKINS, B. J. (1980). Barotropic influences on the growth and decay of nonlinear baroclinic waves. *J. atmos. Sci.*, **37**, 1679–1684.

STONE, P. H. (1972). A simplified radiative-dynamical model for the static stability of rotating atmospheres. *J. atmos. Sci.*, **29**, 405–418.

STONE, P. H. (1978). Baroclinic adjustment. *J. atmos. Sci.*, **35**, 561–571.

STONE, P. H. and CARLSON, J. H. (1979). Atmospheric lapse rate regimes and their parameterization. *J. atmos. Sci.*, **36**, 415–423.

TAYLOR, G. I. (1915). Eddy motion in the atmosphere. *Phil. Trans. R. Soc., Lond.*, A **215**, 1–26.

WHITE, A. A. (1977). The surface flow in a statistical climate model—a test of parameterization of large-scale momentum fluxes. *Qt. Jl R. met. Soc.*, **103**, 93–120.

WILLIAMS, G. P. (1981). On the general character of planetary circulations. Talk presented at the IAMAP Symposium on the Dynamics of the General Circulation of the Atmosphere, Reading.

WILLIAMS, R. T. (1972). Quasi-geostrophic versus non-geostrophic frontogenesis. *J. atmos. Sci.*, **29**, 3–10.

WYATT, L. R. (1981). Linear and nonlinear baroclinic instability of the Northern Hemisphere winter zonal flow. *J. atmos. Sci.*, **38**, 2121–2129.

— 8 —

Transient eddies in mid-latitudes: observations and interpretation

E. O. HOLOPAINEN

8.1 Introduction

One of the main characteristics of the atmospheric circulation in extratropical latitudes is its large variability. This 'large-scale turbulence' is present in any sample of consecutive weather maps and is vividly illustrated in films which show the behaviour of the atmospheric flow pattern as a function of time (e.g., Jenne *et al.*, 1974).

Formally, an arbitrary quantity $s = s(\lambda, \phi, p, t)$ can be written as:

$$s(\lambda, \phi, p, t) = \bar{s}(\lambda, \phi, p, \Delta t_s) + s'(\lambda, \phi, p, t, \Delta t_s) \tag{8.1}$$

where $\bar{s}$ is the time-mean (over sampling duration Δt_s) and s' the transient part of s ($\overline{s'} = 0$). Following a common terminology (e.g., Oort and Rasmusson, 1971; Newell *et al.*, 1972/74), all quantities containing s' will in the following be referred to as transient eddy (TE) quantities. A point to be remembered is that in this terminology TE contributions may arise not only from zonal waves but also from transient axisymmetric (i.e., longitudinally averaged) circulations.

The intensity and characteristics of the transient fluctuations depend upon the sampling duration Δt_s. In many of the pioneering contributions by V. Starr and his collaborators in the 1950s, the transients were defined as deviations from a yearly mean flow ($\Delta t_s = 1$ year) and thus contained a contribution not only from synoptic eddies but also from the seasonal variation. Some recent studies (e.g., Savijärvi, 1978; Youngblut and Sasamori, 1980) have considered transients defined with respect to a one-month mean ($\Delta t_s = 1$ month). Many of the observational studies described or referred to later in this article use data from many winters. In this case the transient eddies

Table 8.1. Different approaches (with examples) used in recent observational studies of transient eddies in the extratropics.

A. Studies of the total contribution of all transients ('Starr-type studies')

1. Zonally averaged statistics
   (Oort and Rasmusson, 1971; Newell *et al.*, 1972/74; Speth, 1974, 1978; van Loon, 1979; Lau and Oort, 1982)

2. Regional studies
   (Tucker, 1979; Holopainen *et al.*, 1980; Alestalo, 1981)

3. Geographical (λ, ϕ) patterns of some statistics of the transients
   (Speth, 1974; Blackmon, 1976; Blackmon *et al.*, 1977, 1979; Lau, 1978; Savijärvi, 1978; Lau, 1979a,b; Lau and Wallace, 1979; van Loon, 1980; White, 1980; Holopainen and Oort, 1981a,b; Lau and Oort, 1982; Lau *et al.*, 1981)

B. Studies on the space-time spectral structure and/or dynamics of the transients

1. Zonal harmonic analysis
   (Willson, 1975; Fraedrich and Böttger, 1978; Tsay and Kao, 1978; Gambo, 1978; Schäfer, 1979; Tomatsu, 1979; Kanamitsu, 1980; Speth and Kirk, 1981; Böttger and Fraedrich, 1980)

2. Spherical harmonic analysis
   (Baer, 1972; Blackmon, 1976; Burrows, 1976; Chen and Wiin-Nielsen, 1978; White, 1980)

3. Eof analysis
   (ECMWF, 1977; Rinne *et al.*, 1981)

4. Filtering
   (Blackmon, 1976; Blackmon *et al.*, 1977, 1979; Holopainen and Nurmi, 1980)

5. Special analyses
   (Lorenz, 1979; van Loon, 1979)

C. Case studies

(Palmén and Newton, 1969; Reed, 1979; Oerlemans, 1980b)

---

include, in addition to intraseasonal fluctuations of different type, contributions from the interannual variability.

Various approaches have been used in the observational studies of transient eddies (Table 8.1). One approach (Class A in Table 8.1) deals with the geographical variation of quantities describing the total effect of all transients (e.g., $\overline{u'u'}$, $\overline{T'v'}$) without any spectral considerations. These 'Starr-type studies' have provided, among other things, documentation of the zonally averaged characteristics of the transient eddies (e.g., Oort and Rasmusson, 1971; Newell *et al.*, 1972/74). Furthermore, in data-rich areas, regional budgets (including the effects of transients) have been worked out for some quantities in a detail which is impossible to achieve on a hemispheric or global scale.

Within Class A studies, three-dimensional distributions of general circulation statistics are also now becoming available, based either on long-term series of synoptic hemispheric analyses or an analyses of world-wide statistics of station data. Examples of such three-dimensional general circulation statistics are the 'GFDL data' and the 'NMC data', most recently described by Lau and Oort (1981a). Differences between these two data sets, which are essentially based upon the same observations, account for a certain measure of the uncertainty currently associated with the 'observed' statistics in the northern extratropics.

Another approach (Class B in Table 8.1) involves spectral decomposition of the transients either in time, space or both. In addition to the classical zonal harmonic (Fourier analysis) approach, expansion of data in spherical harmonics or empirical orthogonal functions (eof) have also been used in recent years. Filtering (in time, space or both) can also be considered as a rough spectral analysis technique, which has its own advantages.

Still another approach (Class C in Table 8.1) involves case studies of representative synoptic phenomena. This approach is important because it may bring physical understanding of the statistics produced in approaches A and B.

In the following, recent studies on the intensity and characteristic structure of the transient disturbances in the northern extratropics are first reviewed (Section 8.2). Section 8.3 reviews the basic problems and some recent findings concerning the energetics of TE in the Northern Hemisphere in winter. Much space is given in this paper to the discussion of the effect of TE on the time-mean flow (Section 8.4); in this connection some new results on the budget of (quasi-geostrophic) potential vorticity are presented.

8.2 The intensity and characteristic structure of the transient eddies in the extratropical latitudes of the Northern Hemisphere

In the 1950s and 1960s the emphasis in the general circulation studies was mainly in the zonally averaged statistics of the atmospheric flow. In the 1970s the longitudinal variation of the flow statistics also began to receive more attention.

Figure 8.1 shows some statistical quantities relating to the 500 mb height in winter according to Blackmon (1976) and commented on in Chapter 3. One measure of the intensity of the transient disturbances, the total rms of Z', is seen in Fig. 8.1(b). In addition to a general increase from south to north up to about 50°N, three maxima are noticed in this quantity roughly downstream of the troughs seen in the $\bar{Z}$ field of Fig. 8.1(a). Figure 8.1(c) shows that these maxima are essentially due to 'low-pass' fluctuations (periods between 10 and

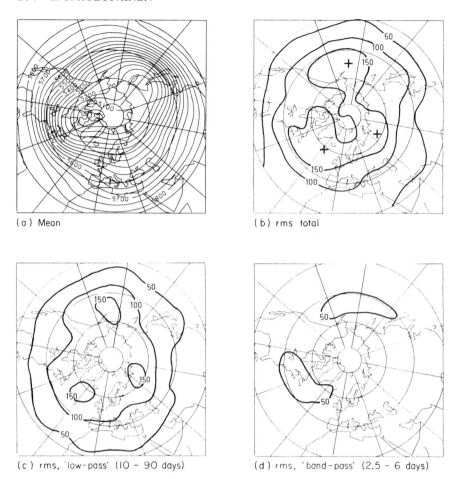

(a) Mean

(b) rms total

(c) rms, 'low-pass' (10 - 90 days)

(d) rms, 'band-pass' (2.5 - 6 days)

Fig. 8.1. Some statistics of the 500 mb geopotential height (m) in a 10-year sample (b, c, and d are redrawn from Blackmon, 1976, Figs. 3a, 4a and 5a). Lines of latitude are drawn every 20° from 20°N to 80°N and those of longitude every 30°.

90 days). The 'band-pass' fluctuations (period 2.5–6 days) which reflect the activity of developing cyclones, have their largest intensity along the major storm tracks off the eastern coast of the two continents (Fig. 8.1d); their relative importance in the time variation of 'weather' (precipitation, cloudiness, etc.) is undoubtedly larger than in the 500 mb height variability seen in Fig. 8.1.

Geographical differences occur not only in the variance fields but also in other TE quantities. Lau (1978, 1979a) and Blackmon et al. (1979) have shown, for example, that the vertical structure and the transfer properties of the band-pass eddies in the storm track areas are typical of developing baroclinic waves whereas those of the low-pass eddies in the areas of the three maxima (Fig.

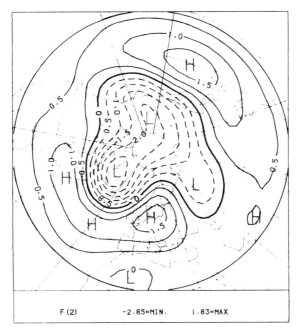

F (2)   -2.85=MIN.   1.83=MAX.

Fig. 8.2. The pattern of the 2nd eof of the 500 mb height, computed by Rinne *et al.* (1981) from a 20-year time series of twice-daily analyses north of 20°N.

8.1c) indicate a more equivalent-barotropic behaviour. (See also Figs. 1.3 and 3.1 of this volume.)

An extensive observational study on the horizontal structure of the transient 500 mb height fluctuations in the northern extratropics is that reported, e.g., in Rinne and Karhila (1979) and Rinne *et al.* (1981). In this study a 20-year series of twice-daily 500 mb height analyses north of 20°N have been converted into empirical orthogonal functions (eof) of the form:

$$Z'(\lambda, \phi, t) = \sum_{v} c_{v}(t) f_{v}(\lambda, \phi) \tag{8.2}$$

where $c_{v}(t)$ is a time-dependent coefficient and $f_{v}(\lambda, \phi)$ the spatial pattern of the vth component.

The first eof pattern and its coefficient basically describes the seasonal variation of Z and is not of interest here. The pattern of the second component, which gives the largest nonseasonal contribution to the total variance, is shown in Fig. 8.2. (This may be compared with Fig. 5.1, which shows the first eof for winter.) An interesting feature is that this component is (contrary to what one might have expected on the basis of most textbooks on dynamic meteorology), not a zonal wave type oscillation, but rather a meridional wave ('zonal index') type fluctuation. Patterns, which to some extent resemble zonal waves, naturally appear as higher components. The time spectra of the eof coefficients (not shown) demonstrates, as can be expected on the basis of Fig. 8.1, the predominance of the long-period fluctuations in the total variance $\overline{Z'^2}$. Furthermore, no gaps in the spectrum can be found, implying that the choice of Δt_s in Eqn. (8.1) is, to a large extent, arbitrary.

The technique most often applied in the studies of the horizontal structure of the atmospheric circulation in the extratropical latitudes is the zonal harmonic (Fourier) analysis. Figure 8.3 shows the spectrum of the 500 mb height variance at $50°N$ in winter as a function of the zonal wave number and period. The total variance has a rather smooth distribution, the form of which indicates that, typically, the period of the fluctuation increases as the zonal wavelength increases. Nevertheless, three broad relative maxima are visible. These can be identified as quasi-stationary ultra-long waves and eastward-propagating long and synoptic waves and, as shown by Fraedrich and Böttger (1978), can be traced on synoptic maps.

Figure 8.3 shows that a large fraction of the total transient variance of the 500 mb height is associated with the ultra-long waves, i.e., with zonal wavenumbers 1-3. By application of suitable filters on the data it is possible to show that part of the observed transient variability on these large scales has properties that can be explained in terms of travelling planetary Rossby waves (Madden, 1979; Böttger and Fraedrich, 1980; Madden and Labitzke, 1981). See Chapter 3 for further discussion of this point.

Since the introduction of the concepts of 'two-dimensional turbulence' (Kraichnan, 1967; Leith, 1968) and of 'quasi-geostrophic turbulence' (Charney, 1971), the laws relating the spectral density to the spatial scale have been the subject of many observational studies. The ' − 3' law for kinetic energy as a function of horizontal wavenumber appears to provide a reasonable representation for wavelengths smaller than about 4000 km but large enough to be described within the present observational system.

8.3   Energetics of the transient disturbances

8.3.1   'Lorenz boxes'

The basic reference concerning the observed energetics of the atmosphere in the form of a 'box diagram', originating in the theory Lorenz (1955), has long been Oort (1964). More recent updatings in this area, e.g., due to Newell et al. (1974), Oort and Peixoto (1974), Peixoto and Oort (1974), Tenenbaum (1976), Tomatsu (1979) and Lau and Oort (1982). Only a few aspects of the energy questions will be taken up here.

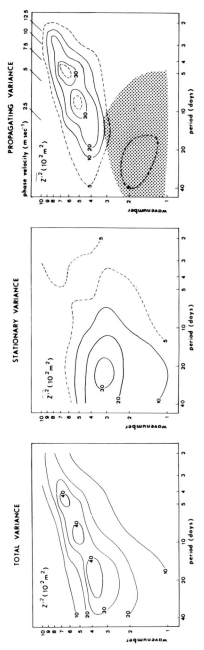

Fig. 8.3. Power spectrum (average of 5 winters) of the 500 mb geopotential height at 50°N in winter according to Fraedrich and Böttger (1978, Fig. 2). (Note: the term 'stationary' refers here to nonpropagating waves whose amplitude oscillates in time, not to the waves seen on the time-mean maps.)

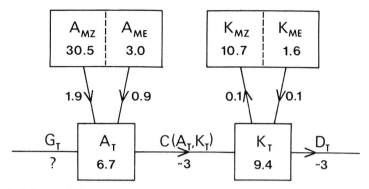

Fig. 8.4. Scheme of the main factors affecting the available potential energy (A) and kinetic energy (K) of the transient eddies in the Northern extratropics ($20°N-80°N$) in winter. The subscripts T, MZ and ME refer, respectively, to transient eddies, time-mean zonally averaged flow and time-mean eddies. The numerical values [except those for $C(A_T, K_T)$ and D_T] are averages of the results obtained from 'GFDL data' and 'NMC data' by Lau and Oort (1982). Unit of energy, 10^5 J m^{-2}; of energy change factors, W m^{-2}.

The basic features of the energetics of the transient disturbances (Fig. 8.4) are those associated with baroclinically unstable disturbances [conversion of $A_M (= A_{MZ} + A_{ME})$ to A_T and of A_T to K_T] and dissipation of K_T. A lot of uncertainty still exists concerning the numerical values. The uncertainties in the evaluation of $C(A_T, K_T)$ are due to the well known difficulties of getting reliable estimates of the vertical velocity (or ageostrophic wind components). Estimates of the energy dissipation outside the atmospheric boundary layer exist only for certain regions (e.g., Kung and Baker, 1975; Holopainen and Eerola, 1979) and differ from each other so much that it is hardly possible to infer from them a representative hemispheric mean value. For these reasons only an 'order of magnitude' value has been inserted in Fig. 8.4 both for $C(A_T, K_T)$ and for D_T.

Equally uncertain is G_T, the generation of A_T by diabatic heating processes. Even the sign of this term is not known for sure. Its numerical value and contributions to it by radiation, latent heat release and subgrid fluxes of sensible heat can probably be established only after sufficiently reliable schemes are available for the parameterization of these processes in the numerical analysis/prognosis models.

One significant new finding concerning the energetics of the transient eddies is the large dissipative effect of TE horizontal heat fluxes on the time-mean eddies: A_{ME} is converted into A_T at the rate of 0.9 W m^{-2}. (The associated time scale is about four days.) The average conversion between $K_M (= K_{MZ} + K_{ME})$ and K_T can be seen from Fig. 8.4 to be small, in agreement with results reported by Holopainen and Oort (1981b). The net dissipative effect of the horizontal TE fluxes of heat and momentum on the time-mean flow, apparent from Fig. 8.4, will be discussed in terms of potential vorticity in Section 8.4.

The above discussion concerns the whole extratropics. For this large region, the relative role of the boundary fluxes appears to be not very large (Lau and Oort, 1982). In local energetics the boundary terms (flux divergence terms) become important.

Geographical distribution of some of the terms affecting K_T (and K_M) have been published for annual statistics by Holopainen (1978b) and for winter statistics in more detail by Lau (1979a). The results show that, in the mid-latitude free atmosphere in winter, K_T is generated (destroyed) over the western oceans (eastern oceans and western continents) by ageostrophic velocities, and that the horizontal flux divergence (convergence) of K_T is important. In general, Lau (1979a) shows that the geographical variation in the energetics of the transient disturbances in winter is, so far as it can be inferred from data, consistent with the corresponding variation of the time-mean flow and the structural statistics of the transient disturbances. It has to be emphasized, however, that we do not have as yet a good enough diagnostic picture for example about the interaction between the cyclogenetic regions and their surroundings, or about the local net effect of the transient disturbances on the local time-mean flow.

8.3.2 Spectral energetics

The available potential energy and kinetic energy can be written as a sum of contributions coming from different spatial scales. If this is done, interactions between the different scales and the spectral transfer of energy enter the picture.

The spectral energetics of the atmosphere has been mostly studied using the zonal harmonic (Fourier) analysis. After Saltzman's (1970) review paper new diagnostic results in this area have been reported, e.g., by Tenenbaum (1976), Tsay and Kao (1978), Kao and Chi (1978), Tomatsu (1979) and Kanamitsu (1980). An observational scale interaction study in terms of a two-dimensional wavenumber has been done for the Northern Hemisphere by Burrows (1976) for a summer period and by Chen and Wiin-Nielsen (1978) for a winter period.

Let $F_K(\mu)$ denote the upscale (from high to low wavenumber) transfer of kinetic energy across wavenumber v. Formally:

$$F_K(v) = \sum_{\mu=1}^{v} L(\mu) \qquad (8.3)$$

where $L(\mu)$ is the gain of kinetic energy at the wavenumber μ due to nonlinear wave–wave interactions. This interaction, which can be approximately determined if the horizontal wind field is known, is the internal kinetic energy exchange between the waves and in a closed region gives no net effect when all the waves are considered together: $F_K(\infty) = \sum_{\mu=1}^{\infty} L(\mu) = 0$. Similarly let $F_A(v)$ denote the corresponding upscale transfer of available potential energy:

$$F_A(v) = \sum_{\mu=1}^{v} S(\mu) \qquad (8.4)$$

where $S(\mu)$ is the gain of available potential energy at the wavenumber μ due to wave–wave interaction; this term can be approximately determined if the horizontal fields of temperature and wind are known. In a closed region it satisfies the condition $F_A(\infty) = S(\infty) = 0$.

Figure 8.5 shows $\bar{F}_K(m)$ and $\bar{F}_A(m)$ as worked out from those observational studies, which use the zonal wavenumber m as a scale index. The curves should be interpreted as a time average of instantaneous transfer functions which, if available, would most likely exhibit a large day-to-day variability.

The basic feature in Fig. 8.5 is the upscale (downscale) transfer of kinetic energy (available potential energy). The recent diagnostic calculations are seen to be basically in agreement with Saltzman (1970), except that they do not show the downscale transfer of kinetic energy at high wavenumbers. (A point that has to be born in mind in this connection is that in diagnostic calculations for open regions such as extratropics one cannot usually exactly satisfy the conditions $\sum_{\text{all}} L(v) = 0$, $\sum_{\text{all}} S(v) = 0$, the summation being over all resolved wavenumbers.)

Figure 8.6 shows, on the basis of data reported by Chen and Wiin-Nielsen (1978), $\bar{F}_K(n)$ and $\bar{F}_A(n)$ for the two-dimensional wavenumber n, which is theoretically more suitable for the study of the spectral energy transfer than the zonal wavenumber m. The curves in Fig. 8.6 are seen to agree qualitatively with those in Fig. 8.5. Note, however, the much larger magnitude of $\bar{F}_A(n)$ compared with that of $\bar{F}_A(m)$. This difference is due to the fact that most of what occurs as a large conversion of the zonal available potential energy into eddy available potential energy in the one-dimensional case, enters in the two-dimensional case into the spectral transfer function $F_A(n)$.

All the waves with n larger than, e.g., 10, are essentially transient waves. To give an example of the interpretation of the results, on the basis of Fig. 8.6, we can thus say that the effect of transient waves with $n > 10$ on the larger (transient or time mean) waves is effectively a thermal (baroclinic) dissipation $[F_A(10) < 0]$ and a barotropic enhancement $[F_K(10) > 0]$, the net time mean effect being definitely dissipative.

Figure 8.5 and 8.6 show the time mean spectral transfer functions and show nothing about the importance of the non-linear wave interaction processes for the growth and decay of the amplitude of any particular transient fluctuation, which occurs at a certain wavenumber and certain frequency. Observational

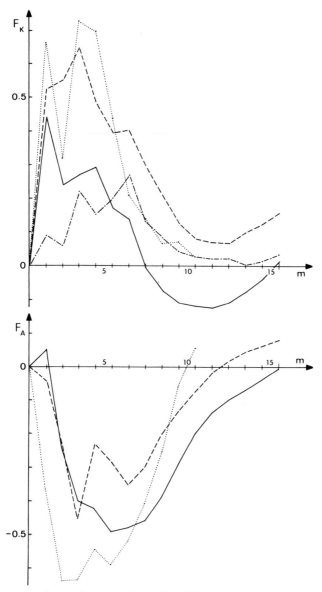

Fig. 8.5. The observed upscale spectral transfer of kinetic energy (F_K) and available potential energy (F_A), in W m^{-2}, in the Northern Hemisphere in winter as a function of the zonal wave number m. The curves have been determined from Eqns. (8.3)–(8.4) and the following data: ——, Saltzman (1970), (20°N–90°N; Feb. + Dec. 63, Jan. 64; 100–1000 mb); –––, Tomatsu (1979), (25°N–75°N; Dec.–Feb. 64/65; 10–925 mb); ····, Kao and Chi (1978), (30°N–60°N; Dec.–Feb. 75/76; 500 mb); –·–·–, Tenenbaum (1976), (0°N–90°N; Jan. 73; 110–1000 mb). $L(0)$ which represents the conversion from K_E to K_Z is not included in the summations.

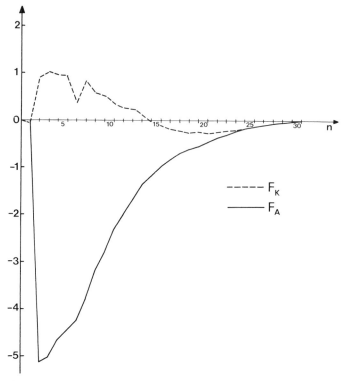

Fig. 8.6. The observed upscale spectral transfer of kinetic energy (F_K) and available potential energy (F_A) in the Northern Hemisphere in winter as a function of a two-dimensional wave number n. Unit, W m^{-2}. (From the data published by Chen and Wiin-Nielsen, 1978.)

studies dealing with this latter problem have been reported by Tsay and Kao (1978), and Kao and Chi (1978). They show that the evolution (growth and decay) of kinetic energy and available potential energy in the long and synoptic-scale waves is to a large extent controlled by the non-linear wave interactions discussed here. However, it is important to bear in mind in this connection that the spectral representation is to some extent an artificial way of describing the changes in the atmospheric circulation pattern. For example, a baroclinic cyclone development over the eastern coast of North America (a more or less regional phenomenon!) may in the spectral presentation be shown as changes in all wavenumbers (including the planetary ones), most of these changes being due to complicated non-linear wave interactions.

8.3.3 Case studies

Case studies of the kinetic energy budget of extratropical cyclones have been the subject of many investigations (see, e.g., Palmén and Newton, 1969; Smith, 1973; Kung and Baker, 1975; Alpert, 1981). One common feature of all cyclone developments seems to be that, if the region for which budget calculations are made includes areas of both ascending and descending motion, there is at some stage of the cyclone development a negative correlation between the specific volume and the vertical velocity ω, i.e., a positive contribution to the conversion of the available potential energy to kinetic energy.

Another common feature in the case study budgets of kinetic energy is that in the boundary layer there is a quasi-balance between the work done by the horizontal pressure gradient force and the work done by the frictional force; this balance results from the approximate Ekman balance in the boundary layer.

Beyond these two somewhat obvious features there seems to be no unique way in which the budget of kinetic energy is fulfilled in the area of developing extratropical cyclones. In the free atmosphere the terms representing the horizontal flux divergence of kinetic energy and the work done by the horizontal pressure gradient force have in most cases the largest magnitude and to a large extent counterbalance each other. These terms can, however, be of any sign depending upon the larger-scale flow pattern, superimposed upon which the cyclone develops.

In the observational studies, in which the case study approach is used, attention should in future be paid to devising schemes, with the aid of which the 'disturbance' can be separated from the larger-scale flow. One scheme could possibly be a spatial filtering technique, which so far has not been applied in budget studies of synoptic-scale phenomena.

8.4 The interplay between the transient disturbances and the time-mean flow in the atmosphere

8.4.1 General

The mutual interaction between the (zonal and/or time) mean flow and the (respective) eddies is one of the fundamental problems concerning the general circulation of the planetary atmospheres. Until the 1920s the eddies in the Earth's atmosphere were considered as a secondary part of the circulation, the primary one to be explained being the axisymmetric (zonally uniform) part. Defant (1921) suggested that the eddies could be important in the required meridional transport of heat and Jeffreys (1926) that they may be important also in the required meridional transport of zonal angular momentum. Observational studies made on the basis of data that became available after World War II showed these hypotheses to be essentially correct.

The basic theoretical framework relating to wave-mean flow interaction is now well established and is presented in the preceding chapter.

In the middle latitudes the meridional TE momentum flux convergence tends to accelerate the mean zonal flow. On the other hand, the meridional TE flux of heat tends to smooth out the mean meridional temperature gradient maintains the shear of the mean zonal flow. The heat transfer and momentum transfer effects of the transient eddies thus oppose each other. An idea about the sign of their net effect is obtained from Fig. 8.4 showing a net drainage of mean-flow energy $(A_{MZ} + A_{ME} + K_{MZ} + K_{ME})$ into energy of transient eddies $(A_T + K_T)$. In this chapter we consider the complicated interplay between the time-mean flow and the transients from another point of view.

The balance equation for a time mean of an arbitrary quantity s (per unit mass) has the form:

$$\partial \bar{s}/\partial t = -\bar{\mathbf{v}} \cdot \nabla \bar{s} - \bar{\omega}\, \partial \bar{s}/\partial p + \bar{S} + S_T \tag{8.5}$$

where $\bar{S}$ is the time-mean intensity of a source term and S_T the three-dimensional transient eddy flux divergence of s:

$$S_T = -\nabla \cdot \overline{s'\mathbf{v}'} - \partial/\partial p\, \overline{s'\omega'} \tag{8.6}$$

However, the transient eddies may be important in inducing part of $\bar{\omega}$ (and also a part of $\bar{\mathbf{v}}$). Furthermore, in addition to the 'local' TE forcing which arises from the different eddy-related parts of the rhs terms in Eqn. (8.5), TE effects may also arise from the vertical boundary conditions. [See also Eqns. (7.8)–(7.16).] Hence, when considering the total tendency (or rate of change) of $\bar{s}$ due to the transients, attention can not only be put on the term S_T but more general considerations are needed.

In the following, some recent observational studies of the geographical distribution of the transient eddy fluxes of heat and momentum, and their effects on the budgets of mean temperature, momentum and vorticity are first reviewed. The main emphasis is, however, put on the net local effect of the horizontal transient eddy fluxes of heat and momentum. For the zonally averaged conditions this is done in terms of Eliassen–Palm fluxes (see Section 7.3.1), for the full geographical distribution by presenting new results derived from the budget of (quasi-geostrophic) potential vorticity.

8.4.2 The eddy heat flux

Since the pioneering works of Starr and his collaborators (e.g., Starr and White, 1954; Peixoto, 1960) many observational studies of the meridional transient eddy flux of temperature, $\overline{T'v'}$, have been made. Zonally averaged values of $\overline{T'v'}$, based on large samples of data, have been published, e.g., by Oort and Rasmusson (1971), Newell *et al.* (1972/74) and Speth (1974). (It should be borne in mind that the zonally averaged TE heat flux is different from the transient variation of the zonally averaged heat flux, studied recently by Stone, 1978 and Lorenz, 1979.) In particular, the data by Oort and Rasmusson (*loc.cit.*) have often been referred to and used in more detailed analyses (e.g., Hantel, 1976; Hantel and Baader, 1978). The meridional component of the transient eddy heat flux has been dealt with in many other studies (e.g., van Loon, 1979, 1980; Opstegh and van den Dool, 1979; Tomatsu, 1979; Tucker, 1979). However, it is only rather recently that studies have appeared which deal not only with $\overline{T'v'}$ but also with $\overline{T'u'}$ (e.g., Lau and Wallace, 1979; Alestalo and Holopainen, 1980); both of these flux components are, of course, needed for estimation of $\nabla \cdot \overline{T'\mathbf{v}'}$.

The wintertime horizontal transient eddy fluxes of temperature in the Northern Hemisphere at 850 mb are shown in Fig. 8.7 (Lau and Wallace, 1979. Compare also Figs. 1.7 and 1.8.) The flux vector $\overline{T'\mathbf{v}'}$ is seen to be clearly downgradient (from high to low mean temperature) but, as pointed out by the authors, the relationship between the downgradient component of the flux and the mean temperature gradient cannot be described by one constant exchange coefficient.

At 300 mb and 200 mb, $\overline{T'\mathbf{v}'}$ (not shown) has, according to Lau and Wallace (1979), a large nondivergent component that arises as a result of the quasi-geostrophic and equivalent barotropic character of the transient fluctuations and does not affect the mean heat budget. Only the irrotational part, which was shown by Lau and Wallace to be clearly downgradient also at 300 mb, is important to $\nabla \cdot \overline{T'\mathbf{v}'}$.

Upgradient (total) transient eddy heat fluxes at 300 mb were reported by Alestalo and Holopainen (1980) for western Europe, which, like western North America, is a region where disturbances are more often in a decaying rather than growing phase of their life cycle. In such regions upgradient heat fluxes are understandable from the arguments given in Section 7.2.

The convergence of the horizontal transient eddy heat flux, $-\nabla \cdot \overline{T'\mathbf{v}'}$, at 700 mb in winter is shown in Fig. 8.8(a). One of the most conspicuous features in the distribution of this term, which in general has a somewhat smaller magnitude than $-\bar{\mathbf{v}} \cdot \nabla \bar{T}$ (Fig. 8.8(b)), appears off the eastern coasts of North America and Asia, where elongated dipoles of cooling to the south and warming to the north are observed. The zero lines associated with each of these two dipoles corresponds roughly to one of the two major cyclone tracks in winter. The developing travelling cyclones thus cause a heat flux roughly perpendicular to their track with the associated cooling (heating) effect on the southern (northern) side.

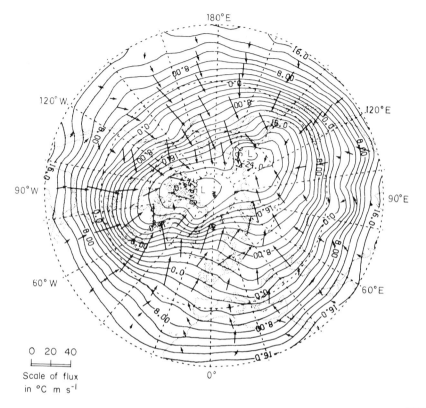

Fig. 8.7. Mean temperature (contours) and the transient eddy flux of sensible heat $\overline{T'v'}$ (arrows) in the northern extratropics at 850 mb in winter according to Lau and Wallace (1979). This figures may be compared with Fig. 1.7 in which is shown the 700 mb picture for one month.

The eastern North Pacific and, in particular, the eastern North Atlantic appear in Fig. 8.8(a) as regions of transient eddy flux divergence (cooling). Fig. 8.8(c) shows that these divergences arise mainly due to long-period (10–90 days) fluctuations in contrast to the storm track features discussed above, which arise (as seen from Fig. 8.8(d)) mainly due to the synoptic eddies (period 2.5–6 days; for filters used see, e.g., Blackmon, 1976). The two eastern oceanic regions are well known for the relatively frequent occurrence of warm blocking highs which undoubtedly are important in determining the characteristics of the transients in these regions.

Transient eddies in the middle latitudes troposphere transport heat not only horizontally but also upwards; associated with this vertical heat transport is the conversion of available potential energy into kinetic energy. Estimates of $\overline{T'\omega'}$, based upon vertical velocities obtained from operational numerical weather prediction models, have been recently reported, e.g., by Lau (1979a), Oerlemans (1980a,b) and Tomatsu (1979). This quantity appears to be well correlated with the meridional flux $\overline{T'v'}$ (e.g., Oerlemans, 1980a) in agreement with the slantwise convection (warm air northwards and upwards, cold air southwards and downwards) predicted by the baroclinic instability theories. Both $\overline{T'v'}$ and $\overline{T'\omega'}$ have largest magnitude in the cyclogenetic regions off the eastern coasts of the two continents.

The average upward flux of heat by transient eddies across, say, 500 mb north of 30°N in winter is of the order of 1×10^{15} W.

The cooling (warming) of the lower (upper) troposphere associated with $\overline{T'\omega'}$ is, on average, of the order of 0.3 K d^{-1}. This flux, the importance of which is discussed, e.g., by Palmén and Newton (1969), is crucial for the maintenance of the positive static stability in the middle latitudes.

8.4.3 Eddy momentum and vorticity fluxes

Studies of the time-mean momentum budget, including a discussion on the effect of the transients, have, on a hemispheric scale, been recently reported by Holopainen (1978a) and Lau (1978); regional budgets of zonal momentum of this type have been published by Holopainen et al. (1980) for North America and by Holopainen (1982) for Europe. These studies show, for example, that the relative importance of the transient eddy flux divergence of zonal momentum varies from place to place; close to the climatological jet stream maxima its effect is, however, small compared with the mean-flow advection of zonal momentum.

The horizontal motion in the extratropical atmosphere is dominated by its rotational component. The vorticity equation combines information from the two equations of horizontal motion and is therefore formally well suited for a study of the forcing of the time-mean flow by the transient eddies. Recent studies in this area are those by Holopainen (1978a), Lau (1979b), Holopainen and Oort (1981a, 1981b). Perhaps the most important finding from these investigations to be mentioned in this connection is that this forcing, which has its largest magnitude in the upper troposphere and lower stratosphere, tends to shift the mean subtropical jetstream polewards and appears to be important for counterbalancing the effect of mean vorticity sources and sinks due to surface friction.

8.4.4 The net local effect of the horizontal TE fluxes on the time-mean flow

(a) *Eliassen–Palm cross-section*

The net local effect of the horizontal TE fluxes of heat and momentum on the zonally averaged time-mean flow has been recently discussed by Edmon et al.

218

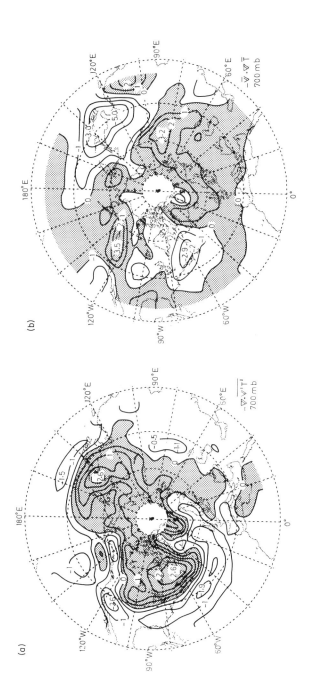

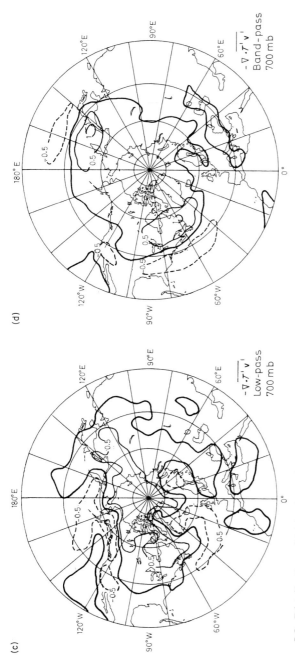

Fig. 8.8. Distribution of (a) horizontal convergence of the transient eddy heat flux; (b) horizontal temperature advection by the mean wind; (c) horizontal convergence of the heat flux due to the low-pass eddies (10–90 days); and (d) horizontal convergence of the heat flux due to the band-pass eddies (2.5–6 days) at 700 mb in winter. Unit, $K \, d^{-1}$. (The panels (a) and (b) are from Lau (1979b), and (c) and (d) courtesy of N-C. Lau.)

(1980). Based upon the theory originally presented by Andrews and McIntyre (1976) they show that it is possible to write the Eulerian set of equations for the zonally averaged flow in such a transformed form, that the meridional eddy transfer of heat disappears from the thermodynamic energy equation and appears in the transformed equation of zonal motion. The latter then has a force term which represents the net local effect of horizontal TE fluxes on the mean zonal flow. On a β-plane, the quasi-geostrophic equations are (7.14) with (7.12) and (7.18). On the sphere, in the quasi-geostrophic framework, this zonal force $[F_\lambda]$ has the form:

$$[F_\lambda] = \frac{1}{a \cos \phi} \nabla \cdot \mathbf{E} \tag{8.7}$$

where the brackets denote a zonal average and:

$$\nabla \cdot \mathbf{E} = \frac{1}{a \cos \phi} \partial/\partial\phi(E_\phi \cos \phi) + \partial/\partial p \, E_p \tag{8.8}$$

$$\begin{cases} E_\phi = -a \cos \phi[\overline{u'v'}] & (8.9) \\ E_p = fa \cos \phi[\overline{\theta'v'}]/(\partial\tilde{\theta}/\partial p) & (8.10) \end{cases}$$

(f is the Coriolis parameter and ($\tilde{\ }$) denotes an area average at an isobaric surface). E_ϕ and E_p are the meridional and vertical components, respectively, of the Eliassen–Palm (EP) flux due to transients. As discussed in Chapter 7, when $\mathbf{E} = E_\phi \mathbf{j} + E_p \mathbf{k}$ is plotted in the meridional plane and the values of $\nabla \cdot \mathbf{E}$ are contoured, the result is an Eliassen–Palm cross-section, which is a convenient way of illustrating the wave–mean flow interaction. The flux itself can be regarded as a measure of the net transfer of wave activity from one latitude and height to another (Eqn. 7.20) and its divergence (equal to the meridional flux of quasi-geostrophic potential vorticity) indicates where and how the zonally averaged time-mean flow experiences the local net effect of transients apart from the residual Coriolis term $f_0[\tilde{v}]$ in Eqn. (7.14).

Figure 8.9 shows the EP cross-section for the transient eddies as determined by Edmon et al. (1980) for the Northern Hemisphere winter. The direction of the flux vectors in such a diagram shows the relative role of the transient eddy heat and momentum forcing (heat fluxes determine the vertical, momentum fluxes the horizontal EP flux component. Hence, it is immediately seen from Fig. 8.9 that the heat flux effect dominates. In the free atmosphere $\nabla \cdot \mathbf{E}$, largely determined by $\partial/\partial p \, E_p$, is seen to be negative in most places. As described by Eqn. (7.14) this implies a tendency to produce easterly acceleration. But, as discussed in Section 7.3.1, the vertical integral of the residual Coriolis term $f_0[\tilde{v}]$ exactly balances that of $\partial E_p/\partial p$. The Coriolis term must therefore give a westerly acceleration which will dominate at low levels. The westerly

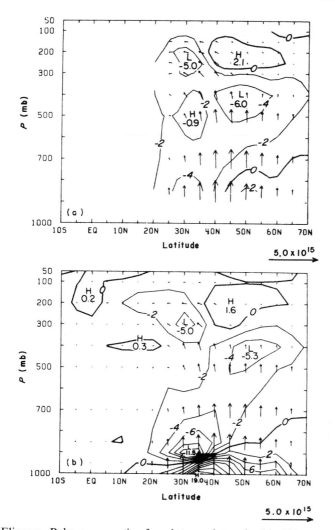

Fig. 8.9. Eliassen–Palm cross-section for winter as determined by Edmon *et al.* (1980) from NMC data (top) and the data from Oort and Rasmusson (1971) (bottom). E_ϕ is multiplied by $2\pi a \cos\phi/g$ and E_p by $2\pi a^2 \cos\phi/g$. The scale for the former in units of m^3 is shown at bottom right. The isolines (every 2.0×10^{15} m^3) are for the EP flux divergence or the net force which the zonally averaged time-mean flow experiences due to horizontal transient eddy flux of heat and momentum.

acceleration implied by the vertical integral of the horizontal convergence of **E** is balanced by surface friction.

As discussed above, $[F_\lambda]$ is equal to the meridional flux of (quasi-geostrophic) potential vorticity. Diagnostic calculations for this latter quantity were reported by Wiin-Nielsen and Sela (1971). They found that in the main part of the troposphere this flux was southward (negative). These independent early calculations are thus in agreement with the findings of Edmon *et al.* (*loc. cit.*).

(b) *Geographical patterns of the transient forcing determined from the budget of (quasi-geostrophic) potential vorticity*

In order to get a picture of the geographical distribution of the forcing due to horizontal transient eddy fluxes of heat and momentum, we may consider the geographical distribution of relevant terms in the budget of time-mean (quasi-geostrophic) potential vorticity:

$$\bar{q} = f + \zeta - f \frac{\partial}{\partial p} \left(\frac{\bar{\theta}}{\tilde{s}} \right)'' \tag{8.11}$$

where ζ is the relative vorticity and $s(= -\partial\Theta/\partial p)$ the static stability. ( )'' denotes a deviation from an isobaric area average ($\tilde{\ }$). From the time-mean equations of motion and the time-mean thermodynamic energy equation we then get for $(\partial\bar{q}/\partial t)_T$, the net rate of change of $\bar{q}$ due to the transients, an equation:

$$(\partial\bar{q}/\partial t)_T = D = D^H + D^V \tag{8.12}$$

where:
$$D^H = f \frac{\partial}{\partial p} \left(\frac{\overline{\nabla \cdot \theta' \mathbf{v}'}}{\tilde{s}} \right) \tag{8.13}$$

is the forcing due to the horizontal transient eddy flux of heat, and:

$$D^V = \mathbf{k} \cdot \nabla \times \mathbf{A}_H \tag{8.14}$$

where

$$\mathbf{A}_H = A_\lambda \mathbf{i} + A_\phi \mathbf{j}$$

and

$$\left\{ \begin{aligned} A_\lambda &= -\frac{1}{a \cos\phi} \frac{\partial}{\partial\lambda} \overline{u'u'} \\ &\quad - \frac{1}{a \cos\phi} \frac{\partial}{\partial\phi} \overline{u'v'} \cos\phi + \frac{\overline{u'v'}}{a} \tan\phi \\ A_\phi &= -\frac{1}{a \cos\phi} \frac{\partial}{\partial\lambda} \overline{u'v'} \\ &\quad - \frac{1}{a \cos\phi} \frac{\partial}{\partial\phi} \overline{v'v'} \cos\phi - \frac{\overline{u'u'}}{a} \tan\phi \end{aligned} \right.$$

is the forcing due to the horizontal transient eddy flux of momentum (compare

Section 7.4.1). D^H and D^V can be called the baroclinic and barotropic TE forcing of the time-mean flow, respectively.

D^V is, approximately, the convergence of the horizontal transient eddy flux of vorticity (e.g., Holopainen, 1978a). At the same level of approximation the term $D (= D^H + D^V)$ is the corresponding flux convergence of (quasi-geostrophic) potential vorticity (Wiin-Nielsen and Sela, 1971). In the quasi-geostrophic potential vorticity equation, D represents the net local effect of the TE horizontal fluxes. (If 'Ertels potential vorticity' were used, the mean budget equation would contain, in addition to a term corresponding to D, a vertical advection term that implicitly depends upon the eddy activity).

The forcing terms D^H and D^V were evaluated for different layers between 100 mb and 850 mb using two long-term statistical data sets for Northern Hemisphere winter (for description of these 'GFDL data' and 'NMC data' see, e.g., Lau and Oort (1981a). Here, only the zonal averages for the layer 300–700 mb are shown (Fig. 8.10). Both data sets give essentially the same patterns which are consistent with Fig. 8.9.

D^H is positive (stabilizing effect) in the subtropics and negative (destabilizing effect) in the higher latitudes. This distribution is natural, owing to the fact that the magnitude of the eddy heat flux typically decreases with height in the middle troposphere and that there is a general divergence (convergence) of the flux in the subtropics (higher latitudes). D^V shows vorticity flux divergence ($D^V<0$) in the subtropics and convergence ($D^V>0$) around 45°–60°N in agreement with the expected vorticity sources and sinks due to surface friction (Holopainen and Oort, 1981b).

In the middle latitudes of both hemispheres some counterbalancing of the terms D^H and D^V is noticed as expected on the basis of the EP arrows in Fig. 8.9. In the more polar latitudes, D^H is seen to have a much larger magnitude than D^V.

One way of picturing the geographical distributions of D^H and D^V is to find out the nondivergent forces $\mathbf{F}_\psi^H \ (= \mathbf{k} \times \nabla \psi^H)$ and $\mathbf{F}^V \ (= \mathbf{k} \times \nabla \psi^V)$, curls of which are D^H and D^V, respectively:

$$D^H = \mathbf{k} \cdot \nabla \times \mathbf{F}_\psi^H = \nabla^2 \psi^H \qquad (8.15)$$

$$D^V = \mathbf{k} \cdot \nabla \times \mathbf{F}_\psi^V = \nabla^2 \psi^V \qquad (8.16)$$

Knowing D^H and D^V from observations, Eqns. (8.15) and (8.16) can be solved for ψ^H and ψ^V and thus also for $\mathbf{F}_\psi^H$ and $\mathbf{F}_\psi^V$. In terms of Eqn. (7.43′), we view the transient eddy potential vorticity flux convergence as equivalent to the curl of an apparent force. (The irrotational parts of the forces, even if possibly of the same order of magnitude as the nondivergent parts are of no real importance because they enter only into the mean divergence or balance equation (7.44), where their relative contributions are very small compared with those of the pressure gradient force and the Coriolis force.)

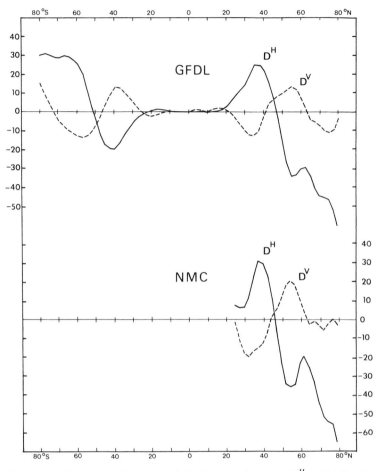

Fig. 8.10. Zonally averaged values of the forcing function D^H (solid line) and D^V (broken line) in the 300–700 mb layer during the northern winter, as determined from GFDL data (upper part) and the NMC data (lower part). Unit, 10^{-12} s^{-2}.

Figure 8.11 shows ψ^H and ψ^V for the 300–700 mb layer in the Northern Hemisphere, obtained from the (global) field of D^H and D^V as calculated from the GFDL data. In most places, $\mathbf{F}_\psi^H$ is seen to be directed westwards. It thus causes a decelerating effect on the westerly time-mean flow. The maximum magnitude of $\mathbf{F}_\psi^H$ is found in Alaska, where it is about 1×10^{-4} m s^{-2}.

From the point of view of $\mathbf{F}_\psi^H$ the layer 300–700 mb is representative of the free atmosphere as a whole. The $f_0\, \partial\bar{\omega}/\partial p$ term in Eqn. (7.43') cancels the curl of $\mathbf{F}_\psi^H$ in the vertical average. The sense of this term must therefore be the opposite to that of $\mathbf{F}_\psi^H$ and will tend to dominate it at low levels. (Alternatively, following Bretherton (1966), the heat flux can be taken as zero at the lower

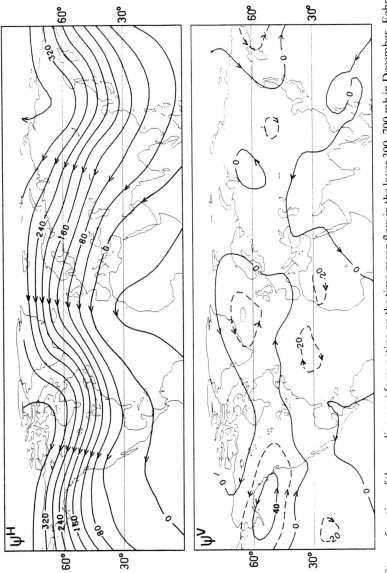

Fig. 8.11. Stream function of the nondivergent force acting on the time-mean flow in the layer 300–700 mb in December–February due to transient eddy horizontal flux of heat (upper part) and momentum (lower part). Unit, $m^2 s^{-2}$.

boundary provided that a delta function in potential vorticity flux is included there. Therefore, $\int D^H \, dm = 0$ and thus also $\int F_\psi^H \, dm = 0$ when the integration is through the whole mass of the air column.) F_ψ^H thus creates a tendency to decrease the shear (baroclinicity) of the time-mean flow.

The 'barotropic' forcing term F_ψ^V is on average directed eastwards in the middle latitudes and westwards south of about 30°N as one might expect on the basis of classical momentum considerations. This force has its largest magnitude $(0.3 \times 10^{-4} \text{ m s}^{-2})$ over the extratropical oceans, where its direction is almost opposite to that of F_ψ^H. From Fig. 8.11, the transient eddies can be seen to cause, via F_ψ^V, cyclonic forcing over the North Atlantic and North Pacific. They thus maintain the local 'centres of action' against the dissipative effects of surface friction; this aspect was discussed in more detail by Holopainen and Oort (1981b).

The sum of F_ψ^H and F_ψ^V in the 300–700 mb layer can be seen in the upper part of Fig. 8.12. The zonally averaged zonal component of this sum force turns out to be approximately equal to the net zonal force obtained from EP flux divergence (see Eqn. 8.7 and Fig. 8.9). Comparing the geographical pattern of $(F_\psi^H + F_\psi^V)$ with the stream function of the time-mean flow (lower part of Fig. 8.12) one notices that not only is the time-mean zonally averaged flow decelerated in this layer by the net local effect of horizontal TE fluxes, but so also the flow in the time-mean waves.

(c) *The effect of transient eddies on the stationary disturbances*

One measure of the intensity of the time-mean waves (stationary disturbances) is the potential enstrophy:

$$B = \tfrac{1}{2}[\bar{q}^{*2}]$$

associated with them. A measure of the net local effect of transients on B is the covariance $[D^*\bar{q}^*]$:

$$(\partial B/\partial t)_T = [D^*\bar{q}^*].$$

Figure 8.13 shows, on the basis of the GFDL and NMC data sets, the meridional distribution of B and $[D^*\bar{q}^*]$ in the 300–700 mb layer. The net effect of the horizontal TE fluxes of heat and momentum on the enstrophy of the stationary disturbances is seen to be negative at all latitudes. The associated dissipation time scale, $B/(\partial B/\partial t)_T$, is of the order of 4 days in the extratropics.

The large thermal dissipative effect of the transients on the stationary disturbances was pointed out by Lau (1979b). That the net local effect of the horizontal TE fluxes on these disturbances is also a dissipative one was indicated by a recent investigation by Youngblut and Sasamori (1980), who studied the budget of quasi-geostrophic potential vorticity for January 1963. In their study, the net local effect of the transient eddies on the monthly-mean

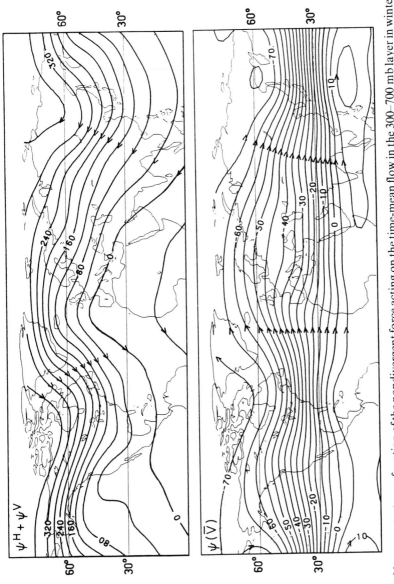

Fig. 8.12. Upper part: stream function of the nondivergent force acting on the time-mean flow in the 300–700 mb layer in winter due to the horizontal transient eddy flux of heat and momentum; unit, m² s⁻². Lower part: stream function of the time-mean flow in the 300–700 mb layer in winter; unit, 10⁶ m² s⁻¹.

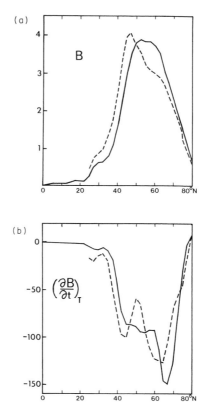

Fig. 8.13. (a) Potential enstrophy of the stationary disturbances, $B = \frac{1}{2}[\bar{q}^{*2}]$, in the 300–700 mb layer in winter: unit, $10^{-10}\,s^{-2}$. (b) The rate of change of B due to transients, $(\partial B/\partial t)_T = [\bar{q}^*D^*]$, in the 300–700 mb layer in winter; unit, $10^{-17}\,s^{-3}$. Solid (broken) line is calculated from the GFDL (NMC) data.

flow was evaluated as $-\overline{v'\cdot\nabla q'}$ ($\approx -\nabla\cdot\overline{q'v'}$) without considering separately the effect of the heat and momentum fluxes.

It was shown by Holopainen and Oort (1981b) that D^V tends to increase the enstrophy ($\frac{1}{2}[\zeta^{*2}]$) of the stationary disturbances in the middle latitudes and to dissipate it in the subtropics the net hemispheric effect being rather small. In Fig. 8.13 these features of D^V are not visible because of the dominance of D^H in the budget of potential enstrophy.

(d) *Discussion*

As a summary of this chapter we can thus say that the primary effect of the transient eddies, which ultimately owe their existence to the instability of the time-mean flow, is the horizontal downgradient heat transfer, which creates a tendency to eliminate the horizontal temperature gradient (and thus the baroclinic component) of the time-mean flow. The main effect of the transient eddy momentum fluxes in the middle latitudes is to enhance the barotropic component of the time-mean flow and to compensate for the effects of surface friction. The net local effect of the horizontal TE fluxes in the free atmosphere appears to be dissipative, not only on the zonally averaged time-mean flow but also on the stationary waves, which ultimately owe their existence to the presence of mountains and the longitudinal differences in heating. This picture concerning the interplay between the time-mean flow and transient eddies is consistent with that obtained in Section 8.3 from considerations of energetics.

The fact that the baroclinic aspects (heat fluxes) of the transient eddies appear to dominate the barotropic ones (momentum and vorticity fluxes) in the interplay between the transient eddies and the time-mean flow tells nothing about the relative importance of the baroclinic and barotropic processes in the internal dynamics of the transient disturbances and thus, for example, in the day-to-day local changes of the atmospheric flow. Perhaps the best evidence that the barotropic processes are important in this latter respect is the fact that it took a long time to beat the barotropic model for short-range forecasts for the middle troposphere in the extratropics.

8.5 Concluding remarks

In the coming few years, our understanding of the transient eddies in the atmosphere will significantly increase when the data from the FGGE year have been thoroughly analysed. Comparative observational studies of the circulation in the Northern and Southern hemispheres with their different forced-flow components will undoubtedly throw further light on the interplay between time-mean flow and the transient disturbances.

Acknowledgements

The author is very much indebted to Drs A. H. Oort and N-C. Lau for allowing their data to be used in this investigation. The main part of the programming was done by L. Rontu and the drafting by P. Nurmi of the Department of Meteorology, University of Helsinki. The first version of the manuscript was written in the Department of Meteorology, University of Wisconsin, where the author worked as a visiting professor during the spring term 1981; the excellent working conditions provided by Prof. J. Kutzbach, Chairman of the Department, are gratefully acknowledged.

References

ALESTALO, M. (1981). The energy budget of the earth-atmosphere system in Europe. *Tellus*, **33**, 360–371.

ALESTALO, M. and HOLOPAINEN, E. (1980). Atmospheric energy fluxes over Europe. *Tellus*, **32**, 500–510.

ALPERT, J. C. (1981). An analysis of the kinetic energy budget for two extratropical cyclones: the vertically averaged flow and the vertical shear flow. *Mon. Weath. Rev.*, **109**, 1219–1232.

ANDREWS, D. G. and McINTYRE, M. E. (1976). Planetary waves in horizontal and vertical shear: The generalized Eliassen-Palm relation and the mean zonal acceleration. *J. atmos. Sci.*, **33**, 2031–2048.

BAER, F. (1972). An alternative scale representation of atmospheric energy spectra. *J. atmos. Sci.*, **29**, 649–664.

BLACKMON, M. L. (1976). A climatological spectral study of the 500 mb geopotential height of the Northern Hemisphere. *J. atmos. Sci.*, **33**, 1607–1623.

BLACKMON, M. L., WALLACE, J. M., LAU, N-C. and MULLENS, S. L. (1977). An observation study of the Northern Hemisphere wintertime circulation. *J. atmos. Sci.*, **34**, 1040–1053.

BLACKMON, M. L., MADDEN, R. A., WALLACE, J. M. and GUTZLER, D. S. (1979). Geographical variation in the vertical structure of geopotential height fluctuations. *J. atmos. Sci.*, **36**, 2450–2466.

BÖTTGER, H. and FRAEDRICH, K. (1980). Disturbances in the wavenumber-frequency domain observed along 50°N. *Contr. atmos. Phys.*, **53**, 90–105.

BRETHERTON, F. B. (1966). Critical layer instability in baroclinic flows. *Q. Jl R. met. Soc.*, **92**, 325–334.

BURROWS, W. R. (1976). A diagnostic study of atmospheric spectral energies. *J. atmos. Sci.*, **33**, 2308–2321.

CHARNEY, J. (1971). Geostrophic turbulence. *J. atmos. Sci.*, **28**, 1087–1095.

CHEN, T.-C. and WIIN-NIELSEN, A. (1978). On nonlinear cascades of atmospheric energy and enstrophy in a two-dimensional spectral index. *Tellus*, **30**, 313–322.

DEFANT, A. (1921). Die Zirkulation der Atmosphäre in den gemässigten Breiten der Erde. *Geograf. Ann.*, **3**, 209–266.

EDMON, H. J., HOSKINS, B. J. and McINTYRE, M. E. (1980). Eliassen-Palm cross sections for the troposphere. *J. atmos. Sci.*, **37**, 2600–2616.

ECMWF (1977). Proceedings of the ECMWF Workshop on the use of empirical orthogonal functions in meterology, 2–4 November, 1977. European Centre for Medium Range Weather Forecasts, Reading, UK.

FRAEDRICH, K. and BÖTTGER, H. (1978). A wavenumber-frequency analysis of the 500 mb geopotential at 50°N. *J. atmos. Sci.*, **35**, 745–750.

GAMBO, K. (1978). A characteristic feature of ultra-long waves at the 500-mb level in the winter season. *J. met. Soc. Japan*, **56**, 435–442.

HANTEL, M. (1976). On the vertical eddy transports in the Northern atmosphere. 1. Vertical eddy heat transport for summer and winter. *J. Geoph. Res.*, **81**, 1577–1588.

HANTEL, M. and BAADER, H-R. (1978). Diabatic heating climatology of the zonal atmosphere. *J. atmos. Sci.*, **35**, 1180–1189.

HOLOPAINEN, E. O. (1978a). On the dynamic forcing of the long-term mean flow by the large-scale Reynolds' stresses in the atmosphere. *J. atmos. Sci.*, **35**, 1596–1604.

HOLOPAINEN, E. O. (1978b). A diagnostic study of the kinetic energy balance of the long-term mean flow and the associated transient fluctuations in the atmosphere. *Geophysica*, **15**, 125–145.

HOLOPAINEN, E. O. (1982). Long-term budget of zonal momentum in the free atmosphere over Europe in winter. *Q. Jl. R. met. Soc.*, **108**, 95–102.

HOLOPAINEN, E. O. and EEROLA, K. (1979). A diagnostic study of the long-term balance of kinetic energy of atmospheric large scale motion over the British Isles. *Q. Jl R. met. Soc.*, **105**, 849–858.

HOLOPAINEN, E. O. and NURMI, P. (1980). A diagnostic scale interaction study employing a horizontal filtering technique. *Tellus*, **32**, 124–130.

HOLOPAINEN, E. O. and OORT, A. H. (1981a). Mean surface stress curl over the ocean as determined from the vorticity budget of the atmosphere. *J. atmos. Sci.*, **38**, 262–269.

HOLOPAINEN, E. O. and OORT, A. H. (1981b). On the role of large-scale transient eddies in the maintenance of the vorticity and enstrophy of the time-mean atmospheric flow. *J. atmos. Sci.*, **38**, 270–280.

HOLOPAINEN, E. O., LAU, N-C. and OORT, A. H. (1980). A diagnostic study of the time-averaged budget of atmospheric zonal momentum over North America. *J. atmos. Sci.*, **37**, 2234–2242.

JEFFREYS, H. (1926). On the dynamics geostrophic winds. *Q. Jl R. met. Soc.*, **52**, 85–104.

JENNE, R. L., WALLACE, J. M., YOUNG, J. A. and KRAUS, E. B. (1974). Observed long-period fluctuations in 500-mb heights. Supplementary text to NCAR films J-4 and J-6. NCAR Technical Note NCAR-TN/STR-94.

KANAMITSU, M. (1980). Some climatological and energy budget calculations using the FGGE III-b analysis during January 1979. Proceedings of the ECMWF 1980 Seminar on 'Data Assimilation Methods'.

KAO, S. K. and CHI, C. N. (1978). Mechanism for the growth and decay of long- and synoptic-scale waves in the mid-troposphere. *J. atmos. Sci.*, **35**, 1375–1387.

KRAICHNAN, R. H. (1967). Inertial ranges in two-dimensional turbulence. *Physics of Fluids*, **10**, 1417–1423.

KUNG, E. C. and BAKER, W. E. (1975). Energy transformations in middle-latitude disturbances. *Q. Jl R. met. Soc.*, **101**, 793–815.

LAU, N-C. (1978). On the three-dimensional structure of the observed transient eddy statistics of the Northern Hemisphere wintertime circulation. *J. atmos. Sci.*, **35**, 1900–1923.

LAU, N-C. (1979a). The structure and energetics of transient disturbances in the Northern Hemisphere wintertime circulation. *J. atmos. Sci.*, **36**, 982–995.

LAU, N-C. (1979b). The observed structure of tropospheric stationary waves and the local balances of vorticity and heat. *J. atmos. Sci.*, **36**, 996–1016.

LAU, N-C. and OORT, A. H. (1981). A comparative study of observed Northern Hemisphere circulation statistics based on GFDL and NMC analyses. Part I: The time-mean fields. *Mon. Weath. Rev.*, **109**, 1380–1403.

LAU, N-C. and OORT, A. H. (1982). A comparative study of observed Northern Hemisphere circulation statistics based on GFDL and NMC analyses. Part II: Transient eddy statistics and the energy cycle. *Mon. Weath. Rev.*, **109**, 889–906.

LAU, N-C. and WALLACE, J. M. (1979). On the distribution of horizontal transport by transient eddies in the Northern Hemisphere wintertime circulation. *J. atmos. Sci.*, **36**, 1844–1861.

LAU, N-C., WHITE, G. and JENNE, R. L. (1981). Circulation statistics for the extratropical Northern Hemisphere. NCAR/TN-171 + STR. National Center for Atmospheric Research, Boulder, Colorado.

LEITH, C. E. (1968). Diffusion approximation for two-dimensional turbulence. *Physics of Fluids*, **11**, 671–673.

VAN LOON, H. (1979). The association between the latitudinal temperature gradient and eddy transport. Part I: Transport of sensible heat in winter. *Mon. Weath. Rev.*, **107**, 525–534.

VAN LOON, H. (1980). Transfer of sensible heat in the atmosphere on the Southern Hemisphere: an appraisal of the data before and during FGGE. *Mon. Weath. Rev.*, **108**, 1774–1781.

LORENZ, E. N. (1955). Available potential energy and the maintenance of the general circulation. *Tellus*, **7**, 157–167.

LORENZ, E. N. (1979). Forced and free variations of weather and climate. *J. atmos. Sci.*, **36**, 1367–1376.

MADDEN, B. (1979). Observations of large-scale travelling Rossby waves. *Rev. Geoph. Space Phys.*, **17**, 1935–1949.

MADDEN, R. A. and LABITZKE, K. (1981). A free Rossby wave in the troposphere and stratosphere during January 1979. *J. Geoph. Res.*, **86**, C2, 1247–1254.

NEWELL, R. E., KIDSON, J. W., VINCENT, D. G. and BOER, G. J. (1972/1974). *The General Circulation of the Tropical Atmosphere and Interactions with Extratropical Latitudes*. Vols. 1/2. The MIT Press, 258 and 371 pp.

OERLEMANS, J. (1980a). An observational study of the upward sensible heat flux by synoptic-scale transients. *Tellus*, **32**, 6–14.

OERLEMANS, J. (1980b). A case study of subsynoptic disturbance in a polar outbreak. *Q. Jl R. met. Soc.*, **106**, 313–326.

OORT, A. H. (1964). On estimates of the atmospheric energy cycle. *Mon. Weath. Rev.*, **92**, 483–493.

OORT, A. H. and RASMUSSON, E. M. (1971). Atmospheric circulation statistics. NOAA Prof. Pap. No. 5. US Dept. of Commerce, 323 pp.

OORT, A. H. and PEIXOTO, J. P. (1974). The annual cycle of the energetics of the atmosphere on a planetary scale. *J. Geoph. Res.*, **79**, 2705–2719.

OPSTEEGH, J. D. and VAN DEN DOOL, H. M. (1979). A diagnostic study of the time-mean atmosphere over northwestern Europe during winter. *J. atmos. Sci.*, **36**, 1862–1879.

PALMÉN, E. and NEWTON, C. W. (1969). *Atmospheric Circulation Systems: Their Structure and Physical Interpretation.* International Geophysics Series 13. New York, Academic Press, 603 pp.

PEIXOTO, J. P. (1960). Hemispheric temperature conditions during the year 1950. Contract AF19(604)-6108, Scientific Report No. 4, Planetary Circulations Project, Dept. of Meteorology, Massachusetts Institute of Technology, 211 pp.

PEIXOTO, J. P. and OORT, A. H. (1974). The annual distribution of atmospheric energy on a planetary scale. *J. Geophys. Res.*, **79**, 2149–2159.

REED, R. (1979). Cyclogenesis in polar air streams. *Mon. Weath. Rev.*, **107**, 38–52.

RINNE, J. and KARHILA, V. (1979). Empirical orthogonal functions of the 500 mb height in the Northern Hemisphere determined from a large data sample. *Q. Jl R. met. Soc.*, **105**, 873–884.

RINNE, J., KARHILA, V. and JÄRVENOJA, S. (1981). The EOFs of the 500 mb height in the extratropics of the Northern Hemisphere. Report No. 17, Dept. of Meteorology, Univ. of Helsinki.

SALTZMAN, B. (1970). Large-scale atmospheric energetics in the wave-number domain. *Rev. Geophys. Space Phys.*, **8**, 289–302.

SASAMORI, T. and YOUNGBLUT, C. E. (1981). The nonlinear effects of transient and stationary eddies on the winter mean circulation. Part II: The stability of stationary waves. *J. atmos. Sci.*, **38**, 87–96.

SAVIJÄRVI, H. (1978). The interaction of the monthly mean flow and large-scale transient eddies in two different circulation types. Part III: Potential vorticity balance. *Geophysica*, **15**, 1–16.

SCHÄFER, J. (1979). A space-time analysis of tropospheric planetary waves in the Northern Hemisphere. *J. atmos. Sci.*, **36**, 1117–1123.

SMITH, P. J. (1973). Midlatitude synoptic scale systems: their kinetic energy budgets and role in the general circulation. *Mon. Weath. Rev.*, **101**, 757–762.

SPETH, P. (1974). Horizontale Flüsse von sensibler und latenter Energie und von Impuls für die Atmosphäre der Nordhalbkugel. *Meteorol. Rundsch.*, **27**, 65–90.

SPETH, P. (1978). The global energy balance of the atmosphere. Part I: The annual cycle of available potential energy and its variability through a ten-year period (1967–1976). *Contr. Atmos. Physics*, **51**, 153–165.

SPETH, P. and KIRK, E. (1981). A one-year study of power spectra in wavenumber-frequency domain. *Contr. Atmos. Physics*, **54**, 186–206.

STARR, V. P. and WHITE, R. M. (1954). Balance requirements of the general circulation. Geophys. Res. Paper No. 35, Geophys. Res. Directorate, Air Force Cambridge Research Center, 57 pp.

STONE, P. (1978). Baroclinic adjustment. *J. atmos. Sci.*, **35**, 561–571.

TENENBAUM, J. (1976). Spectral and spatial energetics of the GISS model atmosphere. *Mon. Weath. Rev.*, **104**, 15–30.

TOMATSU, K. (1979). Spectral energetics of the troposphere and lower stratosphere. *Adv. in Geophys.*, **21**, 289–405.

TSAY, C.-Y. and KAO, S.-K. (1978). Linear and nonlinear contributions to the growth of the large-scale atmospheric waves and jet stream. *Tellus*, **30**, 1–14.

TUCKER, B. (1979). Transient synoptic systems as mechanisms for meriodional transport: an observational study in the Southern Hemisphere. *Q. Jl R. met. Soc.*, **105**, 657–672.

WHITE, G. (1980). On the observed spatial scale of Northern Hemisphere transient motions. *J. atmos. Sci.*, **37**, 892–895.

WIIN-NIELSEN, A. and SELA, J. (1971). On the transport of quasigeostrophic potential vorticity. *Mon. Weath. Rev.*, **99**, 447–459.

WILLSON, M. A. G. (1975). A wavenumber-frequency analysis of large-scale tropospheric motions in the extratropical Northern Hemisphere. *J. atmos. Sci.*, **32**, 478–488.
YOUNGBLUT, C. E. and SASAMORI, T. (1980). The nonlinear effects of transient and stationary eddies on the mean winter circulation. Part I: The diagnostic analysis. *J. atmos. Sci.*, **37**, 1944–1957.

– 9 –

Large-scale structure of the tropical atmosphere

PETER J. WEBSTER

9.1 Introduction

Over the last few decades it has been a common practice to view the atmosphere from a zonally averaged perspective. Although such a procedure may have some advantages for diagnostic or budget studies of the general circulation, it has a distinctly blanketing effect on the appreciation of features that possess strong longitudinal variability. The geographic locations to suffer most from the zonally averaged technique are the equatorial regions. A zonally averaged view of the tropics produces a rather bland and relatively featureless panorama. Compared to the seemingly more vigorous mid-latitudes where baroclinic activity appears well marked in a zonal sense, the low latitudes have gained a distinctly benign reputation.

If we remove the zonal average constraint so that the tropical atmosphere may be viewed with its full longitudinal variability displayed, a markedly different picture evolves. Rather than a zonally averaged meridional circu-lation (Hadley Cell) with magnitudes sufficiently small that it is sometimes difficult to isolate the sign of the circulation with a significant degree of certainty, relatively vigorous regional circulations emerge.

The vigour of the regional tropical circulations is not surprising as they are forced by the largest magnitude heating functions that exist anywhere on the globe — the heating associated with the warm sea-surface tempera-tures and the equatorial land areas. It is the form and complexity of the response to these heating fields that have rendered the understanding of the tropical circulation both a fascinating and frustrating problem. Observations readily verify this complexity. On one hand, it appears that there is a distinct mode preference in response for the largest scale, especially in the low-frequency domain. Such scale preference would seem to point towards simple dynamic relationships. But these motions are composed of a myriad of scales or groups of scales with time scales ranging from seasons to minutes. Thus to understand even the macroscale structure of the tropics we have several involved and interrelated problems to unravel. Besides determining why there is a scale preference for the largest scale motions in the tropics, we must also discern how the smallest scale, the convective element, rectifies in order to drive the macroscale and, in so doing, understand why the convective elements themselves appear to group at intermediate synoptic scales and mesoscales.

Some major relationships have been implied between the low latitudes and the extratropical atmosphere. Particular reference is made to 'tropical-extratropical teleconnections' through which events in the tropics are communicated to higher latitudes and to the reverse problem where the forcing of tropical events is initiated by processes originating in the middle latitudes. The zonally averaged meridional cell may be thought of as the primary, albeit weak, teleconnection between low and high latitudes. However, the intense regional circulation patterns, such as those associated with the monsoons, are probably of considerably greater importance. Indeed, the baroclinic activity at higher latitudes shows similar regionality which may not be unrelated to the low latitude asymmetric structures. Blackmon *et al.* (1977), for example, relate the wintertime maximum in cyclogeneses in the western North Pacific Ocean to the outflow region of the Asiatic jetstream, which, in turn is related to the intense equatorial convection of the winter monsoon.

There appears to be substantial observational evidence to suggest that other teleconnection chains exist between the tropics and higher latitudes. Bjerknes (1966, 1969), Namias (1976), Horel and Wallace (1981), and many others have shown that variations in the macroscale features of the tropical atmosphere, forced by variations in large-scale forcing fields, may be one of the primary causes of interannual variability in the climate of the higher latitudes. Such observations, which correlate the El Niño events of the Pacific Ocean with anomalous location of the higher latitude long waves, have been simulated by general circulation models (e.g., Rowntree, 1972) and have been provided theoretical explanations by a number of studies (Opsteegh and van den Dool, 1980; Webster, 1981, 1982; and Hoskins and Karoly, 1981).

The reverse influence, that of the higher latitudes on the tropics, is less well understood. From a purely zonally averaged perspective, the equatorial influence should be slight or even negligible because zonally averaged weak easterlies dominate the low latitudes. According to Charney (1969), a critical latitude for waves originating on the extratropics will exist at the latitudes corresponding to either weak easterlies or at the transition between the easterlies and westerlies. The theory, which evolved from the more general study of Charney and Drazin (1961) and is described in Chapter 6, suggests an effective barrier for wave propagation between middle latitudes and the tropics.

Observations show that the mean seasonal zonal wind possesses a distinct longitudinal component with regions of westerlies immersed in the equatorial easterlies which in some locations are sufficiently strong to form a bridge between the extratropical westerlies of the two hemispheres. The changes in sign of the equatorial zonal wind component may be tied to the existence of slowly varying forced planetary modes (Webster, 1972, 1973b; Gill, 1980). Furthermore, the observational study of Murakami and Unninayer (1977) shows that the regions of weak equatorial westerlies possess anomalously large transient kinetic energy. Such energy may be a manifestation of the weak westerlies acting as ducts for propagating wave disturbances (Webster and Holton, 1982; Simmons, 1982).

The basic essence of all of the studies cited above which consider the full longitudinal and latitudinal structure of the tropics, is rather clear. It is that a view of the tropics in which the tropical atmosphere is relatively independent of higher latitudes, except for communication via a zonally averaged Hadley cell, has little merit. Indeed it would seem that this parochial view of the tropical atmosphere must be replaced by a perspective of a much grander participator in the global circulation.

In the paragraphs that follow we will emphasize the large-scale slowly varying structure of the tropical atmosphere. The task will be tackled from two fronts. The first constitutes an observational approach, through which we will attempt to isolate the features of the various scales of motion. A second section will attempt to apply a theoretical framework to the observed structure of the tropical atmosphere.

9.2 The observed structure of the tropical atmosphere

9.2.1 The planetary scale

(a) *Zonally averaged structure*

The zonally averaged mean state of the tropical atmosphere is shown in Fig. 9.1 for June, July and August (JJA) and December, January and February (DJF). Overall, the tropical atmosphere is characterized by weak zonally averaged easterlies throughout the troposphere, a weak latitudinal temperature gradient and a small meridional velocity field which indicates net upper level divergence. The major differences between the seasons at low latitude is the reversing of the sense of the meridional circulation, a slight broadening of the surface easterlies in the Northern Hemisphere summer and a sharpening of the latitudinal temperature gradient in the winter subtropics.

These rather bland fields illustrate the traditional view of the tropical general circulation. The lower troposphere convergence in the summer hemisphere marked the ascending leg of the Hadley Circulation which was

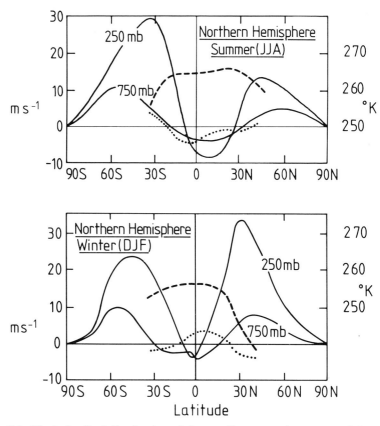

Fig. 9.1. The latitudinal distribution of the zonally averaged structure of the atmosphere for summer and winter. Solid lines refer to the 250 and 750 mb zonal wind fields, dashed curves to the 500 mb temperature field and dotted lines to the 250 mb meridional component of the wind.

termed the intertropical convergence zone (ITCZ). In this model, the mean rising motion at the ITCZ transported heat upwards to move poleward as part of the Hadley Cell. However, the simplicity of this argument was questioned by Riehl and Malkus (1958) who showed that the stability of the tropical atmosphere was such that these simple transports would cool the upper troposphere and thus destroy potential energy. In order to satisfy the heat budget of the general circulation, they argued that only motions that followed a moist adiabatic ascent could actually heat the upper atmosphere. This could only be achieved by upward motion taking place through deep penetrative convection cells. Riehl and Malkus have estimated that at any one time something in excess of 1500 cells in the ITCZ region would be necessary to achieve the appropriate vertical transports for the general circulation heat balance. Thus was established the importance of deep convective processes in the tropical atmosphere.

The character of the ITCZ as a zone of intense convection is evident from the mean seasonal (JJA and DJF) satellite infrared effective temperature distributions shown in Fig. 9.2 (adapted from Liebman and Hartmann, 1981). The IR effective temperature is a keen indicator of deep penetrative convection in the tropics as it demarks the cirrus canopy that accompanies convective processes in the tropics. It also appears to be an indicator of precipitation (Kilonsky and Ramage, 1976). The ITCZ appears as a meandering band of deep convection (cold IR effective temperatures) girdling the tropics. Its mean seasonal position generally coincides with the location of the zonally symmetric upper level divergence pattern noted in Fig. 9.1.

(b) *Longitudinally varying structure*

Other large-scale features of the tropical atmosphere are also evident in Figs. 2.21, 2.22 and 9.2 most of which possess distinct variations in longitude. Perhaps the most notable features in the distribution of tropical cloudiness and precipitation (Fig. 9.2) are the regions of intense equatorial convection (indicated by C) in Indonesia, Africa and South America which show little seasonal variability, and the monsoon convection located in the summer hemispheres (M).

Of all tropical phenomena, the large-scale convective features (C) were the last to be identified as physical entities. This is somewhat surprising as the gravest scales of equatorial motion were the subject of the first systematic global scale data investigation which was undertaken by Sir Gilbert Walker who found significant correlations between various atmospheric quantities existing over space scales measured in terms of multiples of earth radii (see, e.g., Walker, 1923; Troup, 1965). Of the many regions over which strong correlations were identified perhaps the clearest existed between Indonesia and the Western Pacific Ocean and the eastern Pacific Ocean. Thus, although it was known that spatial and temporal relationships did exist apparently no dynamic relationship was conceived. Perhaps the best known of the Walker correlations is the so-called 'southern oscillation index', which is effectively a measure of the pressure gradient between Tahiti and Darwin (Troup, 1965). The index is generally aperiodic with time scales of years and appears to be a strong function of sea-surface temperature variations in the eastern Pacific Ocean, particularly El Niño (Trenberth, 1976).

Figures 9.3(a) and 9.3(b) show the longitudinal seasonal distribution of the deviations from the zonal average of the zonal and meridional velocity components averaged over a 10° latitude strip spanning the equator at 200, 500 and 1000 mb. (One can place these circulations in the context of the three-

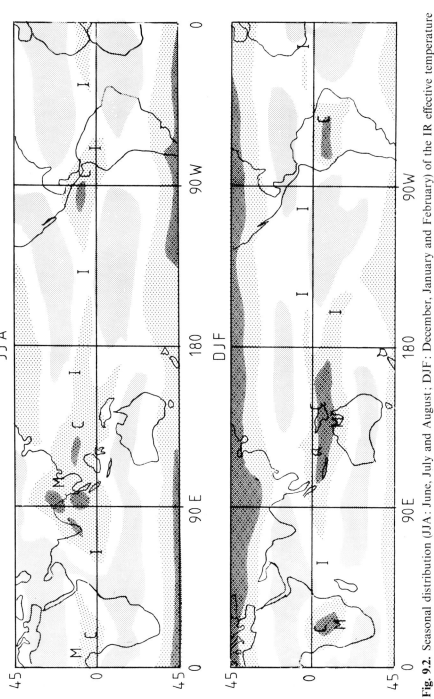

Fig. 9.2. Seasonal distribution (JJA: June, July and August; DJF: December, January and February) of the IR effective temperature averaged over the period 1974–78. Darkest shading denotes temperatures <230 K, stippled <250 K, clear regions <270 K and grey regions >280 K. Labels I, C and M show the location of the intertropical convergence zone, the semi-permanent equatorial convective

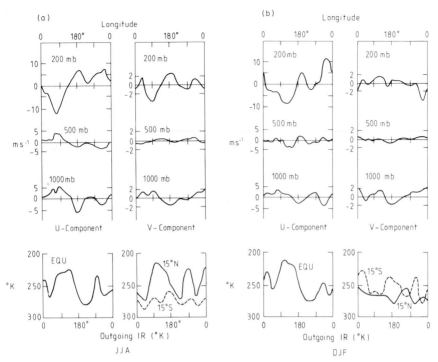

Fig. 9.3. Seasonal distribution for JJA and DJF of the longitudinal structure of the deviations from the zonal average of the zonal and meridional components of the wind field. Distributions are indicated at 1000, 500 and 200 mb. Bottom panels show the longitudinal structure of the IR effective temperature (scale inverted) for the equator, 15°N and 15°S. Note the scale magnification for the meridional velocity component.

dimensional circulation by referring back to Figs. 2.21 and 2.22.) Relatively strong equatorial easterlies dominate the eastern hemisphere, whereas westerlies dominate the western hemisphere. In the lower troposphere the reverse structure is true, indicating an almost perfect anticorrelation between 200 and 1000 mb. We also note that the horizontal flow at 500 mb is relatively weak and that the meridional flow at all levels is small compared to the corresponding zonal component indicating a compensatory divergence pattern existing through the whole atmospheric column. We also note that the regions of upper level divergence (lower level convergence) appear to be associated with mean precipitation (cold areas of Fig. 9.2) This is emphasized by plotting the IR effective temperature along the equator, 15°N and 15°S in the lower parts of Figs. 9.3(a) and 9.3(b). Noting that the low-latitude meridional wind speeds are substantially smaller than their zonal counterparts the planetary scale motions take on the appearance of east–west oriented cells with ascending legs in the regions marked C in Fig. 9.2 and descent elsewhere.

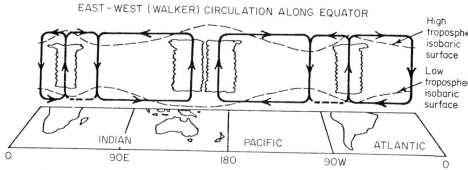

Fig. 9.4. Schematic view of the equatorial symmetric planetary scale features. Regions of ascent correspond to locations C in Fig. 9.2. Note the dominance of the Pacific Ocean–Indonesian cell which is referred to as the Walker Circulation. (From J. Zillman, private communication.)

Such cells appear confined to the equatorial plane and possess a vanishingly small meridional velocity component.

Figure 9.4 shows a schematic diagram (from Madden and Julian, 1971) of the location of the planetary scale motions described above. Several cells are shown, but the dominating structure appears as rising motion over the Indonesian region and the warm sea-surface temperatures of the West Pacific Ocean with descent to the east and west. The Pacific cell associated with Indonesian convection, often called the 'Walker circulation', spans the breadth of the Pacific Ocean and provides a physical manifestation of Walker's southern oscillation index.

Figure 9.3 also shows that there are, despite the relative smallness of the meridional velocity components, regions of substantial cross-equatorial flow. For example, a northward flux may be seen during JJA in the Indian Ocean region towards the heated Asian continent. In the upper levels a return flow is evident. In winter the sense of the circulation is reversed, although removed slightly eastward. Thus in addition to the equatorial flow that appears oriented in the east–west direction the planetary scale flow also possesses a substantial circulation that is asymmetric about the equator. From Fig. 9.3 it seems that the scale of the asymmetric modes is somewhat smaller than the symmetric motions.

The planetary scale motions possess variations on other time scales besides annual or seasonal. Two classes of variation are apparent. The first is the synoptic scale weather event which feeds back to the planetary scale motions. The second class of variation is a modulation of the planetary scale features on time scales longer than synoptic but less than seasonal. For example, in the monsoon regions the established summer monsoon vacillates between extremely active periods and distinct lulls. The latter are referred to as 'breaks'

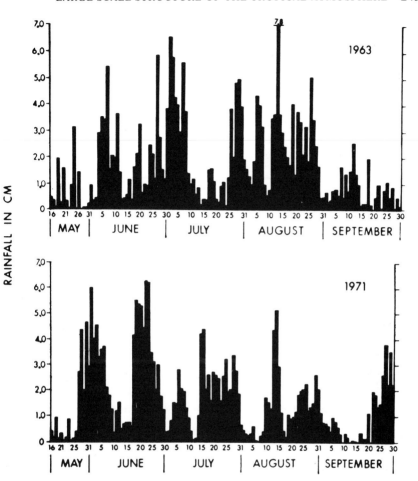

Fig. 9.5. Daily rainfall (cm day^{-1}) along the western coast of India incorporating the districts of Kunkan, Coastal Mysore and Kerala for the summers of 1963 and 1971. Periods of precipitation are referred to as the 'active' monsoon. The lulls in between are called monsoon 'breaks'.

in the monsoon during which time the usual disturbed and precipitating periods of the monsoon over South Asia are replaced by long periods of dry and undisturbed weather. Figure 9.5 summarizes the vacillation between these two phases of the summer monsoon for the years 1963 and 1971 shown by a plot of the daily rainfall along the western coast of India (Kinkon, Coastal Mysore and Kerala). Such low-frequency variability appears characteristic of the precipitating regions of the summer and winter phases of the Asian Monsoon and the African Monsoon. The *active periods* appear to be associated with groups of disturbances and the *lulls* with an absence of disturbance over large areas of the monsoon region.

Extremely long period variations are also observed in the planetary scale features. With changes in the sea-surface temperature structures in the Pacific Ocean during El Niño events, a readjustment of the Walker Cell (see Fig. 9.4) appears to occur with the major convective region moving to the mid-Pacific Ocean (see, e.g., Horel and Wallace, 1981). With the maximum precipitation to the east of Indonesia during El Niño, tropical Australasia, New Guinea and Indonesia fall into a period of relative drought.

9.2.2 The synoptic scale

Viewed on a seasonal or annual basis, the planetary scales appear as coherent structures with well defined regions of steady convection and ascent and associated regions of subsidence. However, viewed on a day-to-day basis, the mean convective regions of the planetary scale are made up of propagating smaller scale disturbances with life times measured in days. The synoptic scale motions at low latitudes possess some unique properties.

The propagating nature of the synoptic scale disturbances may be seen from the time–longitude sections of satellite imagery compiled by Wallace (1970) for the $10°-15°N$ latitude band of the tropics. The section is shown in Fig. 9.6. On such diagrams, propagation is manifested as a coherent line of cloudiness sloping in one direction or another. In the tropical Pacific and Atlantic Oceans westward propagation may be seen for periods of longer than 10 days. A typical length scale for these westward propagating disturbances appears to be about 10^6 m with propagation speeds of order $7-10$ m s^{-1}. It is much more difficult to see examples of such well defined propagation in the geographical regions of low latitudes where continental effects become important. The interpretation of satellite imagery in these regions becomes less clear because of the effect of convective activity associated with the diurnal variation over the moist tropical land areas. In Fig. 9.6, the Indonesian, African and Indian regions all appear as solid white regions indicating high cloud amount. However, westward disturbance propagation is evident in these regions as evidenced by the paths of disturbance emanating from Central Africa during GATE and the Bay of Bengal disturbances which periodically move westward up the Ganges Valley.

To examine the structure of the propagating disturbances, we will make use of data collected during December 1978 which was the first month of the First GARP Global Experiment. (All analyses shown in Figs. 9.7 and 9.8 were adapted from the Australian Numerical Meteorology Research Centre tropical meteorology group. Early interpretation of the analyses appear in the proceedings of *The International Conference on the Early Results of FGGE and*

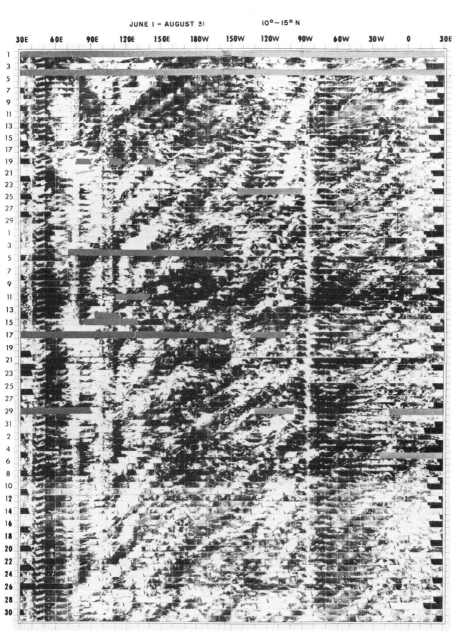

Fig. 9.6. Time–longitude section of visible satellite imagery for the latitude band 10°–15°N of the tropics. Cloud streaks from right to left with increasing time denotes westward propagation. (After Wallace, 1970.)

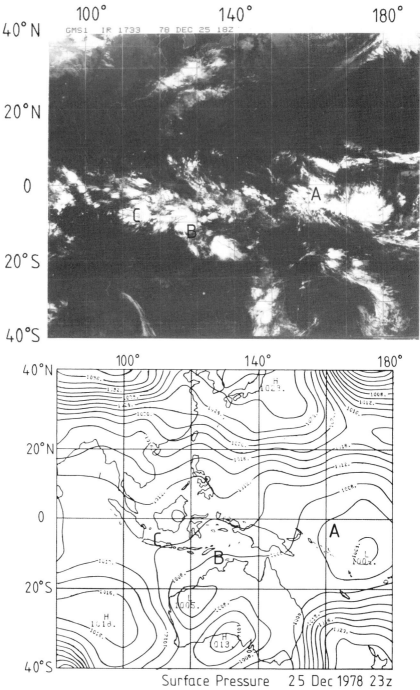

Fig. 9.7. The winter MONEX region of 25 December 1978. Upper panel shows the GMS IR satellite picture with the surface-pressure pattern shown on lower panel. Both panels are on the same projection. Pressure analysis after McAvaney *et al.* (1981). Letters A, B and C identify synoptic-scale disturbances referred to in the text.

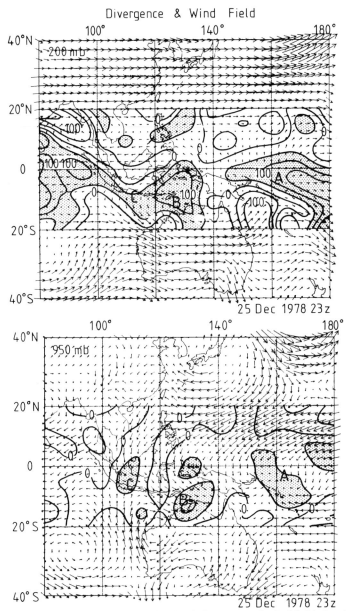

Fig. 9.8. The 250 mb (upper panel) and 950 mb (lower panel) wind fields for the winter MONEX region of 25 December 1978 with the horizontal wind divergence superimposed in the $20°$N–$20°$S latitude strip. In the upper troposphere areas of *divergence* are stippled whereas in the lower troposphere areas of *convergence* are stippled. Stippled areas denote divergence magnitudes greater than $\pm 50 \times 10^{-7}$ s^{-1}. Analyses from McAvaney *et al.* (1981).

Large Scale Aspects of Monsoon Experiments, Tallahassee, Florida, USA, 12–17 January 1981; see, e.g., McAvaney *et al.*, 1981.) Figure 9.7 shows the surface pressure distribution at $2300Z$ and the satellite IR picture for $1800Z$ on the 25 December 1978. Except for the low-pressure region in the northwest of Australia and the broad trough that spans the near equatorial region of the southern hemisphere, the tropical portion of the chart is relatively featureless. In fact, there appears to be little correspondence between regions of intense cloudiness and the pressure field, except that the major cloud regions appear to reside about the axis of a broad equatorial trough. The satellite picture itself reveals significant structures. Three major regions of deep high cloudiness are discernible and are demarked by A, B and C in the diagrams. From the scale of the cloud features it would appear that A, B and C are synoptic scale disturbances.

As the cloudy regions A, B and C of Fig. 9.7 are not as strongly evident in the surface pressure field to the same degree as propagating extratropical disturbances, we resort to an examination of the wind field. The 200 and 950 mb wind fields are shown in Fig. 9.8 in the form of wind vectors with the respective divergence fields (units $10^{-7}\,s^{-1}$) superimposed in the $20°N-20°S$ latitudinal strip. At 950 mb *convergent* areas with magnitudes greater than $-50 \times 10^{-7}\,s^{-1}$ are shaded. At 200 mb *divergent* areas exceeding $+50 \times 10^{-7}\,s^{-1}$ are shaded. The correspondence of upper and lower shaded areas, as depicted in Fig. 9.8, would indicate deep divergent systems. Such is the case for A, B and C. The properties of these three systems; lower tropospheric convergence, deep penetrative convection and upper level divergence, appear characteristic of the synoptic-scale tropical disturbances of the ITCZ and the major convective zones of the low latitudes.

If we refer back to Fig. 9.2, we note that convection denoted by IR irradiance is not the normal situation for the low latitudes. Indeed, for much of the tropics, the dominant signature is relatively warm IR effective temperatures indicating a lack of high cloudiness. One such region is the eastern South Pacific Ocean. With reference to Fig. 9.4 we note that this region corresponds to the subsident leg of the 'Walker circulation'. Here, synoptic-scale events appear to possess a completely different character to those of, for example, the Western Pacific Ocean. Instead of motions that are strongly coupled in the vertical, disturbances in the upper troposphere appear to have little relationship to events in the lower troposphere. The eastern Pacific and eastern Atlantic Oceans, both areas of subsidence, appear to be a location for maximum penetration of extratropical disturbances (Webster and Curtin, 1975) and correspond to regions of weak equatorial westerlies.

The degree of synoptic-scale activity in a particular region of the tropics may be assessed by computing the kinetic energy of the flow for time scales less than some prescribed period. Such perturbations in kinetic energy were calculated by Murakami and Unninayer (1977) for January and February of

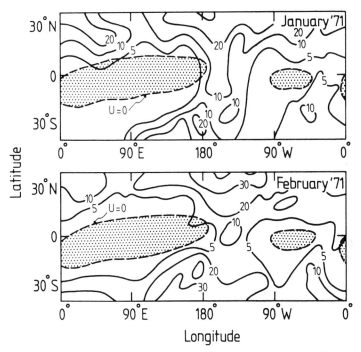

Fig. 9.9. The January and February 1971 perturbation kinetic energy fields calculated on the basis of deviations from a monthly mean. The zero line of the mean DJF zonal component is superimposed. Kinetic energy distribution from Murakami and Unninayer (1977).

1971 for 200 mb and are shown in Fig. 9.9. The zero line of the mean DJF zonal component of velocity (i.e., $\bar{U} = 0$) is superimposed with the region of mean easterlies stippled. The most striking feature is the coincidence of the maximum perturbation kinetic energy at low latitudes with the mean equatorial westerlies.

To illustrate the small surface pressure perturbation associated with synoptic-scale features we present the barograph record for Darwin from 23–28 December 1978, which encompasses the periods of the case study described in Figs. 9.7 and 9.8. The surface pressure trace is shown in Fig. 9.10 with the major variation being associated with the semi-diurnal oscillation which possesses an amplitude of some 4 mb. Little alteration to the semi-diurnal trend is apparent near 25 December 1978, which seems to belie the existence of the disturbance. Indeed at low latitudes only on rare occasions with the passage of a tropical cyclone or hurricane will the synoptic-scale pressure perturbation be larger than the semi-diurnal pressure variation. Thus the two factors, the lack of a notable surface pressure perturbation for most synoptic

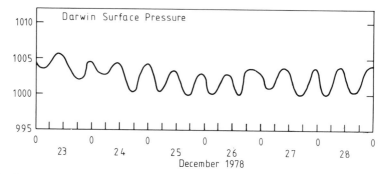

Fig. 9.10. The variation of the Darwin surface pressure for the period 23–28 December 1978. Structure is dominated by the semi-diurnal atmospheric tide.

disturbances and a much larger perturbation being maintained in the tropical atmosphere in the diurnal time scale, require some explanation.

9.2.3 The sub-synoptic scale

The traditional view has the cumulus cloud, occasionally grouped in clusters (the synoptic scale), as the ubiquitous feature of the tropical atmosphere. In this view a scale separation exists between the cumulus scale and the disturbance or synoptic scale. The recent international experiments, in particular the Line Island Experiment, GATE (the GARP Atlantic Tropical Experiment) and MONEX (the Monsoon Experiment) have resulted in data that have done much to alter that simple picture. Whereas convection still remains the dominant process, a new scale, the mesoscale, appears to be the dominant sub-synoptic scale (see, e.g., Zipser, 1969, 1977; Houze and Betts, 1981; Houze, 1982. Houze and Betts, 1981, in particular summarize studies relating to the cumulus and the mesoscale of GATE and provide a clear picture of the multiplicity of scales which exist between the cumulus and the synoptic scales.)

A common feature of most observations appears to be that precipitation does not take place only on the cumulus scale but over much larger regions in a fairly uniform pattern. The studies cited above indicate that the precipitation may be tied to the organized mesoscale features of the disturbance which are manifested as an active deep middle troposphere cloud deck extending often hundreds of kilometres from the convective source. Some 40% of total precipitation of the disturbance appears to be associated with these mesoscale features. An example of one such mesoscale system is shown in Fig. 9.11. In this case the system is associated with a GATE squall-line system and indicates radar echoes associated with broad and general precipitation. Examples of similar structures are given by Houze (1981) and Webster and Stephens (1980)

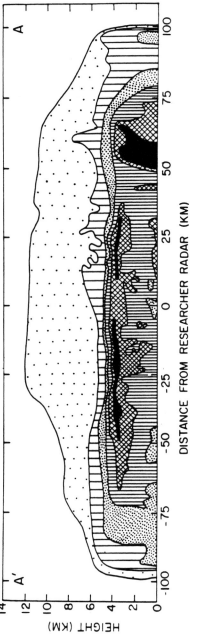

Fig. 9.11. Vertical cross-section through a meso-scale structure showing uniform precipitation between −100 km and 75 km from the GATE ship *Researcher* radar at 2100 GMT 4 September 1974. Various shadings denote intensity of echo. The precipitation may be seen to exist on a much larger scale than the cumulus scale. The black area is the most intense. (From Leary and Houze, 1979.)

in the South China Sea during the Winter MONEX. Webster and Stephens have suggested that the upper tropospheric extended cloud meso-scale system may remain active by radiative destabilization of the cloud deck itself for a considerable period after the convective source has diminished.

Because of the effect of radiation through the tropical column in disturbed regions, it is not surprising to find diurnal variations being a principal component in the synoptic- and the meso-scale. Such variations have been highlighted by Short and Wallace (1980) who compared the distributions of outgoing IR at 0900 and 2100 hours local time for various geographic locations. As may be expected, such variations are largest in the land-ocean mix of Indonesia.

9.3 Aspects of the dynamic structure of the low latitudes

Much of the basic structure of the extratropical atmosphere appears to be described by a unified concept; the quasi-geostrophic theory. Such a unification has been made possible by the considerable effort of a number of researchers who had the benefit of an excellent observational description of the phenomena in question.

No such unified theory has yet emerged for the tropical atmosphere which would have as equally far-ranging implications as the quasi-geostrophic theory of higher latitudes. Furthermore, the data set upon which all theories must ultimately rest has been much poorer than its middle-latitude counterpart. The problem of theory unification is rendered even more difficult by the apparent absence of a significant scale separation as was emphasized in Section 9.2.3.

Despite the lack of a unified theory, considerable progress has been made in understanding the physical processes that dominate the tropical atmosphere by use of comparison with the dynamic processes of higher latitudes. Such comparative scale analyses were first used by Charney (1963). A second approach that has proved fruitful has been the study of the planetary scale motions of the low latitudes using classical tidal theory. Matsuno (1966) and Longuet-Higgins (1968) were among the first to realize the importance of these associations. We will briefly review these two approaches in the following paragraphs, with particular reference to the observational structure described in the last section.

9.3.1 Comparative scales

We choose typical horizontal length and speed and thermodynamic variable scales L, U and χ respectively. In terms of these quantities, the Rossby and

Table 9.1. Typical scales and relative thermodynamic variations (in this case surface pressure) for middle latitude synoptic motions and synoptic and planetary scale tropical motions. Values of $U \sim 10$ m s^{-1} and $H \sim 10^4$ have been assumed.

Length scale	Extratropics 10^6 m	Tropics	
		10^6 m	10^7 m
Froude number	10^{-3}	10^{-3}	10^{-3}
Rossby	10^{-1}	1	10^{-1}
$\delta P/P\vert_{\text{theory}}$	10^{-2}	10^{-3}	10^{-2}
$\delta P/P\vert_{\text{obs}}$	$\frac{30}{1010} \sim 3 \times 10^{-2}$	$\frac{1}{1000} = 10^{-3}$	$\frac{10}{1000} = 10^{-2}$

Froude numbers are defined as U/fL and U^2/gH, where H is a typical height scale. Charney (1963) showed that for *synoptic scale* variations at low latitudes (i.e., $L \sim 10^6$ m, $f \sim 10^{-5}$ s^{-1}) perturbations in the thermodynamic fields scale as:

$$\delta\chi/\chi \sim F_R\vert_{R_0 \sim 1} \tag{9.1}$$

which merely reflects the domination of inertial effects over those of rotation. On the other hand, at higher latitudes ($f \sim 10^{-4}$ s^{-1}) the ratio goes as:

$$\frac{\delta\chi}{\chi} \sim \frac{F_R}{R_0}\bigg|_{R_0 \ll 1} \tag{9.2}$$

reflecting the importance of rotational effects. Thus for the typical values listed above, the scaling arguments predict that the relative variation of thermodynamic quantities, *on the synoptic scale*, should be about an order of magnitude smaller near the equator than in the quasi-geostrophic motions of middle latitudes. The comparative smallness of the low-latitude perturbation may be explained because of the rapidity of the adjustment of the tropical motions to a pressure gradient imbalance; the adjustment being less constrained by rotational effects.

Table 9.1 shows a comparison between observed and theoretical estimates of surface pressure variations for various scales in the tropics and extratropics. Generally there appears to be excellent agreement. However, it is interesting to note that the *planetary scale* features at low latitudes appear to be governed more by Eqn. (9.1) that Eqn. (9.2). Indeed for $L \sim 10^7$ m, $R_0 \ll 1$ even though $f \sim 10^{-5}$ s^{-1}. Thus planetary scale features appear to be quasi-geostrophic even close to the equator, a point noted by Matsuno (1966). Consequently, variations in the thermal structure of planetary scale features such as those in the Pacific Ocean described by Walker (1923) and Madden and Julian (1971), the monsoon circulations and even the atmospheric tides should be about an order of magnitude larger than those associated with the synoptic scale cloud cluster or tropical disturbance.

A consequence of the constraints on the thermodynamic structure of the synoptic scale motions at low latitudes is evident in the balance that determines the vertical velocity. Because of the weak temperature gradients [from Eqn. (9.1) $\delta T \sim F_R T$] the horizontal advective terms in the first law of thermodynamics are considerably smaller than the heating even in undisturbed regions. Thus heating or cooling must be balanced by vertical advection or, in other words, adiabatic effects. In this manner, the slow subsidence over the vast clear region of the tropics is in fact necessary to balance the weak radiative cooling. A radiative cooling rate of order 1°C day^{-1} is effectively balanced by a subsidence of about 0.3 cm s^{-1}. In disturbed regions where latent heating rates exceed 5°C day^{-1} over the synoptic scale, compensatory vertical wind speeds of at least 3 cm s^{-1} are necessary (e.g., Holton, 1979).

A final scale comparison may be made using the vorticity equation. For *extratropical synoptic scale* motions ($L \sim 10^6$ m, $f \sim 10^{-4}$ s^{-1}) or for *low-latitude planetary scale* features of the vorticity equation takes on its divergent form, i.e.:

$$(\partial/\partial t + \mathbf{v}\cdot\nabla)(\zeta + f) + f\nabla\cdot\mathbf{v} \approx 0|_{R_0 \ll 1} \qquad (9.3)$$

whereas for low-latitude synoptic scale motions the form becomes:

$$(\partial/\partial t + \mathbf{v}\cdot\nabla)(\zeta + f) \simeq 0|_{R_0 \sim 1}. \qquad (9.4)$$

In other words, synoptic scale motions should be horizontally non-divergent which would infer that disturbances in the flow field at upper levels may be independent of perturbations at lower levels. Charney (1963) noted that such vertical decoupling only occurs with non-convective disturbances. Thus in regions of strong convection where heating rates exceed 5°C day^{-1} over a large area the local divergence is of sufficient scale to suggest that Eqn. (9.3) is a more descriptive form of the vorticity equation. Within the disturbance balances become exceedingly complex as the cumulus- and meso-scale have to be acknowledged. Holton (1979) points out that in the trough region of a *convective* disturbance both ζ and $\nabla\cdot V$ are large whereas the advective terms remain relatively small. Thus, locally, a rescaling of the vorticity equation is necessary in order to balance the term $(\zeta + f)\nabla\cdot\mathbf{v}$. Probably only vertical momentum transports by the cumulus- or meso-scale may accomplish the balancing.

A further consistency between theory and observations is also apparent. As the planetary scale features appear to be governed by the divergent form of the vorticity equation, strong coupling in the vertical over the entire circulation should occur. This accounts for the observed vertical coherence which characterizes the planetary scale features noted in the wind fields plotted in Fig. 9.3 and shown schematically in Fig. 9.4.

The inherent non-divergent nature of the tropics *before* the generation of a convective disturbance led Charney (1969) and Mak (1969) to consider the possibility of the lateral initiation of disturbances. Based on the relatively simple theoretical states they considered (e.g., the basic states were usually zonally symmetric), it was difficult to relate tropical disturbances with middle latitude forcing. However, it is possible that *in situ* development of disturbances rests with the interactions between the *planetary* and *synoptic scale* motions. We note, for example, that convective synoptical scale disturbance (including tropical cyclones) form in specific geographical regions which correspond to the ascending branches of the planetary scale modes! Such regions are warm ocean regions of the Western Pacific Ocean and Indonesia, tropical Africa and the Western Atlantic and the Central Americas. These locations are regions of low tropospheric divergence (see Fig. 9.4), which is conducive to the development of convective disturbances.

9.3.2 Fundamental modes of the low latitudes

Although the comparative scaling arguments have allowed some insight into the physical processes that maintain the tropical atmosphere and differentiate between the various scales of motion, major advances have emerged from another quarter. The basic advance has been that many of the features of the planetary scale, slowly varying tropical atmosphere may be explained in terms of the classical tidal theory. For example, Matsuno (1966) and Longuet-Higgins (1968) using shallow fluid models on a rotating sphere were able to isolate several distinct modes which, for a parameter range indicative of the large scale and the low frequency, possessed phase speeds and horizontal structures which bore surprising resemblance to the real world. The excitation of these fundamental modes by realistic distribution of forcing functions, such as the sea surface temperature distribution, has been explored in a number of studies (e.g., Webster, 1972, 1973b; Gill, 1980).

The formal development of the governing equations of these fundamental modes exists in many places in the literature (e.g., Matsuno, 1966, and, especially, Longuet-Higgins, 1968, and, for the stratosphere, Holton, 1975). It suffices here to describe briefly the properties of the modes and their basic character as they pertain to the tropical atmosphere.

Figure 9.12 shows schematically the zonal wavenumber one eastward and westward propagating modes identified by Matsuno and Longuet-Higgins plotted against the free parameters of the system which are $E \, (= (2\Omega a)^2/gh,$ where Ω is the rotation rate of the fluid and h its depth) and the frequency λ (normalized by the rotational frequency 2Ω). The letters G_E, G_{II}, K, MY, R_1 and R_2 refer to the eastward and westward propagating gravity wave, the

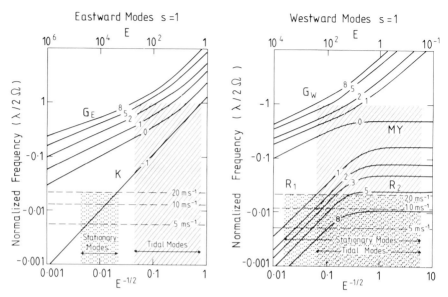

Fig. 9.12. Eigenfunctions of the free inviscid modes of oscillation on a sphere for the gravest longitudinal mode ($s = 1$) propagating (a) eastward and (b) westward, plotted as a function of frequency and E. Horizontal dashed lines refer to Doppler-shifted frequencies corresponding to eastward and westward basic zonal winds of magnitude 5, 10 and 20 m s^{-1}. The region where the low-latitude stationary modes and the tidal modes exist for the parameter range indicative of low latitudes is shaded. The lettering identifies the various classes of modes referred to in the text, and the numbers refer to the eigenfunction of the latitudinal operator. (Adapted from Longuet-Higgins, 1968.)

Kelvin wave, the westward propagating mixed Rossby–gravity wave (denoted MY after Maruyama and Yanai) and the two forms (baroclinic and equivalent barotropic) of the Rossby mode. The horizontal dashed lines correspond to values of the frequency that would have a zero Doppler shifted frequency in basic flows of magnitude 5, 10 and 20 m s^{-1}. Thus interceptions of the dashed lines and the model curves indicate those modes which may exhibit stationary or low-frequency behaviour in the real atmosphere. The region of such intercepts is indicated by stipple. As E becomes larger all the modes become increasingly trapped in the equatorial region. For a basic state of -10 m s^{-1} the 'stationary' Kelvin wave would have a latitudinal e-folding scale of $11.5°$ and extremely small meridional velocities. On an equatorial β-plane ($f = \beta y$) the horizontal structures of the K and MY modes are as illustrated in Fig. 9.13.

For an easterly basic flow only the Kelvin wave possesses a phase speed slow enough to appear as a stationary solution. For a westerly basic flow a range of Rossby solutions may be excited. It is interesting to note that the tidal modes

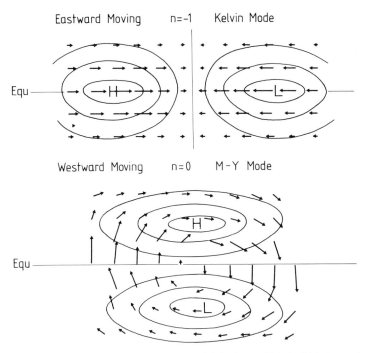

Fig. 9.13. Horizontal structure on an equatorial β-plane of (a) the Kelvin mode and (b) the mixed Rossby–gravity mode. Isobars and isotachs are shown. (After Matsuno, 1966.)

(those modes that correspond to normalized frequencies of order 1) must always be transient for realistic zonal wind speeds (stippled area). All modes shown in Fig. 9.12 are inviscid.

Whereas the horizontal structure of the classical modes explains a considerable part of the variance observed at low latitudes, it should be noted that considerable problems exist with the vertical scales. Inviscid linear solutions predict scales of the Kelvin mode (based on a relative phase speed of order 5 m s^{-1}) of about 1–2 km compared with observations of order 10–15 km (the depth of the troposphere). Figure 9.14 shows the theoretical relationship between the free parameters of a simple isothermal atmosphere and the vertical wavenumber the mode would possess. (Details of the development of relationships between the vertical and horizontal structures of the equatorial modes are given in several studies, e.g., Holton, 1975; Geller, 1970.) In this simple inviscid atmosphere two forms of vertical structure are possible. These are the exponentially decaying Lamb wave and propagating waves. For the frequency range defined by the tidal and planetary scale modes, as discussed in the discussion of Fig. 9.12, a vertical scale separation occurs.

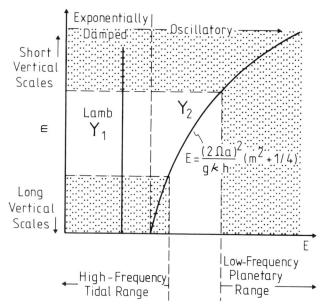

Fig. 9.14. The relationship between the horizontal and vertical structures of the fundamental spherical modes in an isothermal atmosphere. Ranges for the quasi-stationary tropical modes and the tidal motions are shaded.

The tidal motions are either exponentially damped solutions or waves with very large vertical structure; the low-frequency planetary waves, the Kelvin and Rossby modes, possess very small vertical scales according to this theory.

The problem of the difference in vertical scale between the inviscid (Kelvin wave) linear solutions and those observed in the atmosphere was considered by Chang (1977). He argued that the vertical mixing effect of convective activity in the tropical atmosphere was analogous to considering the tropical atmosphere to be strongly dissipative. Chang's mechanism was similar to the concept of 'cumulus friction' formalized by Schneider and Lindzen (1976). Considering the equatorial Kelvin wave Chang found that the effect of strong damping was to produce two forms of vertical wave structures rather than the single form shown in Fig. 9.14. (Formally, two solutions arise because dissipation renders the Doppler-shifted frequency complex as distinct from purely real in the inviscid case: the dispersion relationship for the Kelvin wave thus becomes quadratic.)

The structures may be written in the form $\exp{(i2\pi z/L_z)}\exp{(-\Gamma_i z)}$ and shown in Fig. 9.15 are plots of the vertical wavelength (L_z) against the phase speed and against the attenuation rate (Γ_i) for the Kelvin wave for various values of dissipation rate. The dashed line separates the two vertical mode

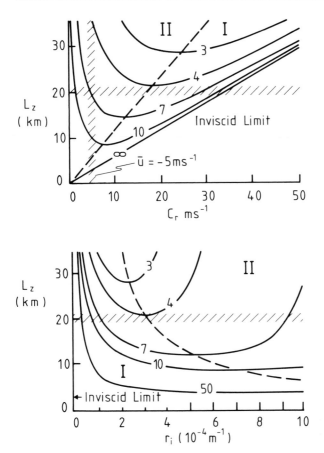

Fig. 9.15. (a) The real Doppler-shifted phase speed c_r plotted as a function of vertical wavelength for various damping timescales. The regimes of modes I and II are clearly evident. The horizontal hatched area denotes the modes excited for a preferred vertical scale whereas the vertical hatched area shows the vertical scale of the excited viscous mode II in a given -5 m s^{-1} basic flow. (b) The imaginary vertical wavenumber (the vertical attenuation rate) as a function of vertical wavelength for various damping timescales. The horizontal hatched line shows the attenuation rates for the two modes for a given vertical scale of excitation. (After Chang, 1977.)

forms. The figure indicates that the vertical wavelength of a viscous mode which can be stationary in an easterly flow of -5 m s^{-1} (i.e., that mode with an equal and opposite phase speed to the basic flow) will be of order 20 km for a 7-day dissipation rate and that it will be trapped in the troposphere, having an e-folding height of about 2 km. Alternatively, Fig. 9.15 indicates that for a forcing with vertical scale of (say) 20 km and the same dissipation rate, two modes will be forced, one with a phase speed of 5 m s^{-1} and e-folding in height 2 km, and the other with a phase speed of 30 m s^{-1} and having an e-folding height of order 10 km. For the latter mode it can be seen that the phase speed is fairly insensitive to dissipation rate. This mode approximates the linear inviscid solution discussed earlier. Chang regards the two modes with different phase speeds and character as indicative of the Kelvin wave behaviour observed in a troposphere which is locally highly convective (and therefore dissipative) and the weakly dissipative stratosphere, respectively.

In summary, any theory that leads to a realistic vertical structure of tropospheric planetary waves in the tropics must incorporate viscous effects. Most importantly, it means that much of the variance of at least the low-frequency circulation of the tropical atmosphere can be understood using simple linear concepts. Furthermore, it explains why waves of the same basic character, the Kelvin wave, appear to possess such different properties in the stratosphere and the troposphere. Finally it resolves several numerical problems. As viscous modes possess scales that are of the same order as the height of the tropopause, a low-order vertical grid resolution may be all that is necessary in the tropics. Unfortunately, this gain in modelling simplicity is cancelled out by the necessity to parameterize adequately the viscous effects of the tropical atmosphere which, in effect, means the proper modelling of convection.

An example of the simulation of the seasonally averaged large scale flow of the tropical atmosphere using a low order vertical representation is shown in Fig. 9.16. Produced by Gill (1980), the diagram represents the superposition of two linear (forced) solutions produced by a symmetric heat source at the equator (providing the Kelvin wave) and an anti-symmetric heat source with maximum amplitude away from the equator (the Rossby wave). Similar reconstructions have been produced by Matsuno (1966) and Webster (1972, 1973a). Compared to the observed mean seasonal state of the atmosphere, the form of the horizontal forced modes is remarkable. For example, the Kelvin wave interpretation of the symmetric steady state circulation about the equator explains the relative magnitudes of u and v shown in Fig. 9.3 as well as the tendency for the symmetric mode to exist on the planetary scale. Holton (1973) and Webster (1973b) have shown that the Kelvin wave is closest to resonance at the largest scale.

9.4 Heating processes

The Earth–atmosphere system is driven by the differences in the distributions of the incoming solar radiation and the outgoing longwave stream. As a large proportion of the shortwave flux is expended in evaporation at the Earth's

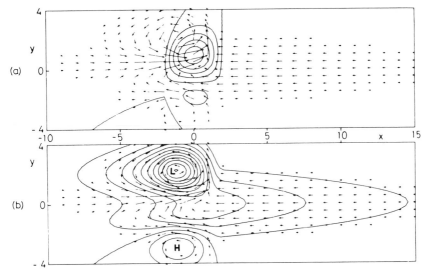

Fig. 9.16. Distribution of the flow obtained by the addition of the response to heating confined about the equator and heating which is concentrated to the north of the equator. Upper diagram shows the distribution of vertical velocity and the lower diagram the perturbation pressure pattern. Vectors indicate the horizontal flow field.
(From Gill, 1980.)

surface and released as latent heat in different regions, a complicated distribution of total heating is produced. In order to minimize the local and regional imbalances a variety of physical processes is invoked.

We can define the total heating as follows:

$$\dot{Q}_T = \dot{Q}_{rad} + \dot{Q}_{cond} + \dot{Q}_{sen} \qquad (9.5)$$

which is given as the sum of the radiational, condensational and sensible heating. It must be emphasized that all of these components are strongly coupled both between themselves and with the dynamics of the system. In the tropical atmosphere, the total heating, especially $\dot{Q}_{cond}$, and the large-scale vertical velocity field are closely related as indicated earlier. Observations point to strong correlations between the ascending regions of the planetary scale features and precipitation. The sign of the vertical velocity provides a cut-off between regions dominated by relatively weak radiational cooling and regions dominated by intense moist-convective heating.

There is a caveat that must be applied to the last statement. On the cumulus and cluster scales there is mounting observational support (e.g., Cox and Griffith, 1979) for the theoretical speculation that radiative effects are of considerable importance in the determination of the local total heating fields (e.g., Stephens and Webster, 1979; Webster and Stephens, 1980).

Even in the relatively strong disturbances described by Leary and Houze (1979) the attenuation of the incoming and outgoing radiation fields by the cloud distribution cannot be neglected if the dynamic structure of the system is to be properly understood. Such suggestions are in sharp contrast to the traditional view of radiation in the tropics. Before GATE and the Monsoon Experiment (MONEX), radiation was supposed to assume a generally passive role throughout the atmosphere with little difference between the disturbed and undisturbed regions.

9.4.1 The distribution of heating

We have noted earlier that the dynamic fields associated with the longitudinally varying planetary scale features of the tropics appeared to be more interesting than the zonally averaged structure. The same is true for the heating fields. The distribution of net radiative flux at the top of the atmosphere, as determined by NIMBUS 3, can be used to illustrate the differences between longitudinal and latitudinal gradients in the total heating. Plots of the zonally averaged net flux (solid line), the latitudinal variation along 90°E (large dashed curve) and the longitudinal variation along 25°N between 0° and 180°E (dashed curve) are shown in Fig. 9.17. The data refers to the mean conditions during July 1969. What is particularly important in the figure is that the *longitudinal gradient* of net flux is of the same magnitude as the *latitudinal gradient*. Most of the variation in the latitudinal profile may be accounted for in the latitudinal gradient of the solar input. However, as the solar input is constant along 25°N, the longitudinal variation in net flux must be due to other effects such as ground albedo and cloud cover; the latter being closely associated with the dynamic systems.

9.4.2 Gross relationship between dynamic structures and heating

Another striking feature of Fig. 9.17 is that the desert regions (10°E to 50°E) appear as net radiative sinks with the outward longwave radiation greater than the net incoming solar flux. On the other hand, the convective monsoon regions (80°E to 180°E) act as net radiative heat sources. In the case shown the radiative heating distribution is probably indicative of the total heating field. The condensational heating will be a maximum in the monsoon regions as Q_{cond} will be strongly tied to the precipitation patterns. In the desert regions the sensible heating will act in a sense opposite to that of the radiational cooling but this component will be smaller and restricted to the lower layers of the atmosphere. Thus diabatically the atmospheric columns above the deserts should be continually cooling and the columns in the monsoon regions

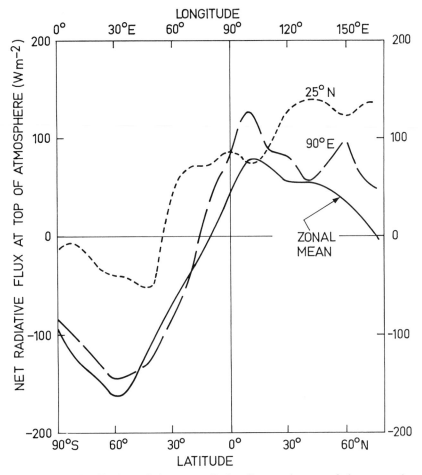

Fig. 9.17. The distribution of the net radiative flux at the top of the atmosphere inferred from NIMBUS 3. Plots of the zonally averaged net flux (solid line), the flux along 90° E (largest dashed line) and along 20° N (dashed curve) for July 1969 are shown. (From Stephens and Webster, 1979.)

continually heating. Obviously, a dynamic response is necessary in order to rectify the imbalance.

To determine the form of the dynamic response data from Newell *et al.* (1972) were used to calculate the heat convergences in the longitudinal section between the arid regions of Saudi Arabia (I), the Arabian Sea (II) and the Bay of Bengal (III). The resultant fluxes, together with the estimation of the vertical profiles of the components of the total heating are shown in Fig. 9.18. The dynamic response to the heating imbalance is such as to converge heat into the upper troposphere of the desert regions and out of the convective

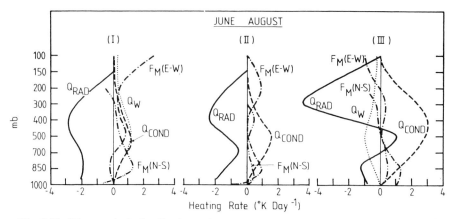

Fig. 9.18. The vertical distribution of the heating components in the atmospheric column above Saudi Arabia (I), the Arabian Sea (II) and the Bay of Bengal (III). Note the relative imbalances in the total heating in I and III. Heat fluxes calculated from the data of Newell *et al.* (1972) are shown.

regions. The net radiative cooling is compensated by adiabatic warming over the deserts and an adiabatic cooling over the Bay of Bengal. In other words, the dynamic response to the longitudinal imbalance of total heating is the generation of a rather vigorous thermally forced dynamic mode.

9.4.3 Cooperative dynamical–heating relationships

In a previous section we considered evidence that suggested that the structure of synoptic scale equatorial disturbances may be considered to be an ensemble of meso-scale systems and convective clouds that appear coherent and organized on the synoptic scale. We also noted that the disturbances appeared to possess distinct geophysical location for their development and followed relatively well defined routes. A great deal of effort has been expended in recent years in attempting to understand the physical processes that promote the development of disturbances and that contribute to their maintenance.

The relative features of the thermal fields on the synoptic scale over most parts of the tropics probably rule out baroclinic instability as a major initiator of tropical disturbances. This may not be universally true as in the Northern Hemisphere summer baroclinicity on the synoptic scale becomes quite large in the Indian and African regions. A more likely and more general candidate is barotropic instability in regions of substantial shear (Bates, 1970; Reed et al., 1977; Holton, 1979). Another possibility is the propagation of disturbances from higher latitudes into the tropics which is discussed below in Section 9.5.1.

Barotropic instability has many attractive properties and probably is an important agent in the initiation of disturbances in the tropics. However, it falls short of explaining all the features of the tropical disturbance. First, disturbances propagate into and through regions of weak shear so that it is improbable that the mature disturbance continues to tap energy from shear flow. Second, the mature disturbance is marked by deep convective processes as we have noted above which is difficult to reconcile with a purely barotropic system.

Early studies by Riehl and Malkus (1958) and Charney and Eliassen (1964) ruled out the possibility that conditional instability alone was a dominating process at low latitudes which could be used to explain the convective nature of tropical disturbances. Even in mature disturbances the boundary layer is far from saturated, which would make penetration of convective elements through the entire depth of the troposphere a difficult feat. Charney and Eliassen proposed a cooperative dynamical–conditional instability process which they termed 'conditional instability of the second kind' (CISK) in the hope of explaining the observed rapid development of tropical storms and hurricanes. They were able to show that an initial low-level convergence field (the dynamic part of the cooperative process) supplies moisture for the cumulus clouds (the convective part) via frictional inflow in the boundary layer. The structure of the tropical atmosphere is such that only convergence occurring in the boundary layer will produce penetrative convection (Ooyama, 1969). Whereas the CISK scheme appears to be quite successful in many respects, it fails to predict a small-scale cut-off for the instability. At the large-scale end, scales greater than about 500–1000 km possess very small growth rates. However, the greatest growth rates predicted by the theory occur for the cumulus scale.

Logical extensions of the CISK mechanism have been made which link the theoretical waveforms noted in the last section with CISK. The low-level convergence is given by the wave structure itself and dissipative effects resulting from the convection were included. For example, Stevens and Lindzen (1978) proposed the concept of wave-CISK. However, despite many appealing features it possesses the same problem as regular CISK: the most unstable mode is the cloud scale itself.

The inability of the CISK theories to produce a small-scale cut-off is disappointing. It would appear at first inspection that all of the elements of the cooperative processes are included in the theory, especially in wave-CISK. However, this may not be the case. If we return to our observational description of tropical convection, we note that the meso-scale structures, as distinct from the individual cumulus elements, may be the dominant sub-synoptic phenomenon. At the time of the development of the CISK theories the importance of the meso-scale phenomena was not appreciated. Perhaps it is their neglect which is the deficiency of the CISK models. But to appreciate their impact we must first understand the physics of the meso-scale processes.

9.5 Tropical–extratropical interactions

9.5.1 Extratropical influences on the tropics

(a) *Uniform zonal flows*

Following the study by Charney (1963) which suggested that the synoptic scale motions at low latitudes were governed by a non-divergent form of the vorticity equation, at least away from convection, the problem of finding a process that initiates an equatorial disturbance became imperative. Charney (1969) posed the hypothesis that mid-latitude influences might drive or initiate events at low latitudes and developed a criterion for the propagation of waves through a zonally averaged basic flow $\bar{u}$. As in Chapter 6 and following Charney (1969) we may write the potential vorticity equation on a β-plane as:

$$(\partial/\partial t + \mathbf{v}\cdot\nabla)q = 0 \tag{45}$$

where the potential vorticity is defined by:

$$q = \nabla^2\psi + \beta y + \frac{f_0^2}{\rho_0}\frac{\partial}{\partial z}\left(\frac{\rho}{N^2}\frac{\partial\psi}{\partial z}\right). \tag{46}$$

If we seek plane wave solutions of the form:

$$\psi = \Psi(y, z)\exp\left[ik(x - ct)\right]$$

we arrive at the governing equation:

$$\frac{\partial^2\Psi}{\partial y^2} + \frac{f_0^2}{N^2}\frac{\partial^2\Psi}{\partial z^2} + \left(\frac{\beta}{\bar{u} - c} - k^2 - \lambda^2\right)\Psi = 0 \tag{47}$$

where:

$$\lambda = \frac{4H^2N^2}{f_0^2}.$$

The criteria for the existence of oscillatory solutions is:

$$0 < \bar{u} - c < \frac{\beta}{k^2 + \lambda^{-2}} \tag{48}$$

which may be interpreted as defining the limit of equatorial propagation of an extratropical wave. Modes with c negative and less than $\bar{u}$ everywhere may

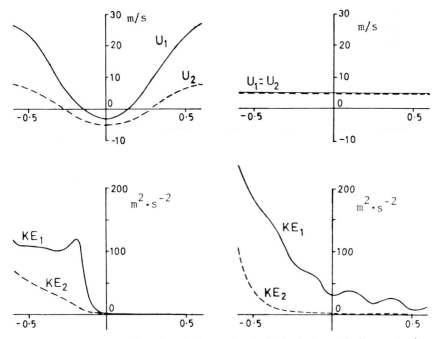

Fig. 9.19. Wave propagation of modes forced in mid-latitudes towards the equator in a viscous atmosphere for (a) a zonally symmetric baroclinic basic state with easterlies at the equator and (b) a zonally symmetric barotropic westerly basic state. Upper panels show the basic states and the lower panels the latitudinal distribution of the perturbation kinetic energy. Subscripts 1 and 2 refer to the upper and lower model troposphere respectively. The abscissa is the sine of the latitude. (After Webster, 1973a.)

propagate deeply into the tropics. Only planetary scale transients come into this category. All other modes will be absorbed or reflected at some latitude poleward of the equator. However, planetary scale transients for which c is sufficiently large and negative for $\bar{u} - c > 0$ at all latitudes possess little energy. With these results Charney (1969) argued that mid-latitude wave disturbances should have little or no effect on the tropical atmosphere. Similar results for zonally symmetric basic states were found by Bennett and Young (1971), Webster (1973a) and Geisler and Dickinson (1974).

An example of the propagation of modes into the tropical regions may be seen in Fig. 9.19 (from Webster, 1973a). Using a linear primitive equation model the response of a basic state in which equatorial easterlies exist and a state that possesses weak westerlies at all latitudes was tested relative to middle latitude forcing. As steady state solutions were sought (i.e., $\partial/\partial t = 0$, so that $c = 0$) the $\bar{u} = 0$ line is the line of transition between oscillatory solutions near the source to exponentially damped solutions near the equator. With westerlies everywhere, cross-equatorial propagation occurs. The decrease in amplitude into the unforced hemisphere is determined by the rates of dissipation assumed in the model.

(b) *Non-uniform zonal flows*

The superposition of the planetary scale stationary modes of low latitudes on the weak zonally averaged flow creates regions of westerlies in the seasonally averaged flow along the equator. The location of these equatorial westerlies in the upper troposphere is shown in Fig. 9.9 together with indications that such regions are occupied by high-frequency transients.

The possibility of mode propagation through such zones of weak westerlies has been shown to exist in analyses of FGGE data by Reed (1981), Paegle *et al.* (1981) and Paegle and Lewis (1981), and also by a number of theoretical studies (Branstator, 1981; Paegle and Paegle, 1981; Simmons, 1982; Webster and Holton, 1982). Both the observation and the theory indicate that middle latitude disturbances could propagate deeply into the low latitudes and occasionally through to the extratropics of the other hemisphere. In all cases the propagation will take place only in the region of equatorial westerlies.

An example of such propagation may be seen in Figs. 9.20 and 9.21 which show two basic non-uniform basic states (upper panels) typified by weak and strong regions of equatorial westerlies, respectively, although the zonal averaged winds in both cases are easterly and of the same magnitude as in Fig. 9.1. The shaded region in the diagrams indicates easterly winds. With a disturbance located at 20°N the free surface non-linear barotropic primitive equation model of Webster and Holton (1982) was allowed to evolve and approached an equilibrium state after about 40 simulated days. The variations in the zonal component of the velocity are shown in the lower panels of Figs. 9.20 and 9.21. In both cases two propagation paths may be seen. The first is in a northeasterly direction away from the source into the middle latitude westerlies and the second into the region of equatorial westerlies. The dual nature of the propagation paths was anticipated by Hoskins and Karoly (1981) from ray-tracing theory. With both basic states the forced modes propagate well into the low latitudes. However, when the equatorial westerlies are strongest, the propagation proceeds through the tropics and into the extratropical region of the adjacent hemisphere. Webster and Holton (1982) showed that the critical parameter that accounted for the differences between the responses was the magnitude of the equatorial westerlies. They were able to show that the meridional component of the group velocity goes as $\bar{u}^2$. Consequently in the strong westerly case greater spatial transmission occurs before dissipation effectively attenuates the magnitude of the response.

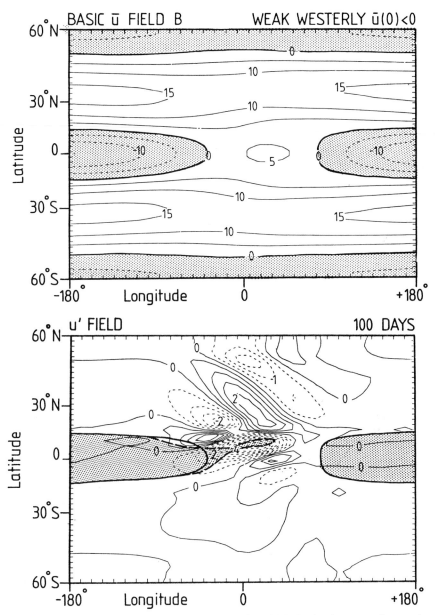

Fig. 9.20. Wave propagation of modes forced in mid-latitudes in a zonally varying basic state possessing a weak westerly maximum at the equator. The zonally averaged state is easterly. Upper panel shows the initial basic state and the lower panel the perturbation zonal velocity component after 100 days of integration. Shaded area denotes the location of the zonal easterlies after 100 days. (From Webster and Holton, 1982.)

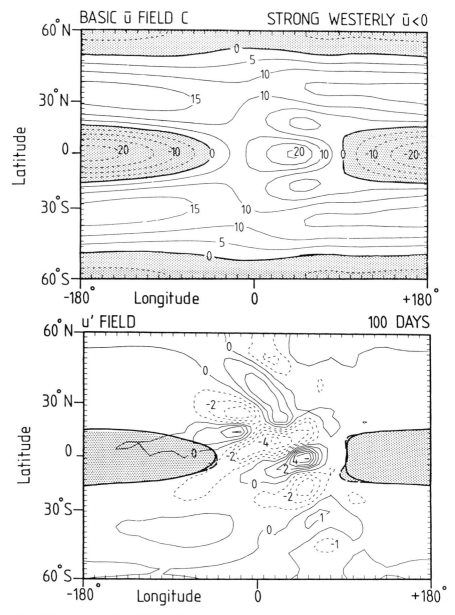

Fig. 9.21. Same as Fig. 9.20 but for a basic state with a strong westerly region at the equator. Zonally averaged state is still easterly. (From Webster and Holton, 1982.)

Webster and Holton point out that at certain times of the year the upper tropospheric westerlies of the eastern Pacific approach the values shown in the strong westerly basic state. We have already noted that the zonally averaged winds of both basic states are easterly. For forced modes of similar or smaller scale to the equatorial westerlies, it would seem that the *zonally averaged* critical latitude has little bearing on the latitudinal propagation.

The consequences of the tropical atmosphere being less insulated from middle latitude wave propagation has considerable implications. For numerical weather prediction it means that a hemispheric model can be expected to degrade because of the omission of transients emanating from the mid-latitudes of the other hemisphere in addition to the usual degradation due to lack of data within the tropics. Thus it may be expected that on occasions degradation may occur quite rapidly especially during periods of active cross-equatorial interaction when the zonal flow at the equator is locally strong and westerly.

As the tropical atmosphere might either transmit, absorb or reflect (see Section 6.3.2) an incident wave train, the location of the equatorial mean easterlies and westerlies would appear a critical factor in determining at least some of the character of the hemisphere from which the train evolved. If the basic zonal flow changes over a period of time, it is quite possible that the transmission characteristics of the incident wave will change. This may occur during El Niño when the phase of the forced planetary scale modes changes (Webster, 1982). Such changes would occur in concert with the poleward propagation of equatorial forced modes resulting directly from the El Niño anomalous forcing.

9.5.2 Tropical influences on the extratropics

In previous sections we have emphasized that the planetary scale modes of low latitudes decay rapidly away from the equator. Such structure would appear to argue against the tropical atmosphere having an appreciable impact on the extratropics. But there is substantial observational evidence to the contrary. Bjerknes (1966, 1969), Namias (1976), Horel and Wallace (1981) and many others have noted that anomalous mean circulations in the extratropics can be linked with large amplitude anomalies in the tropics such as those associated with El Niño (see Fig. 3.10).

Several theoretical studies (Opsteegh and van den Dool, 1980; Webster, 1981, 1982; Hoskins and Karoly, 1981; Simmons, 1982) have tackled the problem of influences emanating from the tropical latitudes. An example from the study of Hoskins and Karoly (*loc. cit.*) is shown in Fig. 9.22. All studies reach similar conclusions. The reason for the poleward mode propagation is

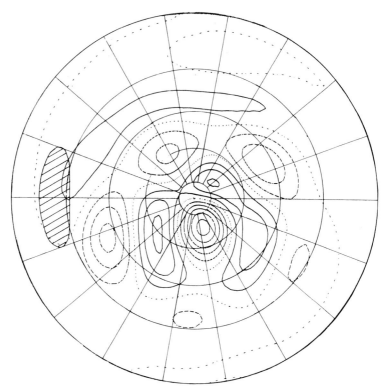

Fig. 9.22. The 300 mb height field perturbation for a steady state, linear solution of a five layer baroclinic model with a heat source in the hatched region perturbing the Northern Hemisphere winter zonal flow. The vertical distribution of the source is of the form $\sin \pi\sigma$. Positive contours are continuous. If the depth averaged heating maximum is 2.5 K day^{-1} then the contour integral is 2 dam. (From Hoskins and Karoly, 1981.)

that the sea-surface temperature anomaly (or rather the latent heating anomaly if one were to result) is occasionally located within the westerlies of the subtropics which allows the excitation of propagating Rossby modes. Once excited, the behaviour of the equivalent barotropic external Rossby waves is well understood and the almost great circle paths known from the ray-tracing techniques of Hoskins and Karoly (1981) (see Section 6.3.2). This explanation does not violate the properties we have found for the planetary scale motions at low latitudes. It also receives some credence from the observation that the anomalies are most noted in the winter hemisphere. It is during winter that the westerly zonally averaged flow is closest to the equator.

References

BATES, J. R. (1970). Dynamics of disturbances on the intertropical convergence zone. *Q. Jl R. met. Soc.*, **96**, 677–701.

BENNETT, J. R. and YOUNG, J. (1971). The influence of latitudinal wind shear upon large-scale wave propagation into the tropics. *Mon. Weath. Rev.*, **99**, 202–214.

BJERKNES, J. (1966). A possible response of the atmospheric Hadley circulation to equatorial anomalies of ocean temperature. *Tellus*, **18**, 824–829.

BJERKNES, J. (1969). Atmospheric teleconnections from the equatorial Pacific. *Mon. Weath. Rev.*, **97**, 163–172.

BLACKMON, M. L., WALLACE, J. M., LAU, N-C. and MULLEN, S. L. (1977). An observational study of the Northern Hemisphere wintertime circulation. *J. atmos. Sci.*, **34**, 1040–1053.

BRANSTATOR, G. (1981). Application of linear barotropic models to the interannual variability of quasi-stationary flow. Paper presented at the IAMAP Symposium on the Dynamics of the General Circulation. I: Emphasis on the mid-latitude troposphere, Reading, UK. 3–7 August 1981.

CHANG, C. P. (1977). Viscous internal gravity waves and low frequency oscillations in the tropics. *J. atmos. Sci.*, **34**, 901–910.

CHARNEY, J. G. (1963). A note on the large-scale motions in the tropics. *J. atmos. Sci.*, **20**, 607–609.

CHARNEY, J. G. (1969). A further note on large-scale motions in the tropics. *J. atmos. Sci.*, **26**, 182–185.

CHARNEY, J. G. and DRAZIN, P. J. (1961). Propagation of planetary-scale disturbances from the lower into the upper atmosphere. *J. Geophys. Res.*, **66**, 83–109.

CHARNEY, J. G. and ELIASSEN, A. (1964). On the growth of the hurricane depression. *J. atmos. Sci.*, **21**, 68–75.

COX, S. and GRIFFITH, K. T. (1979). Estimates of radiative divergence during Phase III of GATE: Part II. Analysis of the Phase I results. *J. atmos. Sci.*, **36**, 586–601.

GEISLER, J. E. and DICKINSON, R. E. (1974). Numerical study of an interacting Rossby wave and barotropic zonal flow near a critical level. *J. atmos. Sci.*, **31**, 946–955.

GELLER, M. A. (1970). An investigation of the lunar semidiurnal tide in the atmosphere. *J. atmos. Sci.*, **27**, 202–218.

GILL, A. E. (1980) Some simple solutions for heat-induced tropical circulations. *Q. Jl R. met. Soc.*, **106**, 447–462.

HOLTON, J. R. (1973). On the frequency distribution of atmospheric Kelvin waves. *J. atmos. Sci.*, **30**, 499–501.

HOLTON, J. R. (1975). *The Dynamic Meteorology of the Stratosphere and Mesosphere*, *Meteorological Monographs*, **15**, No. 37, 218 pp.

HOLTON, J. R. (1979). *An Introduction to Dynamic Meteorology* (2nd Edn), *International Geophysical Series*, *Vol. 23*, Academic Press, 391 pp.

HOREL, J. D. and WALLACE, J. M. (1981). Planetary scale atmospheric phenomena associated with the interannual variability of sea-surface temperature in the equatorial Pacific. *Mon. Weath. Rev.*, **109**, 813–829.

HOSKINS, B. and KAROLY, D. (1981). The steady linear response of a spherical atmosphere to thermal and orographic forcing. *J. atmos. Sci.*, **38**, 1179–1196.

HOUZE, R., JR. (1982). Cloud clusters and large-scale vertical motion in the tropics. *J. met. Soc. Japan*, **60**, 396–410.

HOUZE, R., JR. and BETTS, A. (1981). Convection in GATE. *Rev. Geophys. Space Phys.*, **19**, 541–576.

KILONSKY, B. J. and RAMAGE, C. S. (1976). A technique for estimating tropical open ocean rainfall from satellite observations. *J. appl. Met.*, **15**, 972–975.

LEARY, C. A. and HOUZE, R. A., JR. (1979). Melting and evaporation of hydrometers in precipitation from the anvil clouds of deep tropical convection. *J. atmos. Sci.*, **36**, 669–679.

LIEBMAN, B. and HARTMANN, D. L. (1981). Interannual variation of outgoing IR association with tropical circulation changes during 1974–78. *J. atmos. Sci.*, **39**, 1153–1162.

LONGUET-HIGGINS, M. S. (1968). The eigenfunctions of Laplace's tidal equations over a sphere. *Phil. Trans. R. Soc.*, **262**, 511–607.

274   P. J. WEBSTER

MCAVANEY, B. J., DAVIDSON, N. E. and MCBRIDE, J. L. (1981). The onset of the Australian Northwest Monsoon during Winter MONEX: Broadscale flow revealed by an objective analysis scheme. Proceedings of the International Conference on Early Results of FGGE and large-scale aspects of its Monsoon experiments. Tallahassee, Florida, USA, 12–17 January 1981.

MADDEN, R. D. and JULIAN, P. (1971). Detection of a 40–50 day oscillation in the zonal wind in the tropical Pacific. J. atmos. Sci., 28, 702–708.

MAK, M. K. (1969). Laterally driven stochastic motions in the tropics. J. atmos. Sci., 26, 41–64.

MATSUNO, T. (1966). Quasi-geostrophic motions in the equatorial area. J. met. Soc. Japan, 44, 25–42.

MURAKAMI, T. and UNNINAYER, S. (1977). Atmospheric circulation during December 1970 through February 1971. Mon. Weath. Rev., 105, 1024–1038.

NAMIAS, J. (1976). Some statistical and synoptic relationships associated with El Niño. J. Phys. Oceanogr., 6, 130–138.

NEWELL, R. E., KIDSON, J. W., VINCENT, D. G. and BOER, G. (1972). The General Circulation of the Tropical Atmosphere and Interaction with Extratropical Latitudes. Vol. 1, MIT Press, 258 pp.

OOYAMA, K. (1969). Numerical simulation of the life cycle of tropical cyclones. J. atmos. Sci., 26, 3–40.

OPSTEEGH, J. D. and VAN DEN DOOL, H. M. (1980). Seasonal differences in the stationary response of a linearized primitive equation model: Prospects for long-range weather forecasting. J. atmos. Sci., 37, 2169–2185.

PAEGLE, J. M. and LEWIS, F. P. (1981). Short term effects of tropical convective activity upon higher latitudes. Unpublished manuscript. Paper presented at the IAMAP-ICDM Symposium on the Dynamics of the General Circulation of the Atmosphere. II: Tropical Aspects, IAMAP, Hamburg, 17–19 August 1981.

PAEGLE, J. N. and PAEGLE, J. (1981). The role of barotropic oscillations within atmospheres of highly variable refractive index. Paper presented at the IAMAP Conference on the Dynamics of the General Circulation. I: Emphasis on the mid-latitude troposphere, Reading, UK, 3–7 August 1981.

PAEGLE, J., KALNAY-RIVAS, E. and BAKER, W. E. (1981). The role of zonally symmetric testing in the vertical and temporal structure of the global scale flow fields during FGGE SOP I. Proceedings: International Conference on Early Results of FGGE and Large-Scale Aspects of its Monsoon Experiments. Tallahassee, Florida, USA, 12–17 January, 1981.

REED, R. J. (1981). Upper tropospheric waves in the subtropical Pacific during SOP-I of FGGE and their interaction with the ITCZ. Invited paper presented at the IAMAP Symposium on the General Circulation of the Atmosphere. II: Tropical aspects, Hamburg, FRG, 17–19 August 1981.

REED, R. J., NORQUIST, D. C. and RECKER, E. E. (1977). The structure and properties of African wave disturbances during Phase III of GATE. Mon. Weath. Rev., 105, 317–333.

RIEHL, H. and MALKUS, J. S. (1958). On the heat balance of the equatorial trough zone. Geophysica, 6, 503–538.

ROWNTREE, P. R. (1972). The influence of tropical East Pacific Ocean temperatures on the atmosphere. Q. Jl R. met. Soc., 98, 290–321.

SCHNEIDER, E. K. and LINDZEN, R. S. (1976). A discussion of the parameterization of momentum exchange by cumulus convection. J. Geophys. Res., 81, 3138–3160.

SHORT, D. A. and WALLACE, J. M. (1980). Satellite-inferred morning-to-evening cloudiness changes. Mon. Weath. Rev., 108, 1160–1169.

SIMMONS, A. J. (1982). The forcing of stationary wave motion by tropical diabatic heating. Q. Jl R. met. Soc., 108, 503–534.

STEVENS, D. E. and LINDZEN, R. (1978). Tropical wave-CISK with a moisture budget and cumulus friction. J. atmos. Sci., 35, 940–961.

STEPHENS, G. L. and WEBSTER, P. J. (1979). Sensitivity of radiative forcing to variable cloud and moisture. J. atmos. Sci., 36, 1542–1556.

TRENBERTH, K. (1976). Spatial and temporal variations of the Southern Oscillation. Q. Jl R. met. Soc., 102, 639–653.

TROUP, A. J. (1965). The Southern Oscillation. Q. Jl R. met. Soc., 91, 390, 490–506.

WALKER, G. T. (1923). *Correlation in Several Variations of Weather VIII : A Preliminary Study of World Weather.* **24**, part 4, Superintendent of Government Printing, India, 75–123.

WALLACE, J. M. (1970). Time-longitude section of tropical cloudiness, December 1966–November 1967. *ESSA Technical Report NESC 56* (ESSA TR NESC 56).

WALLACE, J. M. and KOUSKY, V. E. (1968). Observational evidence of Kelvin waves in the tropical stratosphere. *J. atmos. Sci.*, **25**, 900–907.

WEBSTER, P. J. (1972). Response of the tropical atmosphere to local, steady, forcing. *Mon. Weath. Rev.*, **100**, 518–540.

WEBSTER, P. J. (1973a). Remote forcing of the time-independent tropical atmosphere. *Mon. Weath. Rev.*, **101**, 58–68.

WEBSTER, P. J. (1973b). Temporal variation of low-latitude zonal circulations. *Mon. Weath. Rev.*, **101**, 803–816.

WEBSTER, P. J. (1981). Mechanisms determining the atmospheric response to sea surface temperature anomalies. *J. atmos. Sci.*, **38**, 554–571.

WEBSTER, P. J. (1982). Seasonality in the local and remote atmospheric response to sea surface temperature anomalies. *J. atmos. Sci.*, **39**, 41–52.

WEBSTER, P. J. and CURTIN, D. (1975). Interpretations of the EOIE experiment : Spatial variation of transient and stationary modes. *J. atmos. Sci.*, **32**, 1848–1863.

WEBSTER, P. J. and STEPHENS, G. L. (1980). Tropical upper tropospheric extended clouds: Inferences from Winter MONEX. *J. atmos. Sci.*, **37**, 1521–1541.

WEBSTER, P. J. and HOLTON, J. R. (1982). Cross equatorial response to middle-latitude forcing in a zonally varying basic state. *J. atmos. Sci.*, **39**, 722–733.

YANAI, M. and MURAYAMA, T. (1966). Large-scale equatorial waves penetrating from the upper troposphere into the lower stratosphere. *J. met. Soc. Japan*, **47**, 167–182.

ZIPSER, E. J. (1969). The role of organized unsaturated convective downdrafts in the structure and rapid decay of an organized equatorial disturbance. *J. appl. Meteor.*, **8**, 779–814.

ZIPSER, E. J. (1977). Mesoscale and convective downdrafts as distinct components of squall-line circulation. *Mon. Weath. Rev.*, **105**, 1568–1589.

– 10 –

The stratosphere and its links
to the troposphere

JAMES R. HOLTON

10.1 Introduction

In recent years much of the emphasis in stratospheric research has been on the photochemistry of the ozone layer. Comparatively less attention has been given to the role of the stratosphere in weather and climate. Although it is obvious that processes occurring in the troposphere are primarily responsible for weather and climate variability, the stratosphere cannot be totally neglected. Both radiative and dynamical links exist between the troposphere and stratosphere. Experiments with the GFDL 'SKYHI' general circulation model (Fels *et al.*, 1980) indicate that changes in the radiative budget of the stratosphere due to large changes in the concentration of ozone or carbon dioxide may have significant effects on the temperature at the ground. Radiative links between the stratosphere and troposphere have been investigated in detail by Ramanathan and Dickinson (1979) and others (see also the review by Murgatroyd and O'Neill, 1980). In this review we limit the discussion entirely to the dynamics of the stratospheric general circulation and its links to the troposphere. It must be remembered, however, that in a complete treatment the stratosphere must be treated as a coupled system in which radiation, photochemistry, and dynamics all play essential roles.

The plan of this chapter is as follows: In Section 10.2 we review some observational aspects of large-scale stratospheric motions, with emphasis on the influence of the troposphere on the stratosphere. In Section 10.3 we review the so-called transformed Eulerian-mean equations and discuss their utility for diagnostic analysis of stratospheric motions. In Section 10.4 we show how the transformed Eulerian-mean equations can be used to elucidate many dynamical aspects of sudden stratospheric warmings. In Section 10.5 we discuss observational and theoretical aspects of the equatorial stratospheric circulation with emphasis on the quasi-biennial oscillation (QBO). In Section 10.6 we consider the interannual variability of the Northern Hemisphere extratropical stratosphere and its possible links to the troposphere and/or the equatorial QBO. Some conclusions are presented in Section 10.7.

10.2 Observational background

The transient synoptic scale eddies which, together with quasi-stationary planetary waves, are the major features of the general circulation in the troposphere decay rapidly with height above the tropopause. In the stratosphere the circulation is dominated by planetary scale components. During the past decade the observational data base for the stratosphere has increased dramatically due primarily to the availability of global radiance data acquired by satellite borne radiometers (see Barnett, 1980; Gille *et al.*, 1980; and Krueger *et al.*, 1980, for review of various aspects of satellite remote sensing of the stratosphere). Before the satellite era routine meteorological soundings of the stratosphere were limited primarily to the levels between the tropopause and 10 mb (~ 31 km) accessible to radiosonde balloons. Since the radiosonde network is adequate only in the Northern Hemisphere, most climatological studies of the stratosphere have concentrated on the lower stratosphere in the Northern Hemisphere. Climatological studies based on satellite radiance data are just beginning to appear (e.g., Quiroz, 1981), and many aspects of the general circulation above 10 mb remain to be elucidated.

The zonal mean circulation of the stratosphere is driven primarily by differential heating due to the absorption of solar ultraviolet radiation by ozone and the emission of infrared radiation by carbon dioxide and ozone. The radiative equilibrium temperature distribution for the stratosphere and mesosphere computed for the Northern Hemisphere summer solstice (Wehrbein and Leovy, 1981) is shown in Fig. 10.1. In the stratosphere the radiative equilibrium temperatures in the summer hemisphere are close to the observed temperatures shown in Fig. 10.2. But in the high latitudes of the winter hemisphere the radiative equilibrium temperatures are much colder than the observed temperatures, indicating that a strong dynamical heating must exist. In the upper mesosphere (above ~ 65 km) the observed temperature distribution reverses so that the maximum temperature occurs at the winter pole and the minimum at the summer pole. The maintenance of this strong departure from radiative equilibrium in the mesosphere has been discussed recently by Lindzen (1981), Holton (1982) and Matsuno (1982) and will not be pursued here.

The differential heating of the upper stratosphere drives a thermally direct Lagrangian mean meridional circulation with rising motion in the summer hemisphere and strong sinking in the winter polar region. The adiabatic

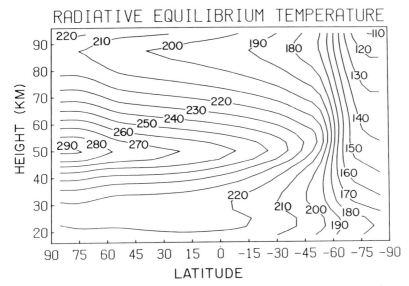

Fig. 10.1. Radiative equilibrium temperature distribution for the stratosphere and mesosphere at solstice (after Wehrbein and Leovy, 1981).

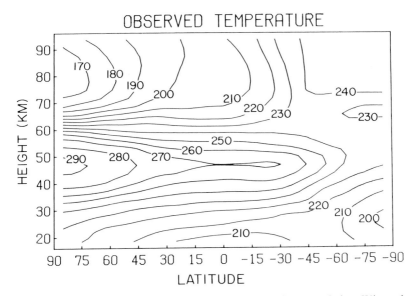

Fig. 10.2. Observed zonal mean temperature distribution at solstice. Winter hemisphere at right (after Murgatroyd, 1969).

cooling (heating) associated with this meridional mass flow maintains the difference between the observed temperature and the radiative equilibrium. The mean zonal flow is in approximate thermal wind balance with the meridional temperature gradient. Thus, the seasonal radiative forcing generates a zonal mean easterly flow in the summer hemisphere, and a westerly flow in the winter hemisphere. The summer easterly vortex is nearly zonally symmetric. But the winter westerly vortex is distorted due to the presence of large amplitude stationary planetary waves superposed on the zonal mean flow.

It is well known that the stationary planetary waves of the winter stratosphere are forced waves generated in the troposphere by orographic forcing and diabatic heating. Charney and Drazin (1961) showed theoretically that such waves can propagate vertically into the stratosphere only when the mean winds are westerly, but less than a critical velocity which is inversely proportional to the total wavenumber squared (see Section 6.4). Hence, stationary planetary waves can propagate vertically into the stratosphere only in winter, when the stratospheric winds are westerly, and even then only the longest waves (zonal wavenumbers 1 and 2) are able to propagate into the stratosphere.

The observed distribution of stationary wavenumber 1 for the Northern Hemisphere winter of 1975–1976 is shown in Fig. 10.3. Note that the maximum geopotential height amplitude that occurs near the stratopause (~ 50 km) at $\sim 65°$N is in excess of 600 geopotential metres and that the phase lines tilt westward with height as well as towards the equator, indicating poleward directed eddy fluxes of momentum and heat.

The situation shown in Fig. 10.3 represents a normal winter mean. However, within any given winter there are irregular variations in the amplitudes and phases of the planetary waves. Occasionally, the amplitudes of the planetary waves increase dramatically over the course of several days, leading to a highly distorted polar vortex as shown in Fig. 10.4. Wave-mean flow interactions occurring as a result of this wave intensification may cause spectacular rises in the temperature of the polar stratosphere—the so-called sudden stratospheric warmings. These will be discussed in some detail in Section 10.4.

The characteristics of the overall circulation in the equatorial stratosphere are rather different from those of the extratropical regions. In the equatorial zone the annual cycle is overshadowed by the quasi-biennial oscillation (QBO) of the zonal mean temperature and wind fields, and planetary Rossby wave modes are overshadowed by the equatorially trapped Kelvin and Rossby-gravity modes (Section 9.3.2).

The QBO is perhaps the most dramatic example in the atmosphere of a quasi-regular variation on the interannual time scale. The QBO consists of an alternating pattern of westerly and easterly mean zonal winds which is symmetric about the equator, with latitudinal half-width of about $12°$. The

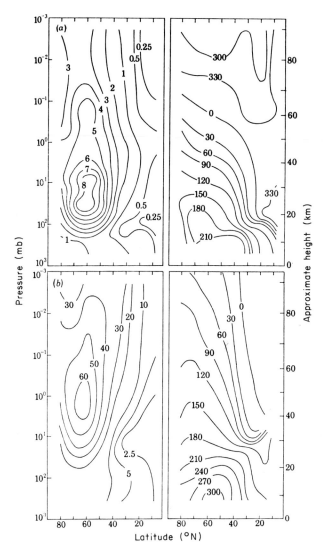

Fig. 10.3. Amplitude and phase of the winter mean (December 1975–February 1976) wavenumber 1 temperature perturbation (upper panels) and geopotential height perturbation (lower panels). Units are K for temperature and dm for geopotential height (after Barnett, 1980).

oscillation has nearly constant amplitude between 35 and 22 km but decays rapidly between the 22 km level and the tropopause. An individual regime of easterlies or westerlies first appears at high levels and propagates downwards at a rate of 1–2 km per month as illustrated in the time–height section of Fig. 10.5.

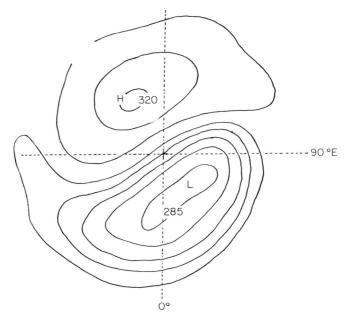

10 mb HT 27 Jan 79

Fig. 10.4. Topography of the Northern Hemisphere 10 mb height field in the presence of a strong planetary wavenumber 1 disturbance (heights in 100s of metres). Figure courtesy of R. S. Quiroz. The diagram extends to latitude 30°N.

According to the theory of Holton and Lindzen (1972) the equatorial QBO is an internally generated oscillation which results from wave-mean flow interactions involving the observed eastward propagating Kelvin waves and westward propagating Rossby-gravity wave. The observed characteristics of these waves have been reviewed by Wallace (1973) and Holton (1975). Both types appear to be forced in the troposphere (presumably by large-scale tropical convective disturbances) and propagate vertically into the stratosphere.

Thus, the equatorial stratosphere and the extratropical winter stratosphere are both coupled dynamically to the troposphere via vertically propagating wave disturbances. These waves provide the major dynamical link between the stratosphere and the troposphere. Hence, a major portion of this chapter is devoted to the dynamics of the interaction between forced waves generated in the troposphere and the mean flow in the stratosphere.

The dynamical influence of the stratosphere on the troposphere is a much more difficult problem. Some would argue that the stratosphere has no significant dynamical impacts on the troposphere. However, there are theoretical reasons to believe otherwise. As will be discussed in Section 10.6

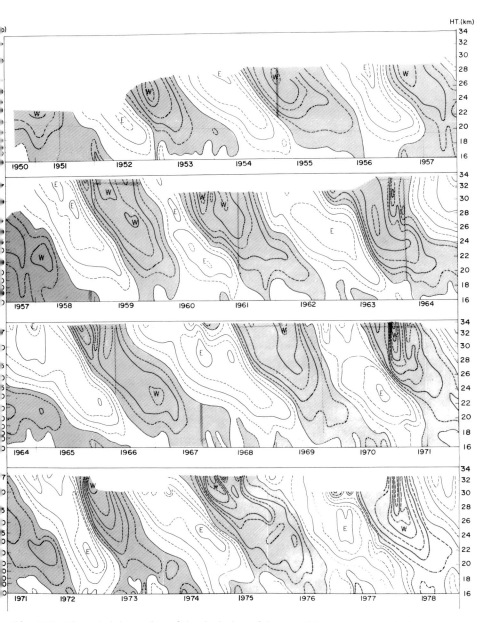

Fig. 10.5. Time–height section of the deviation of the monthly mean zonal wind near 9°N from the long term average. Solid isothachs are at 10 m s^{-1} intervals. Shaded areas are westerlies (after Coy, 1979).

there is also observational evidence which suggests that the stratospheric QBO may have some influence on the interannual variability in the troposphère, and that stratospheric sudden warmings may affect the high latitude regions of the troposphere.

10.3 Dynamical background

The dynamics of the stratosphere is discussed in some detail in Holton (1975). However, since the publication of Holton's monograph several important advances have occurred in stratospheric dynamics. In particular, Andrews and McIntyre (1978a) have shown that a Lagrangian mean formulation of the dynamics equations provides a conceptual description of wave-mean flow interaction which is more fundamental than the conventional Eulerian mean approach. Despite theoretical advantages of the Lagrangian mean formulation, there are severe obstacles to its practical use as a diagnostic for observational data or general circulation model output (McIntyre, 1980). An alternative formulation that has proved to be very useful is the transformed Eulerian mean system introduced by Andrews and McIntyre (1976a, 1978b) and Boyd (1976). Several recent studies have demonstrated the utility of the transformed Eulerian mean equations in diagnosing wave-mean flow interactions. Before discussing these applications we here briefly review the development of the transformed Eulerian mean.

Following Holton (1975) we adopt a log pressure coordinate system in which $z \equiv -H \ln (p/p_s)$. Here p is a local pressure, p_s is a constant reference pressure, and $H \equiv RT_s/g$ is a constant scale height. In an isothermal atmosphere z varies exactly as the geometric height. Departures of z from the geometric height in the stratosphere are small enough to be ignored here.

We shall now manipulate the basic equations in a manner similar to that in Section 7.3.1 but concentrating on the theory relevant to the stratosphere. Following the notation of Holton (1975) the zonal mean momentum equations, thermodynamic energy equation, and continuity equation may be written approximately for quasi-geostrophic motions on a mid-latitude β-plane as follows:

$$\frac{\partial [u]}{\partial t} - f_0[v] = -\frac{\partial}{\partial y}[u^*v^*] \qquad (10.1)$$

$$f_0[u] = -\frac{\partial [\Phi]}{\partial y} \qquad (10.2)$$

$$\frac{\partial}{\partial t}\left(\frac{\partial [\Phi]}{\partial z}\right) + N^2[w] = \frac{\kappa}{H}[Q] - \frac{\partial}{\partial y}[v^*\Phi_z^*] \qquad (10.3)$$

$$\frac{\partial [v]}{\partial y} + \frac{1}{\rho} \frac{\partial}{\partial z} (\rho [w]) = 0 \tag{10.4}$$

where $\rho(z) \propto \exp(-z/H)$ is a basic state density, $N^2 \equiv g \, d\ln \theta_s/dz$ is the buoyany frequency squared [where $\theta_s(z)$ is a basic state potential temperature], and we have used the hydrostatic approximation $\partial [\Phi]/\partial z = R[T]/H$ to write Eqn. (10.3) in terms of the 'thickness', $\partial [\Phi]/\partial z$.

Differentiating Eqn. (10.2) with respect to z we obtain the thermal wind equation:

$$f_0 \frac{\partial [u]}{\partial z} = -\frac{\partial^2 [\Phi]}{\partial y \, \partial z} = -\frac{R}{H} \frac{\partial [T]}{\partial y}. \tag{10.5}$$

The thermal wind balance Eqn. (10.5) imposes a strong constraint on the ageostrophic meridional circulation as is discussed, for example, in Holton (1979a). In the absence of a meridional circulation ($[v] = [w] = 0$) the eddy momentum flux divergence $\partial [u^*v^*]/\partial y$ and the eddy heat flux divergence $\partial [v^*\Phi_z^*]/\partial y$ would tend separately to change the mean zonal wind and temperature distributions and hence would destroy the thermal wind balance. However, the pressure gradient force which results from an infinitesimal departure of $[u]$ from thermal wind balance immediately generates a mean meridional circulation ($[v], [w]$) which adjusts the mean zonal wind and temperature fields so that Eqn. (10.5) remains valid at every instant. In many situations the compensation by the ($[v], [w]$) circulation is so complete that the mean zonal flow remains nearly constant despite the existence of large eddy flux divergences in Eqns. (10.1) and (10.3). Thus, it is rather difficult to diagnose the net eddy forcing of the mean flow using the conventional Eulerian mean equations.

An alternative approach that provides clearer diagnostics of wave-mean flow interaction processes is the transformed Eulerian mean. Following Andrews and McIntyre (1976a) we define a residual mean meridional streamfunction $\tilde{\Psi}$ such that:

$$\tilde{\Psi} \equiv [\Psi] + [v^*\Phi_z^*]/N^2 \tag{10.6}$$

where $[\Psi]$ is the streamfunction for the ($[v], [w]$) mean meridional circulation:

$$[w] = \frac{\partial [\Psi]}{\partial y}; \quad [v] = -\frac{1}{\rho} \frac{\partial}{\partial z} (\rho [\Psi]). \tag{10.7}$$

Thus, by analogy with Eqn. (10.7) we may define a residual mean meridional circulation $(\tilde{v}, \tilde{w})$ as follows:

$$\tilde{v} = [v] - \frac{1}{\rho} \frac{\partial}{\partial z} \left(\frac{\rho [v^*\Phi_z^*]}{N^2} \right) \tag{10.8a}$$

$$\tilde{w} = [w] + \frac{\partial}{\partial y}\left(\frac{[v^*\Phi_z^*]}{N^2}\right). \qquad (10.8b)$$

Substituting from Eqn. (10.8) into Eqns. (10.1) and (10.3) we obtain the transformed Eulerian mean zonal momentum and thermodynamic energy equations:

$$\frac{\partial[u]}{\partial t} - f_0\tilde{v} = -\frac{\partial}{\partial y}[u^*v^*] + \frac{f_0}{\rho}\frac{\partial}{\partial z}\left(\rho\frac{[v^*\Phi_z^*]}{N^2}\right) = \frac{1}{\rho}\nabla\cdot\mathbf{E} \qquad (10.9)$$

$$\frac{\partial}{\partial t}\left(\frac{\partial[\Phi]}{\partial z}\right) + N^2\tilde{w} = \frac{\kappa}{H}[Q] \qquad (10.10)$$

where following Andrews and McIntyre (1976a) we have defined the flux vector in the $y-z$ plane:

$$\mathbf{E} \equiv \rho(-[u^*v^*], f_0[v^*\Phi_z^*]/N^2). \qquad (10.11)$$

Andrews and McIntyre coined the name 'Eliassen–Palm' (EP) flux for $\mathbf{E}$ defined in Eqn. (10.11) in recognition of the earlier work by Eliassen and Palm (1961) who showed the relation of $\mathbf{E}$ to the wave energy flux.

In the transformed equations (10.9)–(10.10) the eddy forcing is expressed entirely in terms of the divergence of $\mathbf{E}$ in Eqn. (10.9). Eddy quantities do not appear at all in the thermodynamic energy equation. For steady, conservative linear waves $\nabla\cdot\mathbf{E} = 0$ (Eliassen and Palm, 1961; Andrews and McIntyre, 1976a; Boyd, 1976) so that the eddy forcing of the residual mean meridional circulation $(\tilde{v}, \tilde{w})$ vanishes. Under such conditions the residual circulation is driven only by zonal mean diabatic heating and may be identified with the diabatic circulation (i.e., $\tilde{w}$ is just that vertical velocity for which the adiabatic heating/cooling balances the diabatic cooling/heating). Under such conditions the residual mean meridional circulation is approximately equal to the Lagrangian mean meridional circulation. The residual circulation is thus more relevant to the motions of trace species in the stratosphere than is the conventional Eulerian mean meridional circulation (Dunkerton, 1978; Holton, 1981).

The eddy forcing $\nabla\cdot\mathbf{E}$ has a useful interpretation in terms of potential vorticity. For linear waves the zonal mean and eddy components of the quasi-geostrophic potential vorticity equations are (Holton, 1975):

$$\frac{\partial[q]}{\partial t} = -\frac{\partial}{\partial y}[q^*v^*] - [S] \qquad (10.12)$$

$$\left(\frac{\partial}{\partial t} + [u]\frac{\partial}{\partial x}\right)q^* + v^*\frac{\partial[q]}{\partial y} = -S^* \qquad (10.13)$$

where:

$$[q] = f_0 + \beta y + \frac{1}{f_0}\frac{\partial^2[\Phi]}{\partial y^2} + \frac{f_0}{\rho}\frac{\partial}{\partial z}\left(\frac{\rho}{N^2}\frac{\partial[\Phi]}{\partial z}\right) \tag{10.14}$$

$$q^* = \frac{1}{f_0}\nabla^2\Phi^* + \frac{f_0}{\rho}\frac{\partial}{\partial z}\left(\frac{\rho}{N^2}\frac{\partial\Phi^*}{\partial z}\right). \tag{10.15}$$

$[S]$ and S^* are the zonal mean and eddy components of the net potential vorticity sink. With a little manipulation it can be shown that when u^* and v^* are evaluated geostrophically:

$$[q^*v^*] = \frac{1}{\rho}\nabla\cdot\mathbf{E} \tag{10.16}$$

where $\mathbf{E}$ is the EP flux defined in Eqn. (10.11). Thus from Eqn. (10.9) we see that the net eddy forcing of the mean flow can alternatively be expressed in terms of the potential vorticity flux $[q^*v^*]$.

Multiplying Eqn. (10.13) by q^* and averaging zonally we find that:

$$[q^*v^*] = \left(\frac{\partial[q]}{\partial y}\right)^{-1}\left(-\frac{1}{2\partial t}[q^{*2}] - [q^*S^*]\right). \tag{10.17}$$

Thus if eddy sources and transience vanish ($S^* = 0$, $\partial/\partial t = 0$) in Eqn. (10.17) the potential vorticity transport by the eddies vanishes. Equations (10.16) and (10.9) then show that there is no forcing of the mean flow by the eddies. This is the 'non-acceleration' theorem of Charney and Drazin (1961), later generalized by Andrews and McIntyre (1976a) and Boyd (1976). If the eddy sources consist of Newtonian cooling and Rayleigh friction with constant and equal decay coefficients, S^* takes the simple form αq^* where $\alpha > 0$. Equation (10.17) then shows that for growing or damped waves $[q^*v^*] < 0$ provided that $\partial[q]/\partial y > 0$. Thus for typical conditions with $[q]$ increasing towards the North Pole the potential vorticity transport by growing or damped eddies is equatorward and from Eqn. (10.16) the EP flux is convergent. Referring back to Eqn. (10.9) we see that for $[q^*v^*] < 0$ the eddies produce a net easterly mean flow acceleration. Such an acceleration will of course tend to destroy the thermal wind balance in the mean flow so that a compensatory ($\tilde{v}$, $\tilde{w}$) residual mean meridional circulation must arise to keep the thermal field in thermal wind balance with the changing $[u]$ field. However, unlike the situation for the conventional Eulerian mean formulation, the compensation cannot be complete. Studies based on numerical simulations (Dunkerton et al., 1981) and observational data (Palmer, 1981a) show that the actual mean flow acceleration tends in general to parallel the time evolution of the eddy forcing $\nabla\cdot\mathbf{E}$ while the eddy momentum and heat fluxes separately can provide little clue as to the actual net forcing. As stressed in Chapter 7, the situation is less clear in the troposphere.

The quasi-geostrophic form of the non-acceleration theorem does not apply in equatorial regions and so cannot be used to diagnose wave-mean flow interaction processes in the equatorial QBO. An approximate form of the theorem valid in equatorial regions was derived by Holton (1974). Andrews and McIntyre (1976a) and Boyd (1976) have shown that the theorem applies for the linearized primitive equations in spherical coordinates. For zonally symmetric motions in equatorial regions the transformed Eulerian mean momentum, Eqn. (10.9) is still approximately valid. But the quasi-geostrophic EP flux must be replaced by its primitive equation form which has the components:

$$E^y = -\rho[u^*v^*] + \rho[u]_z[v^*\Phi_z^*]/N^2$$
$$E^z = -\rho[u^*w^*] + \rho(f - [u]_y)[v^*\Phi_z^*]/N^2. \tag{10.18}$$

In particular, the vertical momentum flux $\rho[u^*w^*]$ plays a crucial role since for both Kelvin and Rossby-gravity modes this term represents a substantial portion of the total flux of wave activity. For such equatorially trapped disturbances, where the wave fields decay away from the equator, we can integrate Eqn. (10.9) meridionally across the equatorial region to get the approximate channel average momentum equation:

$$\frac{\partial[\langle u \rangle]}{\partial t} \simeq \frac{1}{\rho}\frac{\partial \langle E^z \rangle}{\partial z} \tag{10.19}$$

where $\langle \ \rangle$ denotes a meridional average, and we have neglected the term involving the residual circulation. The non-acceleration theorem still applies so that $\partial \langle E^z \rangle / \partial z = 0$ for steady conservative waves. Equation (10.19) will be used in Section (10.5) in our discussion of the equatorial QBO.

10.4 The extratropical winter stratosphere

Dynamical forcing from the troposphere is the major source of disturbances in the extratropical winter stratosphere. The circulation of the winter stratosphere is strongly influenced by vertically propagating forced planetary waves. Such waves are excited in the troposphere by topographic forcing and diabatic heating (see Chapter 6 and references). The relative importance of these two types of wave forcing is still uncertain. However, linearized steady state wave calculations by Lin (1982) indicate that topographic forcing alone can account for most of the observed characteristics of the stationary planetary wave perturbations in the stratosphere.

Charney and Drazin (1961) showed that planetary waves can propagate vertically into the stratosphere provided that the mean zonal flow is westerly, but less than a critical value, U_c, referred to as the 'Rossby critical velocity'. For an idealized model in which the latitudinal dependence of the wave and mean flow fields is neglected, the perturbation potential vorticity, Eqn. (10.13), has solutions of the form:

$$\Phi^*(x, z, t) = \Psi(z) \exp [z/2H] \exp [ik(x - ct)]. \tag{10.20}$$

Substitution of Eqn. (10.20) into Eqn. (10.13) yields the standard one-dimensional wave equation:

$$d^2\Psi/dz^2 + m^2\Psi = 0 \tag{10.21}$$

where:

$$m^2 \equiv \frac{N^2}{f_0^2} \left[\frac{\partial [q]/\partial y}{[u] - c} - k^2 - \frac{f_0^2}{4H^2N^2} \right].$$

Vertical propagation requires $m^2 > 0$. Thus for stationary waves ($c = 0$) vertical propagation is possible only for $[u]$ satisfying the inequality

$$0 < [u] < U_c = \frac{\partial [q]}{\partial y} (k^2 + f_0^2/4H^2N^2)^{-1}. \tag{10.22}$$

For typical conditions in the winter stratosphere only the longest waves (zonal wavenumbers 1 and 2) satisfy the condition (10.22). Although the relationship (10.22) has been derived for a highly idealized flow situation it appears to be qualitatively valid for more realistic flows although inclusion of meridional variability in both the waves and mean flow may enhance the wave propagation (Simmons, 1974).

The criterion (10.22) also shows why the extratropical summer stratosphere is relatively undisturbed. Stationary waves cannot propagate vertically beyond the $[u] = 0$ 'critical surface' that exists in the lower stratosphere. Thus, planetary waves forced by topography and diabatic heating are trapped in the troposphere in the summer. It is known that the summertime planetary wave regime of the troposphere is distinctly different from that of the winter season (see Chapter 3). However, it is not clear to what extent these differences are due simply to changes in the tropospheric forcing due to the altered positions and intensities of tropospheric mean winds and heat sources, rather than to any possible influence of the stratospheric critical surface.

In winter, quasi-stationary, ultralong vertically propagating planetary waves create a considerable distortion of the zonal mean polar vortex in the stratosphere (cf. Fig. 10.3). However, due to the fairly long dissipation time scales for stationary planetary waves in the winter stratosphere, the time mean wave perturbations are relatively unimportant for wave-mean zonal flow interactions when compared with the subseasonal time scale oscillations in wave amplitude. As shown in the last section, wave transience (i.e., time rate of change of wave amplitude) produces a net eddy potential vorticity flux. For amplifying waves the potential vorticity flux is negative, as it also is for waves subject to thermal and mechanical dissipation. Thus, amplifying waves will tend to decelerate the mean zonal flow.

There are several possible sources of wave transience in the stratosphere. One obvious source is simply the response in the stratosphere to planetary wave oscillations in the troposphere. Observations suggest that some, though not all, stratospheric wave oscillations are direct responses to tropospheric planetary wave oscillations. The causes of amplitude and/or phase variations of tropospheric planetary waves are, however, not well understood (see Chapter 6). Tung and Lindzen (1979) suggest that anomalous amplification of stationary planetary waves throughout the troposphere and the stratosphere may be due to a linear resonance which occurs when the stratospheric polar night jet core descends to a lower than normal elevation. A somewhat different model involving *non-linear* resonant amplification has been proposed by Plumb (1982). In Plumb's model wave growth is facilitated by a feedback loop in which the mean zonal wind is altered by the growing wave so that the wave becomes closer to resonance and can grow more rapidly.

These resonance models provide a theoretical basis for expecting strong linkages between tropospheric and stratospheric planetary wave variations and, in particular, for the oft cited supposed relationship between sudden stratospheric warmings and tropospheric blocking. However, tropospheric blocking events occur much more frequently than stratospheric warmings, and tend to be regional rather than planetary in scale. Thus, although blocking may possibly be a necessary precursor for sudden stratospheric warmings, it is certainly not a sufficient condition.

Sudden stratospheric warmings have provided a fascinating challenge to dynamic meteorologists for many years. The literature on observational and dynamical aspects of stratospheric warmings is extensive (e.g., see reviews by Quiroz et al., 1975; Schoeberl, 1978; Holton, 1980). The theoretical basis for our present understanding of the dynamics of sudden warmings has been discussed by McIntyre (1982), and a complete review of the climatology of observed warmings has been given by Labitzke (1982).

Observational and modelling work during the past few years (e.g., O'Neill and Taylor, 1979; O'Neill, 1980; Dunkerton et al., 1981; Palmer, 1981a,b; Butchart et al., 1982; O'Neill and Youngblut, 1982) has confirmed beyond reasonable doubt that sudden stratospheric warmings result from wave-mean flow interaction induced by the amplification of vertically propagating forced planetary waves.

The recent modelling results of Dunkerton et al. (1981) and the observational studies of Palmer (1981a,b) and O'Neill and Youngblut (1982) demonstrate that the transformed Eulerian mean equations of Section 10.3 provide a powerful set of diagnostics for the study of sudden stratospheric warmings.

The utility of the transformed equations in the diagnosis of sudden warmings arises partly from the fact that in linear planetary wave theory the

EP flux appears as the flux in a conservation equation for wave activity of the form (Edmon *et al.*, 1981):

$$\frac{\partial A}{\partial t} + \mathbf{V} \cdot \mathbf{E} = D \tag{10.23}$$

where D represents the dissipation. For quasi-geostrophic β-plane motions it may be verified from Eqns. (10.16) and (10.17) that:

$$A \simeq \tfrac{1}{2}[q^{*2}]/[q]_y; \quad D = -[q^*S^*]/[q]_y.$$

A is a measure of the local wave activity that, unlike the wave energy density, is conserved in the absence of dissipation and wave-mean flow interaction. The EP flux can thus be regarded as an indicator of the propagation of wave activity in the meridional plane.

Additional support for the utility of $\mathbf{E}$ as a measure of the flux of wave activity comes from the fact that, as pointed out by Edmon *et al.* (1980), for situations in which the group velocity can be defined it turns out that:

$$\mathbf{E} = \mathbf{c}_g A \tag{10.24}$$

where $\mathbf{c}_g$ is the projection of the local group velocity on the meridional plane. According to the ray theory of geometric optics, for quasi-steady conditions the group velocity will tend to be guided up the gradient of the refractive index (see Karoly and Hoskins, 1982, and O'Neill and Youngblut, 1982, for details). Matsuno (1970) showed that the index of refraction for stationary planetary waves is given approximately as:

$$Q \simeq [q]_y/[u] \tag{10.25}$$

where:

$$[q]_y = \beta - [u]_{yy} - \frac{f_0^2}{\rho} \frac{\partial}{\partial z}\left(\frac{\rho}{N^2} \frac{\partial[u]}{\partial z}\right).$$

Thus, the pattern of EP flux in the meridional plane is, for a given tropospheric wave source distribution, sensitively dependent on the distribution of the refractive index in latitude and height.

From Eqn. (10.25) we see that away from regions of strong flow curvature the magnitude of Q is approximately inversely proportional to $[u]$. Thus, $\mathbf{V}Q$ is directed from the region of mid-latitude westerlies towards the weak wind region of the tropics. As a consequence the EP flux vectors tend to propagate primarily upwards and equatorwards under normal conditions. However, the strong meridional curvature associated with the polar night jet produces a local maximum in $[q]_y$ which may produce a maximum in Q at high latitudes which serves to guide a portion of the wave activity into the high latitude stratosphere. A poleward shift and intensification of the polar night jet can 'precondition' the mean flow by enhancing the high latitude maximum in Q so that the EP flux vectors switch from equatorward to poleward tilts in the polar

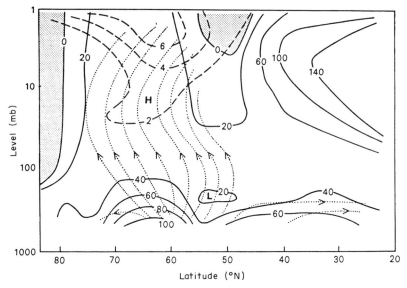

Fig. 10.6. Contours of refractive index Q for 26 February 1980 (solid lines), integral curves of the EP flux vector (dotted lines) and contours of the EP flux divergence (dashed lines, units of m s^{-2}) (after Palmer, 1981b).

region (Palmer, 1981b). The resulting strong flux convergence, as shown for the late February 1980 warming in Fig. 10.6, leads to rapid deceleration of the mean zonal flow in the polar region, and hence by thermal wind considerations a reversal of the normal pole to equator temperature gradient.

Whether the 'preconditioning' of the mean flow mentioned above is a necessary prerequisite for the occurrence of a sudden stratospheric warming is still unclear. Nor is it clear whether resonance and/or critical layers play essential roles in the warming process. Nevertheless, recent model studies (Butchart et al., 1982) clearly demonstrate that the distribution of the refractive index (which is controlled by the mean flow profile) is absolutely crucial in determining the mean flow response to a given tropospheric forcing. The frequent occurrence of periods of enhanced wave activity several weeks prior to major warmings (Labitzke, 1982) suggests that preconditioning of the mean flow is indeed a common occurrence and that a 'sudden' warming is only the dramatic climax to a process which develops over a span of several weeks.

10.5 Dynamics of the equatorial stratosphere: the quasi-biennial oscillation

The observed characteristics of the equatorial QBO have been summarized in Section 10.2. There can be little doubt that the QBO is driven by the interaction between the mean flow and vertically propagating equatorial waves generated in the troposphere (primarily Kelvin and Rossby gravity waves), although laterally propagating stationary Rossby modes of extratropical origin may generate a significant portion of the easterly acceleration. Little is known concerning the source for the Kelvin and Rossby gravity waves. Observational studies by Yanai and Hayashi (1969) and others have confirmed that the Rossby gravity waves are generated in the upper troposphere. It is commonly assumed that convective disturbances in the equatorial zone excite these waves—i.e., that they are diabatically forced. However, the observational data base for tropical winds and latent heat release is insufficient to directly relate observed Kelvin and Rossby gravity waves to their sources. Therefore, despite the fact that such sources probably have substantial interannual variability related to the Southern Oscillation and other sources of equatorial climate anomalies, it has not been possible to directly relate such variability to observed variations in the period of the stratospheric QBO.

Dynamical models of the equatorial QBO have generally specified that the tropospheric wave sources were constant in time so that any variability in period could only result from inherent variability in the stratospheric oscillation. Various aspects of equatorial wave-mean flow interaction have been studied theoretically since the wave-mean flow interaction theory of the QBO was first proposed by Lindzen and Holton (1968). Most theoretical studies of the QBO have utilized one-dimensional models in which the height dependence of the mean zonal wind was computed as a function of time under conditions of steady Kelvin and Rossby gravity wave forcing at the lower boundary. The first such model was that of Holton and Lindzen (1972). In Holton and Lindzen's model the meridionally integrated vertical component, $\langle E^z \rangle$, of the EP flux in Eqn. (10.19) was evaluated by assuming that the Kelvin and Rossby gravity waves were single frequency modes, steady in amplitude, but subject to damping in the form of radiative dissipation modelled by Newtonian cooling. The meridionally integrated EP flux then decays with height as follows:

$$E^z = \sum_{i=1}^{2} A_i \exp\left(-2 \int_{z_g}^{z} \lambda_i(z)\, dz \right) \tag{10.26}$$

where $i = 1$ for Kelvin waves and $i = 2$ for Rossby gravity waves. Here A_i designates the specified meridionally integrated EP flux at the lower boundary level (z_g). A_1 is negative and A_2 is positive. The inverse vertical decay scales λ_i are given by:

$$\lambda_1 = \tfrac{1}{2} N k \alpha \hat{\omega}^{-2} = \frac{\alpha}{2W_1}$$

$$\lambda_2 = \tfrac{1}{2} N \beta \alpha \hat{\omega}^{-3} [1 + k\hat{\omega}/\beta] = \frac{\alpha}{2W_2} \tag{10.27}$$

where N is the buoyancy frequency, k the zonal wavenumber, α the Newtonian cooling coefficient, $\hat{\omega}$ the Doppler-shifted frequency, $\beta \equiv df/dy$, and W_i the vertical component of the group velocity. Thus, for both wave modes the vertical decay scale is just the distance in which energy propagates in one 'e-folding' decay time. Since the group velocities for the Kelvin and Rossby gravity modes depend on the square and cube of the Doppler-shifted frequency, respectively, the EP flux divergence tends to increase as $\hat{\omega} \to 0$. Hence, in the presence of westerly shear the Kelvin waves are able to propagate in the region of easterlies but tend to be rapidly damped and a westerly acceleration occurs ($\partial E^z/\partial z > 0$). In the presence of easterly shear the Rossby gravity waves are able to propagate in the region of westerly winds but tend to be even more rapidly damped and an easterly acceleration occurs ($\partial E^z/\partial z < 0$).

It is this strong dependence of the wave-drain acceleration on the mean wind distribution which provides the basis for the equatorial QBO. As Plumb (1977) showed, it is in principle possible to have a steady solution for the mean wind when the Kelvin and Rossby gravity wave sources are steady. However, for realistic conditions such a solution would be unstable to small perturbations in the mean wind and a self excited non-linear oscillation will result. Dunkerton (1981) showed that in a compressible atmosphere the period of this oscillation is inversely proportional to the EP flux of the waves at the tropopause. The period is in no way controlled by the annual or semiannual cycles. Thus, the *quasi-biennial* character of the oscillation is simply a coincidence.

The non-linear instability model of the QBO has been confirmed in a laboratory analogue by Plumb and McEwan (1978). In their experiment the working fluid is water contained in an annular vessel. The bottom boundary consists of a flexible membrane which is vibrated to produce two waves of equal amplitude and equal but opposite phase speeds propagating eastward and westward, respectively. The waves are damped by viscous dissipation as they propagate vertically. For wave forcing exceeding a certain critical amplitude the Doppler-shifted phase speed dependence of the viscous wave damping amplifies small perturbations in the interior mean flow and leads to alternating downward propagating easterly and westerly shear zones, just as in the atmosphere.

Considering the combined evidence from observational studies, numerical simulations, and the Plumb–McEwan laboratory simulation, there can be little doubt that the equatorial QBO is generated by equatorial wave-mean flow interactions. However, a complete three-dimensional theory has yet to be developed.

That an understanding of the meridional structure of the QBO is an important aspect of any complete theory, is clear from the work of Andrews and McIntyre (1976a). They showed that the meridional distribution of the mean flow acceleration produced by damped Rossby gravity waves is very sensitive to the relative magnitudes of viscous and thermal damping. In particular, for steady waves subject only to thermal damping the mean flow acceleration has two symmetric maxima occurring north and south of the equator, but vanishes at the equator. However, the addition of moderate mechanical damping completely changes the profile. Furthermore, because of the inverse Doppler-shifted frequency dependence of the acceleration, Andrews and McIntyre (1976b) suggested that thermally damped Rossby gravity waves might act to enhance any small cross equatorial mean wind shear and eventually produce barotropic instability.

In a detailed study of equatorial waves in the presence of meridional wind shear, Boyd (1978) showed that, for equal mechanical and thermal dissipation rates, the mean flow acceleration forced by Rossby gravity waves has a maximum at the equator and an approximate Gaussian decay with latitude, consistent with the structure of the observed QBO. In addition, the resulting mean flow acceleration tends to reduce the magnitude of any cross equatorial mean wind shear. This latter tendency tends to preserve the equatorial symmetry of the QBO even in the presence of the perturbing influence of the approximately antisymmetric annual cycle.

Unfortunately, although it is easy to rationalize Newtonian cooling in terms of radiative damping, there seems to be no analogous mechanical dissipation which can rationalize the use of a Rayleigh friction coefficient. However, Andrews and McIntyre (1976a) pointed out that the mean flow acceleration produced by an exponentially growing wave has the same form as that produced by a steady wave subject to Rayleigh frictional damping. Thus, it appears that the observed latitudinal structure of the equatorial QBO can be accounted for provided that wave transience is included in the model.

As we indicated in the previous section, wave transience is essential in driving the mean flow decelerations observed during sudden stratospheric warmings. However, the possible role of wave transience in the equatorial stratosphere has only been recently explored. In a numerical study of equatorial wave-mean flow interaction Holton (1979b) found that even when no mechanical dissipation was included, the mean flow acceleration patterns for both Kelvin and Rossby gravity waves in the presence of cross equatorial shear were very similar to those computed by Boyd (1978) for steady mechanically damped waves. Holton attributed this result to the effects of wave transience. He pointed out that although the wave forcing was specified to be steady at the tropopause level the wave-mean flow interaction process in the descending mean wind shear zone itself generated substantial temporal change in wave amplitude, and hence wave transience effects.

Due to numerical problems Holton was not able to use his model to simulate a complete cycle of the QBO. However, the wave transience effect has been included in a one-dimensional simulation by Dunkerton (1981). He demonstrated that wave transience is probably the primary cause of the

QBO—at least in a chronological sense—because it is transience which initially establishes the easterly and westerly shear zones. Wave absorption due to radiative damping, however, plays an essential role in making the transience induced mean wind changes irreversible. Thus, it appears that future models of the equatorial QBO, whether one or two-dimensional should explicitly take account of wave transience effects.

10.6 Interannual variability in the stratosphere

As we have seen in the previous section, interannual variability in the form of a quasi-biennial oscillation is the predominant dynamical feature of the equatorial stratosphere. Although interannual variations of comparable regularity do not occur in the extratropical stratosphere, there is very pronounced year to year variability in the winter stratosphere, at least in the Northern Hemisphere where adequate data coverage is available.

Quiroz (1981) has used a combination of satellite and conventional data to derive monthly mean meridional cross sections of the mean zonal geostrophic wind from 500–0.4 mb in the Northern Hemisphere winter months for 1975–76 through 1978–79. Strong interannual variability is apparent not only in monthly means but seasonal means as well (Fig. 10.7). Because of the strong sensitivity of planetary wave propagation to the mean wind distribution, the planetary wave climatology should exhibit an equally large interannual variability. Of course, a large fraction of the observed stratospheric variability may be due simply to interannual variability in vertically propagating planetary waves caused by processes originating in the troposphere. However, as argued in Section 10.4, the strong sensitivity of planetary wave propagation to the mean wind distribution suggests that interannual variability in stratospheric means winds might (through its influence on wave reflection and/or wave resonance) generate some planetary wave variability in the troposphere.

The mean wind departure fields of Fig. 10.7 clearly indicate a tendency for a see-saw oscillation between polar and mid-latitudes in which a positive anomaly at high latitudes goes with a negative anomaly in mid-latitudes, and vice versa. Furthermore, this four-year sample suggests the presence of a quasi-biennial oscillation in which alternating years have positive and negative polar anomalies, respectively. Many observational studies have been devoted to the search for extratropical QBOs. However, only a few of these have attempted to relate extratropical QBOs to the equatorial QBO.

Holton and Tan (1980, 1982) used a 16-year record of gridded monthly mean geopotential height and temperature data for several levels in the Northern Hemisphere stratosphere to study the influence of the equatorial QBO on the extratropical circulation. They composited this data with respect

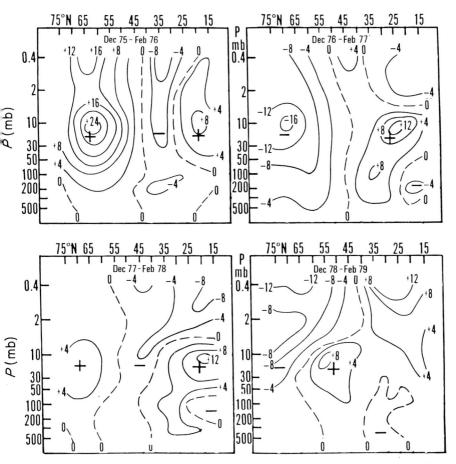

Fig. 10.7. Wintertime (December–February) mean zonal wind anomalies (m s^{-1}) computed by subtracting long period monthly means from monthly means for each year (after Quiroz, 1981).

to the phase of the equatorial QBO by placing each of the monthly means into either a westerly or easterly category depending on the sign of the equatorial mean zonal wind at 50 mb. Their studies indicated that the zonally symmetric see-saw oscillation apparent in Fig. 10.7 remains even when a much larger sample is examined. During the westerly phase of the equatorial QBO at 50 mb the polar heights are lower and mid-latitude heights higher than during the easterly phase (Fig. 10.8). They also found a QBO in the

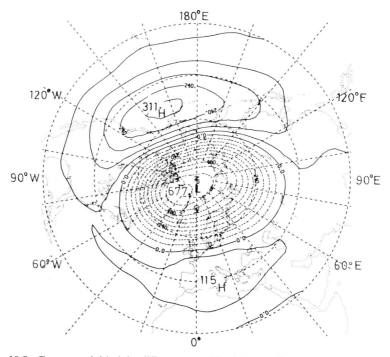

Fig. 10.8. Geopotential height differences at 10 mb (westerly category minus easterly category) for January–March for 16 years of data composited with respect to the phase of the equatorial QBO at 50 mb. Units: geopotential metres. The outer latitudinal circle is 20°N (after Holton and Tan, 1982).

amplitude of stationary planetary wavenumber 1 in the early winter (November–December) with larger amplitude occurring during the easterly phase of the equatorial QBO at 50 mb. However, this signal was not present in the later winter months of January–March. For those months, rather, wavenumber 2 was stronger during the westerly phase of the equatorial QBO. Despite this apparent QBO in the stationary wave amplitudes, Holton and Tan (1982) were unable to detect a significant QBO in the monthly mean stationary wave EP fluxes. Thus, it appears that the extratropical QBO is primarily a zonally symmetric oscillation in which the polar night jet strength is positively correlated with the equatorial mean zonal wind speed at 50 mb.

Despite the strong evidence for a link between the equatorial QBO and observed interannual variability in the Northern Hemisphere winter stratosphere provided by Holton and Tan's studies, other interpretations remain possible. Van Loon *et al.* (1981) have carried out a study in which the winter mean geopotential fields are composited with respect to extremes of the Southern Oscillation in the troposphere (see Section 3.5). For the same 16-year

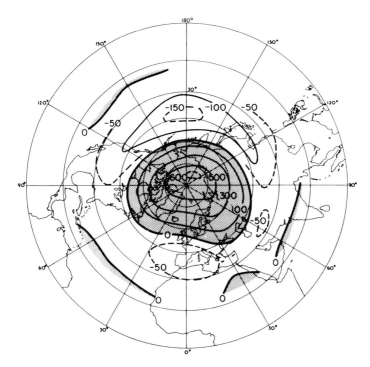

Fig. 10.9. Geopotential height differences at 10 mb (low/wet category minus high/dry category) for December–February for 11 years of data composited with respect to the Southern Oscillation. Units: geopotential metres. The outer latitudinal circle is the equator (after Van Loon *et al.*, 1981).

sample used by Holton and Tan, Van Loon *et al.* find that 4 years were characterized by high sea-level pressure and low rainfall in the equatorial South Pacific and 7 years were characterized by low sea-level pressure and high rainfall. Apparently quite by chance the high/dry years in the Southern Oscillation composite generally correspond to westerly years for the equatorial QBO at 50 mb, and low/wet years correspond to easterly years of the QBO. Thus the zonally symmetric see-saw in geopotential heights reported by Holton and Tan occurs with an almost identical pattern in the Van Loon *et al.* composite (Fig. 10.9).

We are thus left with an uncertainty as to whether the Southern Oscillation or the equatorial QBO (or both) are responsible for causing the observed interannual zonally symmetric see-saw of temperatures and heights in the Northern Hemisphere winter stratosphere. Wallace and Chang (1982) and Labitzke (1982) have attempted to resolve this issue by carefully examining longer data records. Wallace and Chang found that although the strongest (weakest) polar night jets generally occurred for the westerly (easterly) phase of the equatorial QBO at 50 mb and in conjunction with high/dry (low/wet) years for the Southern Oscillation, there were a couple of years that were inconsistent with each of the two compositing schemes. Thus, until much longer observational records are available it seems unlikely that the true source of the observed zonally symmetric interannual variability in the Northern Hemisphere stratosphere will be determined by observations alone.

10.7 Concluding remarks

In this chapter we have stressed dynamical processes that depend on the forcing of stratospheric motions by tropospheric sources—primarily vertically propagating planetary waves and equatorial waves. There can be little doubt that a large fraction of the observed variability in the stratospheric general circulation on both subseasonal and interannual time scales is due directly or indirectly to the influence of tropospheric wave forcing. However, it is much less clear to what extent the stratospheric general circulation can influence the troposphere. At present our understanding of tropospheric planetary waves seems inadequate to determine whether reflections and/or resonances related to the mean wind distribution in the stratosphere are significant factors in their dynamics.

Evidence from unusual events such as the major stratospheric warming of January 1977 suggests that on at least some occasions wave-mean flow interaction processes which first occur in the stratosphere can modify the flow throughout the troposphere. On that occasion the polar warming penetrated all the way to the surface poleward of 60°N (Quiroz, 1980). However, such dramatic evidence of a downward directed linkage between the stratosphere and troposphere is uncommon. It seems clear that observational studies alone will not be adequate to determine the extent of stratospheric dynamical influence on the troposphere. Rather, carefully diagnosed numerical studies would appear to offer the best opportunity to elucidate this problem. The diagnostic techniques discussed in Section 10.2 should prove of continuing value in our search for an understanding of the dynamical links between the troposphere and stratosphere.

References

ANDREWS, D. G. and McINTYRE, M. E. (1976a). Planetary waves in horizontal and vertical shear: The generalized Eliassen-Palm relation and the mean zonal acceleration. *J. atmos. Sci.*, **33**, 2031–2048.
ANDREWS, D. G. and McINTYRE, M. E. (1976b). Planetary waves in horizontal and vertical shear: Asymptotic theory for equatorial waves in weak shear. *J. atmos. Sci.*, **33**, 2049–2053.

ANDREWS, D. G. and McINTYRE, M. E. (1978a). An exact theory of non-linear waves on a Lagrangian-mean flow. *J. Fluid Mech.*, **89**, 609–646.

ANDREWS, D. G. and McINTYRE, M. E. (1978b). Generalized Eliassen-Palm and Charney-Drazin theorems for waves on axisymmetric means flows in compressible atmospheres. *J. atmos. Sci.*, **35**, 175–185.

BARNETT, J. J. (1980). Satellite measurements of middle atmosphere temperature structure. *Phil. Trans. R. Soc. Lond.*, **A296**, 41–57.

BOYD, J. P. (1976). The non-interaction of waves with the zonally averaged flow on a spherical earth and the interrelationships of eddy fluxes of energy, heat and momentum. *J. atmos. Sci.*, **33**, 2285–2291.

BOYD, J. P. (1978). The effects of latitudinal shear on equatorial waves. Part II: Application to the atmosphere. *J. atmos. Sci.*, **35**, 2259–2267.

BUTCHART, N., CLOUGH, S. A., PALMER, T. N. and TREVELYAN, P. J. (1982). Simulations of an observed stratospheric warming with quasi-geostrophic refractive index as a model diagnostic. *Q. Jl R. met. Soc.*, **108**, 475–502.

CHARNEY, J. G. and DRAZIN, P. G. (1961). Propagation of planetary scale disturbances from the lower into the upper atmosphere. *J. Geophys. Res.*, **66**, 83–109.

COY, L. (1979). An unusually large westerly amplitude of the quasi-biennial oscillation. *J. atmos. Sci.*, **36**, 174–176. Also see Corrigendum, 1980, *J. atmos. Sci.*, **37**, 912–913.

DUNKERTON, T. J. (1978). On the mean meridional mass motions of the stratosphere. *J. atmos. Sci.*, **35**, 2325–2333.

DUNKERTON, T. J. (1981). Wave transience in a compressible atmosphere. Part II: Transient equatorial waves in the quasi-biennial oscillation. *J. atmos. Sci.*, **38**, 298–307.

DUNKERTON, T. J., HSU, C.-P. F. and McINTYRE, M. E. (1981). Some Eulerian and Lagrangian diagnostics for a model stratospheric warming. *J. atmos. Sci.*, **38**, 819–843.

EDMON, H. J., HOSKINS, B. J. and McINTYRE, M. E. (1980). Eliassen-Palm cross-sections for the troposphere. *J. atmos. Sci.*, **37**, 2600–2616. (See also Corrigendum, 1981, *J. atmos. Sci.*, **38**, 1115.)

ELIASSEN, A. and PALM, E. (1961). On the transfer of energy in stationary mountain waves. *Geofys. Publ.*, **22**, no. 3, 1–23.

FELS, S. B., MAHLMAN, J. D., SCHWARZKOPF, M. D. and SINCLAIR, R. W. (1980). Stratospheric sensitivity to perturbations in ozone and carbon dioxide: Radiative and dynamical response. *J. atmos. Sci.*, **37**, 2265–2297.

GILLE, J. C., BAILEY, D. P. and RUSSELL, J. M. (1980). Temperature and composition measurements from the l.r.i.r. and l.i.m.s. experiments on Nimbus 6 and 7. *Phil. Trans. R. Soc. Lond.*, **A296**, 205–218.

HOLTON, J. R. (1974). Forcing of mean flows by stationary waves. *J. atmos. Sci.*, **31**, 942–945.

HOLTON, J. R. (1975). *The Dynamic Meteorology of the Stratosphere and Mesosphere*. Boston: Am. Meteorol. Soc., 218 pp.

HOLTON, J. R. (1979a). *An Introduction to Dynamic Meteorology*. Second Edition, Academic Press, New York, 391 pp.

HOLTON, J. R. (1979b). Equatorial wave-mean flow interaction: A numerical study of the role of latitudinal shear. *J. atmos. Sci.*, **36**, 1030–1040.

HOLTON, J. R. (1980). The dynamics of sudden stratospheric warmings. *Ann. Rev. Earth Planet. Sci.*, **8**, 169–190.

HOLTON, J. R. (1981). An advective model for two-dimensional transport of stratospheric trace species. *J. Geophys. Res.*, **86**, 11,989–11,994.

HOLTON, J. R. (1982). The role of gravity wave induced drag and diffusion in the momentum budget of the mesosphere. *J. atmos. Sci.*, **39**, 791–799.

HOLTON, J. R. and LINDZEN, R. S. (1972). An updated theory for the quasi-biennial cycle of the tropical stratosphere. *J. atmos. Sci.*, **29**, 1076–1080.

HOLTON, J. R. and TAN, H-C. (1980). The influence of the equatorial quasi-biennial oscillation on the global circulation at 50 mb. *J. atmos. Sci.*, **37**, 2200–2208.

HOLTON, J. R. and TAN, H-C. (1982). The quasi-biennial oscillation in the Northern Hemisphere lower stratosphere. *J. met. Soc. Japan*, **60**, 140–148.

KAROLY, D. J. and HOSKINS, B. J. (1982). Three-dimensional propagation of planetary waves. *J. met. Soc. Japan*, **60**, 109–123.

KRUEGER, A. J., GUENTHER, B., FLEIG, A. J., HEATH, D. F., HILSENRATH, E., MCPETERS, R. and PRABHAKARA, C. (1980). Satellite ozone measurements. *Phil. Trans. R. Soc. Lond.*, **A296**, 191–204.

LABITZKE, K. (1982). On the interannual variability of the middle stratosphere during the Northern winters. *J. met. Soc. Japan*, **60**, 124–139.

LIN, B-D. (1982). The behavior of stationary planetary waves forced by topography and diabatic heating. *J. atmos. Sci.*, **39**, 1206–1226.

LINDZEN, R. S. (1981). Turbulence and stress due to gravity wave and tidal breakdown. *J. Geophys. Res.*, **86**, 9707–9714.

LINDZEN, R. S. and HOLTON, J. R. (1968). A theory of the quasi-biennial oscillation. *J. atmos. Sci.*, **25**, 1095–1107.

MATSUNO, T. (1970). Vertical propagation of stationary planetary waves in the winter Northern Hemisphere. *J. atmos. Sci.*, **27**, 871–883.

MATSUNO, T. (1982). A quasi-one-dimensional model of the middle atmosphere circulation interacting with internal gravity waves. *J. met. Soc. Japan*, **60**, 215–226.

MCINTYRE, M. E. (1980). An introduction to the generalized Lagrangian-mean description of wave mean-flow interaction. *Pure appl. Geophys.*, **118**, 152–176.

MCINTYRE, M. E. (1982). How well do we understand stratospheric warnings? *J. met. Soc. Japan*, **60**, 37–65.

MURGATROYD, R. J. (1969). The structure and dynamics of the stratosphere. *The Global Circulation of the Atmosphere* (G. A. Corby, Ed.), London, Royal Meteorological Society, pp. 159–195.

MURGATROYD, R. J. and O'NEILL, A. (1980). Interaction between the troposphere and stratosphere. *Phil. Trans. R. Soc. Lond.*, **A296**, 87–102.

O'NEILL, A. (1980). The dynamics of stratospheric warmings generated by a general circulation model of the troposphere and stratosphere. *Q. Jl. R. met. Soc.*, **106**, 659–690.

O'NEILL, A. and TAYLOR, B. F. (1979). A study of the major stratospheric warming of 1976/77. *Q. J. R. met Soc.*, **105**, 71–92.

O'NEILL, A. and YOUNGBLUT, C. E. (1982). Stratospheric warmings diagnosed using the transformed Eulerian-mean equations and the effect of the mean state on wave propagation. *J. atmos. Sci.*, **39**, 1370–1386.

PALMER, T. N. (1981a). Diagnostic study of a wavenumber-2 stratospheric sudden warming in a transformed Eulerian-mean formalism. *J. atmos. Sci.*, **38**, 844–855.

PALMER, T. N. (1981b). Aspects of stratospheric sudden warmings studied from a transformed Eulerian-mean viewpoint. *J. Geophys. Res.*, **86**, 9679–9687.

PLUMB, R. A. (1977). The interaction of two internal waves with the mean flow: Implications for the theory of the quasi-biennial oscillation. *J. atmos. Sci.*, **34**, 1847–1858.

PLUMB, R. A. (1981). Instability of the distorted polar night vortex: a theory of stratospheric warmings. *J. atmos. Sci.*, **38**, 2514–2531.

PLUMB, R. A. and BELL, R. C. (1982). A model of the quasi-biennial oscillation on an equatorial beta-plane. *Q. Jl R. met. Soc.*, **108**, 335–352.

PLUMB, R. A. and MCEWAN, A. D. (1978). The instability of a forced standing wave in a viscous stratified fluid: A laboratory analogue of the quasi-biennial oscillation. *J. atmos. Sci.*, **35**, 1827–1839.

QUIROZ, R. S. (1980). Variations in the zonal mean and planetary wave properties of the stratosphere and links with the troposphere. *Pure Appl., Geophys.*, **118**, 416–427.

QUIROZ, R. S. (1981). The tropospheric-stratospheric mean zonal flow in winter. *J. Geophys. Res.*, **86**, 7378–7384.

QUIROZ, R. S., MILLER, A. J. and NAGATANI, R. M. (1975). A comparison of observed and simulated properties of sudden stratospheric warmings. *J. atmos. Sci.*, **20**, 265–275.

RAMANATHAN, V. and DICKINSON, R. E. (1979). The role of stratospheric ozone in the zonal and seasonal radiative energy balance of the Earth-troposphere system. *J. atmos. Sci.*, **36**, 1084–1104.

SCHOEBERL, M. R. (1978). Stratospheric warmings: Observation and theory. *Rev. Geophys. Space Phys.*, **16**, 521–538.

SIMMONS, A. J. (1974). Planetary scale disturbances in the polar winter stratosphere. *Q. J. R. met. Soc.*, **100**, 76–108.

TUNG, K. K. and LINDZEN, R. S. (1979). A theory of stationary long waves. Part II: Resonant Rossby waves in the presence of realistic vertical shears. *Mon. Weath. Rev.*, **107**, 735–750.

VAN LOON, H., ZEREFOS, C. S. and REPAPIS, C. C. (1981). *Evidence of the Southern Oscillation in the Stratosphere*. Publication No. 3, Academy of Athens, Research Center for Atmospheric Physics and Climatology, Athens, Greece.

WALLACE, J. M. (1973). General circulation of the tropical lower stratosphere. *Rev. Geophys. Space Phys.*, **11**, 191–222.

WALLACE, J. M. and CHANG, F-C. (1982). Interannual variability of the wintertime polar vortex in the Northern Hemisphere middle stratosphere. *J. met. Soc. Japan*, **60**, 149–155.

WEHRBEIN, W. M. and LEOVY, C. B. (1982). An accurate radiative heating and cooling algorithm for use in a dynamical model of the middle atmosphere. *J. atmos. Sci.*, **39**, 1532–1544.

YANAI, M. and HAYASHI, Y. (1969). Large-scale equatorial waves penetrating from the upper troposphere into the lower stratosphere. *J. met. Soc. Japan*, **47**, 167–182.

– 11 –

The oceanic general circulation and its interaction with the atmosphere

DAVID L. T. ANDERSON

11.1 Introduction

Benjamin Franklin as early as 1770 viewed the Gulf Stream as a river of warm water flowing along the eastern coast of the United States and thence to the east well beyond the Grand Banks. With this idea in mind Maury (1844) argued that 'this warm water spreads itself out for thousands of square leagues over the cold waters around and covers the ocean with that mantle of warmth which tends to mitigate in Europe, the rigors of winter'. Even before this time, Sabine in 1825 had suggested that variations in the strength of the Gulf Stream were responsible for variations in the climate of Europe (see Sabine, 1846). He suggested that the strengthening and extension eastwards of the Gulf Stream resulted from a strengthening of the trades winds a month or two earlier, increasing the flow into the Caribbean and thence via the Florida current to the Gulf Stream. It is clear then that the effect of the ocean in modulating climatic conditions was realized quite some time ago, and that the idea of teleconnections is just as old.

Maury's statement has frequently been interpreted to mean that Western Europe's mild winter is a result of the northward heat transport by the Gulf Stream; however, there are several difficulties with this interpretation. First, Maury's statement above deals with the surface circulation, whereas the calculation of northward heat transport across a given latitude requires knowledge of the ocean circulation to all depths. Second, there is a difficulty in defining the location and properties of the Gulf Stream. For example, it is not possible to determine where and at what temperature the water flowing northward in the Gulf Stream returns to the south, and for this reason it is not possible to identify very meaningfully its contribution to the heat transport. Third, although the waters off Ireland are some 5°C warmer than those off the west coast of North America at a similar latitude, this change results from differences in the general circulation of the ocean basins, which in turn is affected by the winds and differential heating, and is not attributable to the Gulf Stream alone.

In fact both near-surface conditions and the response of the water column throughout the ocean basin (and not just the Gulf Stream) are important in understanding the role of the ocean in the climate system, their relative importance depending on the time scale of interest. With a heat capacity of approximately 1000 times that of the atmosphere, the ocean can simultaneously be viewed as having on the one hand a damping effect on climate change and on the other, because of its ability to act as a moisture and heat source, as a cause of climate variability.

In Section 11.2, the ocean heat transport and general circulation will be considered. Knowledge of heat transport is sufficiently inadequate that we cannot even be sure of the sign in some cases. This part of the paper is concerned with processes that take place on long time scales: decades to centuries, though we have no data to show variation in heat transport on this time scale.

Sections 11.3 and 11.4 deal with variability on shorter time scales and are concerned mainly with the near surface circulation. At middle latitudes the role of the ocean in inducing changes to the atmospheric circulation is unclear. Lau (1981) in a 15-year model simulation of the atmospheric circulation with seasonally prescribed ocean temperatures has shown that there is considerable interannual variability (see Chapter 5). Separating the natural variability of the atmosphere from that resulting from changes in the ocean is in general difficult. There is nothing corresponding to the Southern Oscillation in Lau's model calculation, however, and this is at least suggestive that the low-latitude ocean plays an important role in the Southern Oscillation. In Section 11.4, a description and partial explanation of the oceanic changes associated with El Niño is given.

11.2 The oceanic heat transport and general circulation

11.2.1 Calculations of heat transport

The ocean can both store and transport heat. Its large thermal capacity acts to reduce the seasonal extremes of temperature relative to those over land, whereas the ability to transport heat means that it can act to reduce the temperature gradient between the equator and the pole (if the heat transport is polewards). Over the years, there have been various estimates of the importance of oceanic heat transport. Interest has considerably increased in the last few years, with the realization that the ocean heat transport is an important factor in understanding the climate response to increasing CO_2.

Vonder Haar and Oort (1973) and Oort and Vonder Haar (1976) inferred the ocean transport and seasonal variation as a residual from atmospheric measurements. Their estimates were much larger than oceanographers had anticipated, based on calculations from oceanographic data. Regardless of which approach is more accurate, the ocean's contribution is a significant fraction of the total, particularly equatorward of 45°. In the next few years, as part of the World Climate Research Program (WCRP), considerable attention is likely to be devoted to improving our estimates of the ocean contribution to heat transport and the diabatic processes responsible for water mass conversion.

At present four basic methods are available for estimating northward transport. The oldest is to calculate net flux through the surface and then to infer transport out of a region by either calculating storage or assuming it zero when averaged over the annual cycle. The disadvantage of this method is that it relies on dubious parameterizations of surface fluxes that are often applied beyond their range of applicability. The second method is to calculate directly the net radiation at the top of the atmosphere and the divergence of the atmospheric heat flux, to infer first the flux of radiation into the ocean and then the oceanic transport. This method suffers from the obvious drawback of being a residual calculation. The third method is to use numerical models to simulate the ocean circulation and the oceanic heat transport. Deficiencies in the models (for example, inadequate parameterizations of deep convection, improper simulation of the depth of the thermocline, or the method of including diabatic forcing) all prejudice the results.

The final method involves direct oceanic calculations across a section. In this approach the potential temperature and velocity normal to the section are determined. The calculation is easiest if the mass transfer is zero as would be the case for a coast to coast section in a bounded ocean basin. The difficulty is that although $T(x, z)$ may be known reasonably well, the contemporaneous velocity field is not and must be inferred usually using ideas based on a simplified conception of ocean circulation. For example, one might estimate the mass transport by integrating the wind stress curl from the eastern boundary westward using the Sverdrup relation:

$$\beta v = \text{curl } \tau. \tag{11.1}$$

This meridional transport must then be returned in a western boundary current. For the Florida current the transport has been *directly* measured to be 30 Sverdrups $(3 \times 10^7 \text{ m}^3 \text{ s}^{-1})$ which is comfortingly consistent with the value estimated by integrating Eqn. (11.1) across the Atlantic basin at 24°N. The partition of transport with respect to depth and hence temperature can be made more easily. Usually this information is not at hand, so one has to perform this partition somewhat arbitrarily (Bryan, 1962; Bennet, 1978) or diagnostically using an inverse technique (Roemmich, 1980).

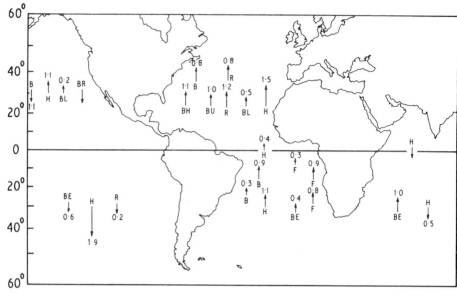

Fig. 11.1. Composite of heat transport measurements in units of 10^{15} W from a variety of sources. In some cases authors give more than one value (giving an indication of error bounds). These have not been included here and only some notional average value has been taken. Not all calculations are really independent in that the same data sources are sometimes used. B, Bryan (1962); BE, Bennet (1978); BH, Bryden and Hall (1980); BL, Bryan and Lewis (1979); BR, Burridge (1982); BU, Bunker (see Bryden and Hall, 1980); F, Fu (1981); H, Hastenrath (1980); R, Roemmich (1980, 1981). At 24°N in the Atlantic there is reasonable agreement at 1.1×10^{15} W; H at 1.5×10^{15} W is high and BL, a model calculation, at 0.5×10^{15} W is low. In the North Pacific, however, the sign is dubious and likewise in the Indian Ocean. All results appear to indicate equatorward heat flux in the South Atlantic. The measurements all represent heat transport across a given latitude from coast to coast in that ocean basin to which they apply.

Figure 11.1 gives a composite of the results of various studies using one or other of the above techniques. Some of the results by different authors may not be totally independent because they use common data. Sometimes authors give information on the sensitivity of their results, e.g., how different the transport is when a different 'model' is used to specify v. Thus Fig. 11.1 does not represent all our knowledge, but does give a feel for the scatter of the results. No indication of conceived reliability of the results has been given. Although the direct oceanic calculation at 24°N in the Atlantic is thought more reliable than the corresponding one in the Pacific (because in the former the details of the western boundary current are better known and the technique less sensitive to errors in v), in general, it is difficult to favour one study over another. Bryden and Hall (1980) discuss this point further.

Figure 11.1 shows that there is modest agreement that the heat flux in the

Atlantic is everywhere northwards and that at 24°N its value is $\sim 1 \times 10^{15}$ W. This value is well short of the 3×10^{15} W required by Vonder Haar and Oort for the Atlantic plus Pacific. One might anticipate that the Pacific being larger could carry the 2×10^{15} W required for consistency, but this would not appear to be the case. No Pacific study suggests a value this large, and in fact there is no consensus in even the sign of the transport. The Pacific could in fact transport heat equatorward as Bryan (1962) and Burridge (1982) would indicate. These results are in agreement with the water mass study of Stommel and Csanady (1980) who noted that cold water is correlated with high salinity; since there is a net northward transport of salt to balance precipitation, there should be an accompanying equatorward heat transport. It is thus by no means clear that the oceans individually need to transport heat poleward and indeed perhaps only in two of the four ocean basins is this so.

11.2.2 Processes involved in heat transport

The ocean is driven by wind stress and density differences. The latter corresponds roughly to the sinking of heavy fluid in high latitudes and an envisaged general upward motion at lower latitudes. This circulation, called the thermohaline circulation, since it depends on both thermal and salinity effects, is undoubtedly a part of the mean meridional circulation of the ocean, and is important in determining the mean stratification of the ocean.

The wind-driven circulation is important at all latitudes, but is most marked in the subtropical gyres, largely horizontal circulations, of depth of order 1000 m. The western boundary currents of the oceans, e.g., the Gulf Stream and Kuroshio, are parts of this circulation. These gyres are thought to be largely Ekman driven in the sense that the subtropical easterlies imply northward transport, the westerlies imply southward transport, and the centre of the gyre then corresponds to the region of convergence of this Ekman flux. Consequently there are downward bowing isopycnals, associated geostrophic currents around the gyres, and very strong currents along the western boundary. The wind driven circulation as pictured above is quasi-horizontal, but there is also a mean meridional circulation associated with it. In the absence of boundaries this circulation would be equal and opposite to that of the atmosphere. With boundaries present, the model of Bryan and Lewis (1979) suggests that this picture is not substantially modified. Although parts of the meridional circulation can be considered as primarily wind driven or thermohaline driven, in general one cannot readily distinguish the two components. For example, the subtropical gyres are displaced poleward with depth, and become more closely related to the convective driven circulations of higher latitudes. The picture is further complicated by the presence of deep circulations which are driven by eddies generated near the surface. The Gulf Stream exhibits baroclinic–barotropic instability. The eddies generated transport momentum vertically and give rise to a mean circulation in the deep water (Holland and Rhines, 1980). This process depends on the properties both of the density field and of the wind field. Finally, further interactions between wind-driven baroclinic motions and topography can also contribute to the mean circulation.

Although it is mathematically possible to subdivide the heat transport into contributions from meridional circulation, gyre circulations, Ekman fluxes, and eddies, the interpretation in terms of physical processes is ambiguous and no attempt is made to do this here. The heat transport by the meridional circulation cannot easily be identified with the thermohaline circulation alone.

Although it is difficult to partition the transport of heat by the mean circulation according to physical processes, it is sometimes possible meaningfully to relate changes in the forcing to changes in the heat transport. Seasonal changes in the heat transport have been studied by Bryan (1982) and Fig. 11.2 shows the heat transport in the Pacific and Atlantic for January and July. The large seasonal change in the Pacific is not *directly* supported by oceanographic evidence, but the *residual* calculations of Oort and Vonder Haar do show a very large cross-equatorial seasonal heat flux. The Atlantic shows seasonal variation also but of much smaller amplitude. The sign of the seasonal variation is consistent with an Ekman-type argument. At 40°N the westerlies are stronger in winter. This means a larger Ekman flux towards the equator. The temperature of the southward flow is probably close to that of the mixed layer. The location or depth of the return flow is not known, but it will be at a lower temperature. Changes in the Ekman transport are therefore associated with increased equatorward heat transport in winter at the latitude of the westerlies, implying a smaller net poleward heat transport there.

11.2.3 Thermohaline circulation and deep convection

The heat transport measurements of Section 11.2.1 suggest a possible difference between the Atlantic and Pacific oceans. This difference could result from the fact that there is deep water formation in the Atlantic but not in the Pacific.

Ellis (1751) was the first to note that the ocean at depth was cold. He used equipment no more elaborate than a rope and a bucket with lids which were open on descent and forced closed on ascent. The significance of Ellis's observation was noted by Count Rumford (1800), who realized that the coldness of the deep water in the tropics suggested a polar origin. From the experiments of Marsigli (1681), it was known that density differences could drive a circulation, but whether in the oceans this was comparable with the wind-driven circulation was not obvious. Indeed, it seems to have been partly the debate concerning this scientific question which led to the *Challenger*

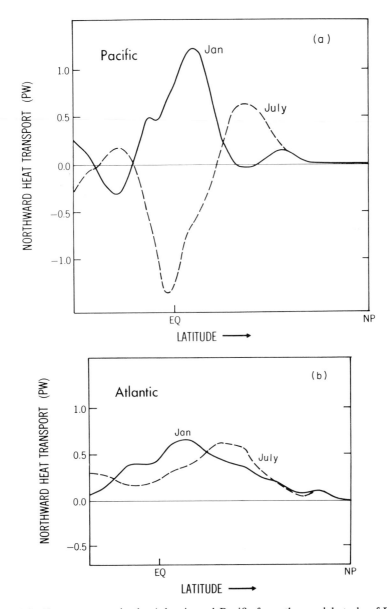

Fig. 11.2. Heat transport in the Atlantic and Pacific from the model study of Bryan (1982) which shows the large seasonal variation in the Pacific. The Atlantic heat transport has less seasonal variation but a larger mean.

expedition of 1872. This expedition lasted $3\frac{1}{2}$ years, charted all the world's oceans apart from the Arctic; took so much data that 76 authors required 23 years to prepare the reports, but still failed to resolve the question. This failure is not surprising since, in a highly interactive system like that of the mean circulation, it is not in general possible to unambiguously separate the two processes.

The seasonal variation in surface heating and wind stirring cause a surface mixed layer to form, whose depth generally increases poleward, where it reaches several hundred metres in winter. In the Greenland sea, and probably the Norwegian sea as well, convection can extend all the way to the bottom. It would appear, however, that the horizontal scale of this convection is rather small (about 50 km) and intermittent. Further north in the Arctic, deep water does not form because of the ice cover and the presence of very low salinity near-surface water. Heat exchange with the atmosphere probably takes place elsewhere mainly in shallow continental shelves.

The cold water formed in the Greenland/Norwegian sea passes through the three gaps in the mid-Atlantic ridge (see Fig. 11.3) (sill depth approximately 700 m) and then flows southward as Count Rumford had suggested it should. Originally, it had been thought that there was a general equatorward spreading but this cannot be the case if a geostrophic vorticity balance applies. The relation:

$$\beta v = f w_z \tag{11.2}$$

implies that if there is to be upwelling (needed to maintain the thermocline against diffusive effects) then the mean value of v in the deep water has to be positive, i.e., northwards. A way round this embarrassment was proposed by Stommel (1958) in which deep water in the North Atlantic would flow equatorward along the western boundary detraining into the interior as it goes. The interior flow could then be northward and slowly upwards consistent with Eqn. (11.2). This general picture has never been confirmed. Vertical velocities are much too small to measure and so, generally, are mean horizontal velocities in the interior ocean. The picture is further complicated by the interleaving of water masses and topographic effects. But there is evidence in the Atlantic to support the existence of a deep, equatorward flowing, density-driven western boundary current (see Jenkins and Rhines, 1980, for example). In addition to open-ocean convection, deep water may be formed near continental margins—the Weddell and Ross seas being examples of areas where this process occurs. The reader interested further in deep water formation can find excellent reviews by Warren (1981) and Killworth (1983).

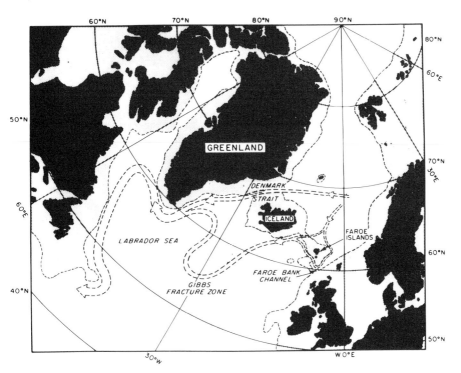

Fig. 11.3. The paths followed by deep water flowing out from the Norwegian sea showing the tendency to form a southward boundary current along the western boundary (from Worthington, 1969).

Figure 11.4 from Ostlund is a graphic illustration of the sinking of dense water in high latitudes.

Deep convection would appear to be of importance to climate on a very long time scale, the time scale of overturning of the ocean (> 100 years), but it may also be so on somewhat shorter time scales in a more indirect manner. Salinity is rather important in high latitudes as density is much more sensitive to salinity changes at low than at high temperatures. There is no atmospheric feedback on salinity as there might be on temperature, suggesting that salinity changes may in fact be cumulative. At the present time (Rooth, personal communication), the ocean between Scotland and the mid-Atlantic ridge is considerably fresher than it was in the Discovery section of 1956 and deep water flow through the Gibbs fracture zone has been cut off. The thermohaline circulation is not yet well understood so the significance of such changes to the baroclinic structure of the ocean have not yet been assessed.

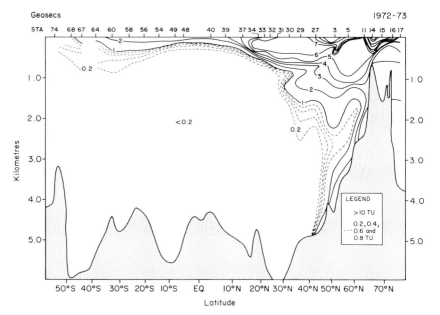

Fig. 11.4. Tritium section in the western North Atlantic suggestive of sinking of northern deep water. Prepared by Rooth and Ostlünd at the University of Miami (see Broecker, 1981).

11.2.4 Climatic changes resulting from increasing atmospheric carbon dioxide

Of concern over the last few years has been the increasing levels of CO_2 in the atmosphere. The time scale of consideration here depends on the rate of increase of CO_2, so we might anticipate changes on the scale of decades to a century, and that for changes on this time scale the atmospheric response is in quasi-equilibrium. It is possible that the ocean will absorb a significant fraction of the extra CO_2 emitted and so directly mitigate the increase. It can also, because of its thermal inertia, reduce or delay the anticipated warming. However, the extent to which it will do so depends on the depth to which the excess heat flux penetrates. It can be anticipated that penetration of thermal anomalies will be inhibited at low latitudes where the ocean static stability is high, and also at high latitudes where the ocean is either buffered from the atmosphere by ice cover or has low-salinity, low-density water near the surface. At intermediate latitudes the penetration depth could be expected to be larger, particularly in the convective regions discussed earlier. These ideas appear to be substantiated in the coupled atmosphere–ocean model of Bryan *et al.* (1982). Figure 11.5 shows a plot of the change in temperature of atmosphere and ocean resulting from a quadrupling of the CO_2, 25 years

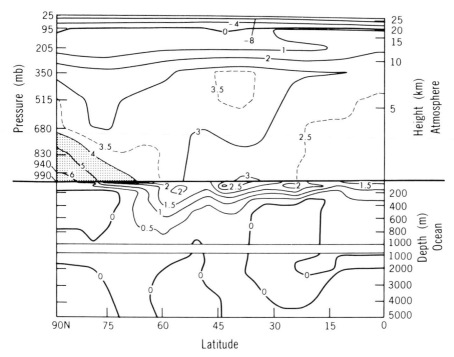

Fig. 11.5. Temperature differences in atmosphere and ocean resulting from a quadrupling of CO_2. The equatorial temperature anomaly is produced quickly and does not penetrate down into the interior because of the shallow strong thermocline (from Bryan *et al.*, 1982).

earlier. One should note, however, that the SST in the tropics increases in a much shorter time than 25 years since only a relatively shallow layer is heated. (The seemingly unreasonably large increase in CO_2 is chosen to try to emphasize changes resulting from CO_2 over those resulting from natural variability in the model.) Studies using an ocean model that transports heat and one that does not, suggest that the changes in the temperature of the atmosphere are substantially less (by more than 10°C in surface temperature in the region around 60°N) where the ocean is permitted to transport heat (Manabe, personal communication). This suggests that previous studies which have not included ocean heat transport may have overestimated the CO_2 effect.

Current knowledge and models of the ocean circulation are not adequate to give definitive answers to the CO_2 problem. Work is required to improve our modelling of sea ice, deep convection and the thermohaline circulation. It is one of the objectives of the WCRP that ocean models be improved to such an extent that they can be used to answer questions related to how the ocean adjusts to changes in external parameters.

11.3 Ocean-atmosphere interaction on short time scales: The extratropics

The ocean appears to be implicated in climatic variations on short (< 10 year) time scales, but separating ocean induced variability from natural variability has been difficult. Some progress, however, has been made in recent years, particularly in implicating the low-latitude ocean. In this section we will discuss the causes and consequences of variability in the extratropics before turning to the dynamics of the equatorial region in Section 11.4.

It has been suggested (Frankignoul and Hasselmann, 1977) that the ocean acts to rectify short time scale (synoptic) atmospheric forcing, generating SST anomalies whose variance grows with time. Such a process can be modelled by an equation of the form:

$$d(hT)/dt = \dot{Q}/\rho C_p \qquad (11.3)$$

where $\dot{Q}$ is the net heating per unit area, T is temperature of the mixed layer, ρ its mean density and h its depth, taken constant for the present. The fact that SST anomalies are limited in magnitude to about $2°C$ suggests the presence of some damping mechanism. One contender is atmospheric *negative* feedback (although the feedback could be positive for shorter periods of time). Negative feedback (used by Frankignoul and Hasselmann) is not the only mechanism for controlling the amplitude of anomalies. A model slightly better than Eqn. (11.3) allows h to vary and augments Eqn. (11.3) with an equation relating changes in potential energy of the mixed layer to wind stirring and convection. Such an equation can be written as:

$$d/dt(\tfrac{1}{2}h^2 T) = M + [h\dot{Q}/\rho C_p]\mathscr{H}(-\dot{Q}) \qquad (11.4)$$

where M describes mechanical mixing by the wind stress that leads to entrainment of fluid into the mixed layer from below, and the second term describes a process whereby potential energy can be destroyed when the ocean is cooled (Gill and Turner, 1976); here $\mathscr{H}$ denotes the Heaviside function. Random heating in this model still induces a runaway SST anomaly, but random stirring does not.

The models represented by Eqns. (11.3) and (11.4) are very simple slab models, that is, a horizontal layer of depth h and temperature excess T overlying a cold deep ocean. In a more elaborate model with a detailed vertical structure, heat can be deposited at depth when the mixed layer is deep, and this process can efficiently control the size of anomalies induced by random heating. In other words, the ocean can restrain the growth of anomalies by internal processes. It follows that while the ocean may respond to high-frequency atmospheric forcing to produce anomalies it is less clear that the atmosphere need respond directly to that anomaly; it does not need to provide negative feedback.

Davis (1976, 1978) has examined SST anomalies and sea level pressure (SLP) anomalies in the North Pacific where Namias (1976) suggests that the ocean causes atmospheric anomalies. Davis's calculations are consistent with the Namias view in that he finds that SST in summer *can* be used as a predictor of SLP in autumn. The problem is that SLP in summer seems to be an equally good predictor. One cannot rule out that both are responses to some other unidentified causal mechanism. At other seasons the correlations are less clear. Using annual means, as opposed to stratifying the data by season, appears to lead to the conclusion that the atmosphere forces the ocean, as the random forcing notion would suggest.

The model of Haney (1980) reproduces much of the structure and evolution of a mid-Pacific ocean temperature anomaly, provided that the dynamics include the atmospheric forcing which produces the anomalies via Ekman currents. Different mixing resulting from the different wind stress might also be important but this process was not investigated. The initial conditions play some part in the evolution in that a subsurface anomaly (up to 400 m) may influence the surface anomaly through advection of mean temperature but bigger changes result from the anomalous forcing. Such experiments indicate the importance of anomalous atmospheric forcing which is in this case coherent rather than random, but do not give any indication as to whether the ocean played any part in producing this anomalous forcing. Because the modelling of the evolution of the SST anomalies requires information on the evolving wind field such studies, like Haney's, cannot be used predictively unless one can predict the evolution of the wind field. Opinion appears to be divided as to whether mid-latitude SST anomalies in meteorological models are associated with significant change in atmospheric circulation. It is at present far from clear how much the mid-latitude ocean affects the atmospheric circulation let alone the extent to which it can be used to predict climate change (Barnett, 1981).

The vertical structure of temperature anomalies appears to be quite varied; some are shallow, some are deep (to about 400 m) but have no surface expression; some are deep with a surface expression. One might anticipate that the random forcing of the mixed layer could produce anomalies only to the depth of the winter mixed layer. Deeper anomalies, perhaps with no surface expression, could result from Ekman pumping or from low-frequency baroclinic wave adjustment (Bryan and Ripa, 1978). Measurements in the North Pacific centred on 40°N suggest that in weak forcing conditions, the thermocline moves in accord with expected Ekman pumping ideas (the depth of the thermocline should decrease under enhanced cyclonic wind stress curl). During the winter months, however, when storm forcing is strongest, the response is opposite to that expected (White *et al.*, 1980). This suggests that horizontal advection may be important and deeper anomalies may move in from higher latitudes where the mixed layer is deeper and hence could have been formed by anomalies in mixed layer depths, perhaps generated stochastically. While the formation of the various kinds of anomalies is unclear, as is their relative frequency of occurrence, one might anticipate that shallower mixed layer anomalies would be the most frequent.

Sufficiently little is known about interannual variations in the gyre structure of either the Pacific or Atlantic ocean to state whether these have any climatic repercussions or not. There are suggestions that the formation of 18°C water in the subtropical gyre of the Atlantic does vary interannually and that the Gulf Stream may respond to this. The winter of 1976–77, which was very cold in the Eastern USA may have been one during which the Gulf Stream transport was consequently increased. The enhanced flow, according to Worthington (1977) is not a result of wind stress forcing, but enhanced buoyancy forcing. During a cold winter, the mixed layer extends deeper in the northern part of the subtropical gyre, perhaps with a difference of some 300 m between cold and mild winter depths, leading to enhanced gradients across the stream. One must caution, however, that calculations of transport in the Gulf Stream are difficult to make and its seasonal changes, even more so its interannual differences, are not well known.

11.4 Ocean–atmosphere interaction on short time scales: The tropics

11.4.1 Introduction

Barnett (1981) has shown that SSTs in the Pacific may be a useful indicator of subsequent changes in surface air temperature anomaly over North America (one to three seasons in advance). The predictability is seasonally as well as geographically variable, with the major predictive ability coming from SSTs in an area near the equator and along the coast of Ecuador. This is an area where SST variations are known to be large and points to a connection with the El Niño phenomenon and the Southern Oscillation. It has been recognized for a long time that the Southern Oscillation has a correlation scale much larger than the Pacific basin, extending westwards into the Indian Ocean and north to the Aleutians and Canada. Studies by Hoskins and Karoly (1981), and Horel and Wallace (1982) have done much to clarify the global extent of the Southern Oscillation, and the teleconnections between the Pacific and distant areas, emphasizing the role of atmospheric Rossby waves in this process. In this section a review will be given of what is known about the role of the ocean in the Southern Oscillation. In Section 11.4.3 the observational description of the cycle will be given. This summary involves aspects of equatorial dynamics and wave theory which may not be familiar to the meteorological community, and so a simple review of relevant aspects of the dynamics is given in Section 11.4.2 together with supporting observations. Section 11.4.4 describbes model results and their contribution to understanding the dynamics of the process.

11.4.2 Simple equatorial dynamics and supportive evidence

The simplest model of equatorial flow is obtained by applying the zonal momentum equation at the equator in the absence of pressure gradients and friction:

$$u_t = \frac{1}{\rho_0} \frac{\partial \tau^x}{\partial z},\tag{11.5}$$

where τ^x is the zonal component of the wind stress. The equatorial thermocline is generally shallow, depth of order 100 m, and sharp, so that the balance implied by Eqn. (11.5) means a layer of 100 m depth can be spun up to ~ 1 m s^{-1} in only 10 days by a windstress of 0.1 N m^{-2}. A more elaborate treatment which includes the dependence of the Coriolis parameter on y shows that for a uniform wind stress the zonal velocity is equatorially confined within a scale which is comparable to that of the equatorial Kelvin wave, roughly 300 km. This jet is commonly referred to as the Yoshida jet.

   If energy can travel westward from a boundary or the edge of the forcing region, it is possible eventually for a different balance to develop, that between the longitudinal pressure gradient p_x and the windstress given by:

$$o = \frac{-p_x}{\rho_0} + \frac{\tau^x}{\rho_0 H},\tag{11.6}$$

where H is the depth of the wind forced layer. Equation (11.6), strictly valid only for a uniform windstress implies that the thermocline develops a slope proportional to τ^x. For a uniform north–south windstress, there is a solution corresponding to Eqn. (11.6) in which the thermocline tilts down to the north for a southerly wind.

   The above balances are unrealistic in some regards because they imply no pressure gradients beneath the wind forced layer and no motion at any depth, not even in the surface layer. If the stress is not uniform in y, the steady solution involves motion with the zonal velocity given by:

$$u_x = -v_y = \frac{1}{\beta H} [\tau_x^y - \tau_y^x]_y\tag{11.7}$$

and this expression, by geostrophy, implies that the pressure gradient varies latitudinally.

The mechanism for changing from Yoshida jet dynamics to the establishment of thermocline tilts, involves baroclinic wave propagation. The most important waves are Kelvin and the lowest equatorial Rossby waves, and we shall see later that such waves play an important role in the remote forcing theory of El Niño. Regrettably, there is at present no compelling *direct* evidence to support Kelvin wave propagation as a precursor to El Niño, but there is a growing amount of circumstantial evidence to support the existence of equatorial waves in other circumstances. First, a notable example was the fortuitous Kelvin wave event in April 1980, when three independent teams of observers [Ericksen, Knox and Halpern (both unpublished) and Ripa and Hayes (1981)] were able to correlate their measurements and to show propagation along the equator over distances of 10 000 km, at a speed of 2.7 m s^{-1}. By the time the signal moved into the shallow thermocline region near Galapagos it did not appear any longer to be a pure Kelvin wave (it was not symmetric about the equator), but was sufficiently similar to greatly increase confidence in the equatorial wave guide as a means of rapid communication in the ocean over long distances. Secondly, Kelvin waves at periods of 1–2 months can also explain the tide gauge records of Luther (1980) who finds coherence over large distances between stations within a radius of deformation of the equator but not with those outside. Thirdly, the existence of significant semi-annual variability in temperature records in the east Pacific where the semi-annual wind variability is small, seems to suggest this signal has propagated from further west where semi-annual variation in the winds is stronger (Kindle, 1979; Meyers, 1979). Finally, the work of Weisberg *et al.* (1979) has documented the existence of antisymmetric wave modes (mixed Rossby gravity waves) in the Atlantic, and there is some evidence for high frequency inertia gravity waves in the Atlantic and Pacific, though some ambiguity in interpretation still remains. There has been considerable work on this end of the spectrum but it cannot be reviewed here. From the above (and from additional evidence to be presented later), it would appear that the equatorial wave guide is real and that Kelvin waves at least can propagate great distances along it (almost the whole width of the Pacific). It is remarkable that this should be so in the presence of dissipative processes, strong shears and tilting thermoclines.

Quantitative verification of the Yoshida jet is not easy (nor strictly to be expected) but there is support for a current in the middle eastern Indian Ocean with a scale of $\sim 3°$ in spring and autumn (Wyrtki, 1973), and its presence has been interpreted as verification of Eqn. (11.5) (O'Brien and Hurlburt, 1974). Because the current exists for a month or two, this is long enough for wave propagation to at least partially establish the conditions expressed by Eqn. (11.7). However, it is likely that it is a mixture of the conditions corresponding to Eqns. (11.5) and (11.7) which is really being observed with the meridional scale set by both the curvature of the wind and the radius of deformation.

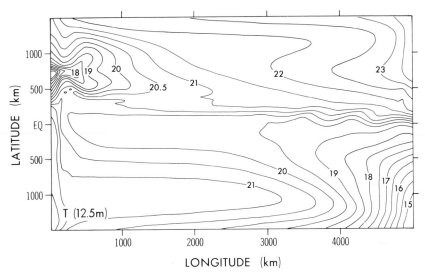

Fig. 11.6. Plot of surface temperature in response to a uniform wind from the south (from Philander and Pacanowski, 1980).

The pressure gradient given by Eqn. (11.6) is strictly valid only for the case of uniform wind stress, but for most wind profiles, provided the wind is relatively uniform within a range of a few degrees of the equator, the thermocline can be expected to slope as suggested by Eqn. (11.6). The time required to establish this slope is about the time it takes a Kelvin mode and the lowest equatorial Rossby mode to propagate across the forcing region. Their respective wave speeds are $c \sim 3 \text{ m s}^{-1}$ and 1 m s^{-1}, and so the time scale for adjustment is typically a few weeks.

There is evidence that the thermocline slopes from east to west in all oceans. In the Indian Ocean the thermocline is tilted downwards to the east during May and June and again in November, and the winds are westerlies at around this time (Wyrtki, 1973; Eriksen, 1979). In the Atlantic the $23°C$ isotherm slopes upwards to the east in July–December, and this slope nearly vanishes from January–June, these properties are consistent with the increased easterly wind along the equator in summer (Merle, 1980). In the Pacific west of $95°W$ the zonal wind stress is westward and strongest during June–August (Tsuchiya, 1979; Wyrtki, 1975; Wyrtki and Meyers, 1976), and the zonal pressure gradient appears to be strongest (Tsuchiya, 1979) in June–July. In addition, the difference in sea level between Canton island ($3°S, 172°W$) and Talara ($5°S$ on the west coast of South America) increases between June and October, decreases between December and February and remains low till the following June. Talara, being a coastal station, will be affected by coastal upwelling induced by the longshore wind. In the far eastern Pacific the wind stress has a mean southerly component, and therefore should induce an

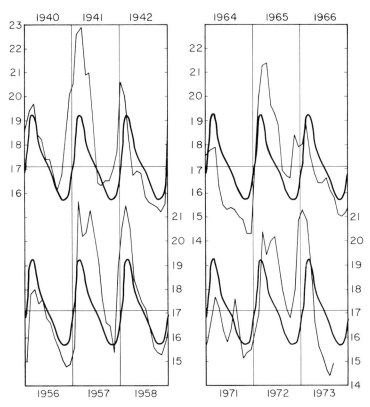

Fig. 11.7. Sea-surface temperature at Puerto Chicama (7°S) on the coast of South America averaged over four El Niño events (thin line). The annual mean cycle is denoted by the heavy line. The time of occurrence of El Niño appears to be related to the annual cycle (from Wyrtki, 1975).

asymmetry in the thermal structure between hemispheres, with isotherms deepening from south to north. Something like this tilt does exist in the Pacific and also occurs in the model results of Philander and Pacanowski (1980) (see Fig. 11.6). This asymmetry will later be invoked to explain the observed asymmetry in the temperature anomalies during El Niño.

11.4.3 Interannual variability in the equatorial Pacific

There are several indicators of El Niño, but attention here will be focused on near surface temperature and its change. A rapid rise in surface temperature occurs in El Niño years over large areas of the Pacific Ocean— along the coast of Ecuador and Peru and in a strip along the equator.

The excessive warming in the SH coastal band occurs during a period when the surface would in any case be warming by solar insolation and thus appears closely tied to the annual cycle (Fig. 11.7), but the excess warming is not likely to be primarily a consequence of this local forcing. Recent El Niño events occurred in 1957–58, 1965, 1972–73 and possibly 1976. Figure 11.8 shows the changes that occurred before, during and after the El Niño event of 1972. The figure illustrates that the scale of the temperature anomalies is comparable with the equatorial radius of deformation, that anomalously warm water can exist for long distances along the equator at least as far west as Canton Island (172°W, 3°S), and that changes in the east are large ($\sim 5°C$) between warm and cold years but smaller in the west.

It is useful to describe El Niño as consisting of three phases. In the initial phase, warm water appears quite suddenly along the coast of Ecuador and Peru. It appears to be of equatorial origin and to advect along the coast. The intermediate phase is one of slow damping of the anomalies in the east but a westward spread along the equator at a speed in the range 50–100 cm s^{-1} (Hickey, 1975; Rasmusson and Carpenter, 1982). In the third phase, warm water quickly reappears along the coast of Ecuador and Peru usually about a year later. This third phase does not always occur, and when it does it occurs only at coastal locations. In addition to the larger events which occur once every 6–10 years, smaller events occur roughly every 2–3 years, at least over the last few decades. It is as if the mechanism is operating on a time scale of about 3 years, but every so often everything acts in concert and a bigger event occurs. The priming mechanism may even operate on the scale of months, but only at certain times, perhaps, does the atmosphere respond sympathetically and develop a strong anomaly. There is an indication (Rasmusson and Carpenter, 1982) that there are changes in the windfield and SST centred on 90°W, 25°S a few months before the coastal warming along Ecuador and Peru, but whether this is significant or not is as yet unclear.

At one time it was thought that the occurrence of high SST along the coastal strip resulted from reduced coastal upwelling, which would occur if the alongshore (equatorward) component of the wind decreased. Wyrtki (1975), however, showed that coastal winds do not decrease, and Enfield (1980) showed that they may even increase creating stronger upwelling. This does not imply a lowering of SST since, as will be shown later, the upwelled water is warmer as a result of changes to the subsurface stratification. Hickey (1975) and Wyrtki (1975) showed that a weakening of the trade winds occurred far to the west (west of 140°W). This observation led to the idea of a remote forcing of El Niño, via an equatorial Kelvin wave.

In fact, Wyrtki (1975) suggests that sea level stands higher in the West Pacific preceding an El Niño, as a consequence of stronger easterly winds. (Stronger winds do often appear to precede El Niño but perhaps not always and are probably not essential to the process.) El Niño appears to

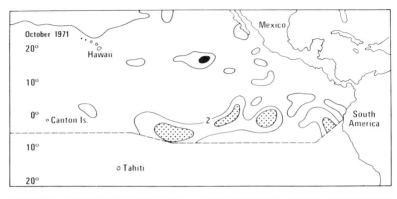

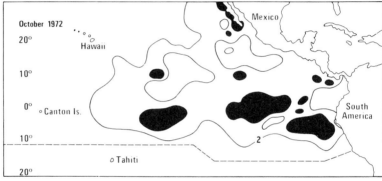

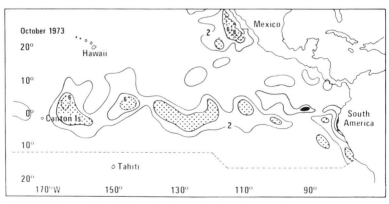

Fig. 11.8. SST anomalies before, during and after the El Niño which started in early 1972, showing the equatorial scale of the anomalies (from Fletcher, 1979). Cold and warm anomalies of more than 4°F are indicated by stippling and shading, respectively.

be triggered when the winds suddenly and mysteriously reduce in the West Pacific. Theory suggests that an equatorial Kelvin wave is locally forced and then propagates rapidly eastward until it reaches the eastern boundary. This wave acts to deepen the thermocline in the eastern equatorial ocean. As noted earlier, the thermocline slopes upward towards the east and at certain times of the year meets or is very close to the surface in the east. This suggests that this area is quite sensitive to changes in thermocline depth—small changes are able to give rise to significant temperature anomalies. The wind is, however, not meridionally uniform and this leads to north–south slopes of the thermocline and isotherms superimposed on the east–west gradient (see, e.g., Tsuchiya, 1972).

A Kelvin wave propagating eastwards along the equator following a relaxation in the Trade winds would therefore be expected to lower the thermocline and so induce surface temperature changes (warm anomalies), but there is no indication of this happening. Warm temperatures are usually first observed at the coast of South America, not along the equator (1982–3 was an exception). Lowering of the thermocline would also be expected to show up in sea level data, but Hickey's data (1975) at Christmas Island (2°N, 157°W) do not show any convincing indication of the expected increase in sea level. (On two occasions there was a rise at about the expected time, but the rise is smaller than might have been expected.) A possible explanation for this for the 1972 El Niño has been given by Gill (personal communication) involving a combination of Kelvin waves and reflected Rossby waves but relies on a coincidence of the waves travelling in opposite directions. This imposes restriction on the time variation of the wind stress which may not hold in general.

When the thermocline is lowered in the eastern ocean, possibly by a remotely forced Kelvin wave, theory suggests that it must propagate polewards along eastern boundary in both hemispheres as a coastal Kelvin wave (Anderson and Rowlands, 1976). Again one would anticipate changes in SST. Because of the asymmetry in the mean thermal structure between hemispheres, one also expects an asymmetry in SST response, and it would be of a sign such that the largest SST changes occur in the Southern Hemisphere. This asymmetry also appears to be consistent with the data (Rasmusson and Carpenter, 1982). Such a depression affects not only the SST but also sea level. Long tide-gauge records exist along the west coasts of both North and South America and these have been examined by Enfield and Allen (1980). These authors found a large-scale correlation between stations in the Northern Hemisphere and in the Southern Hemisphere. The data were monthly means and therefore not very suitable for detection of propagation at speeds comparable to that of Kelvin waves, i.e., $2°$ day^{-1} or $60°$ month^{-1}, but as far as the authors could tell, there was a lag between poleward stations and equatorial ones, consistent with coastal wave propagation. Such a correlation could also be expected if there were any coherent large-scale atmospheric adjustment but Enfield and Allen conclude that this is not so and favour the wave propagation theory. The signal (in sea level) may even be less attenuated in the Northern Hemisphere and propagate further than it does in the

Southern Hemisphere.

Although observations do not show the expected lowering of the thermo-cline at Canton, there are observations which show large changes in subsurface thermal structure near the eastern boundary. This evidence is largely due to the sequence of cruises run in 1971–73 by the Instituto Oceanográfico de la Armada del Ecuador (INOCAR) and Instituto del Mar del Perú (IMARPE). Pre El Niño (November–December 1971), the thermocline was shallow and sharp with a weak thermal gradient beneath. By November and December 1972 the thermocline was much more diffuse, the 15° isotherm having been depressed about 200 m. This broadening of the stratified zone appears to propagate poleward along the eastern boundary, to be most marked in the near surface zone to about 100 m depth and to occur quite rapidly, large changes being observed between November–December 1971 and February–March 1972. This is consistent with Wyrtki's result who used the depth of the 15° isotherm to interpret sea level, assuming that the two fields were correlated. The results of Huyer (1980), however, show deeper changes though at different space and time scales. The data showing poleward propagation of stratification changes are consistent with a similar pro-pagation of sea level changes.

Details of the processes involved have not been worked out: the INOCAR/IMARPE sections are not of sufficient duration to allow an extraction of the seasonal cycle from the data. Nor is it known how the changes in the weak thermal gradient affect or result from changes in the undercurrent. Changes in the east–west slope should influence the undercurrent con-siderably, but there are few data with which to assess this.

11.4.4 Results from models

Models using only one active layer (i.e., neglecting the undercurrent and thermostad) have been used to illustrate how a change in the wind stress in the mid-to-west Pacific can change the thermocline depth basin width. Figure 11.9 from McCreary (1976) shows the adjustment of the model thermocline topography after a weakening of the westward trade winds by an amount equivalent to a stress of $0.05 \, \text{N m}^{-2}$. The dotted lines in the upper left panel of the Figure delineate the region of weakened trade winds; the wind stress change is a maximum in the central portion and weakens markedly in the surrounding areas. After 35 days, a downwelling signal rapidly radiates from the region of the wind patch as a packet of equatorially trapped Kelvin waves and already the equatorial thermocline (and also sea level) begins to tilt to generate a zonal pressure gradient that balances the wind there. In the region of strongest wind curl, the thermocline begins to move closer to the surface, and this upwelling signal propagates westward as a packet of Rossby

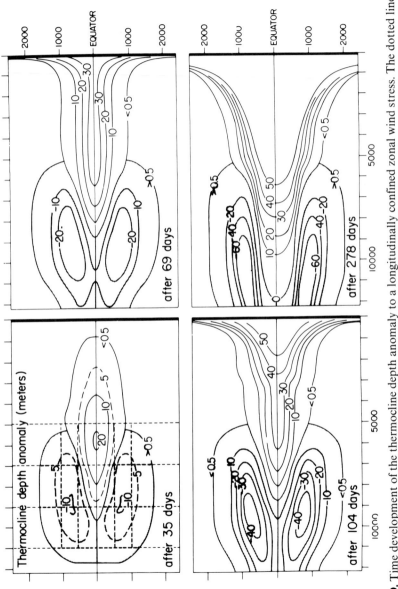

Fig. 11.9. Time development of the thermocline depth anomaly to a longitudinally confined zonal wind stress. The dotted lines in the upper left panel indicate the region of the wind. The solid lines in each panel indicate the presence of an eastern ocean boundary. Horizontal distances are in kilometres (after McCreary, 1976).

waves. After 69 days the packet of Kelvin waves has already reflected from, and spread along, the eastern boundary. As time passes, this downwelling signal propagates back into the ocean interior as a packet of Rossby waves, and more and more upper-layer water piles up in the eastern ocean. As shown in the lower two panels, the source of the water is a broad area of the western and central ocean where the thermocline is uplifted; in some regions the thermocline rises over 65 m. Note that in this case it is the arrival of the Kelvin wave that causes the initial deepening of the pycnocline in the eastern ocean.

Although such calculations are interesting, we must ask whether they are simulating the real course of events. The calculations are predicated on the observation by Wyrtki that the SE trades weaken dramatically before El Niño occurs off Peru. Rasmusson and Carpenter (1982) also find evidence for a weakening of the zonal wind stress in the west before Peruvian warming, as assumed by McCreary. Other models with the assumption of increased easterlies in the year preceding El Niño have been run (Hurlburt *et al.*, 1976; Kindle, 1979). In such calculations, the ocean is first primed by running the model with easterlies prior to switching them off. The development is similar to McCreary's though there is some change in detail. A double peak in the temperature at the coast is detected in the model of Kindle but this relies very heavily on timing between wave pulses and reflection from the western boundary (whose reflective characteristics must be assumed to be uncertain because of the complicated geometry and topography).

Modelling support for Kelvin wave influence on the eastern equatorial region is provided by Busalacchi and O'Brien (1982) who have run a fifteen-year simulation using Wyrtki–Meyers wind data for the period 1961–77. Figure 11.10(a) shows the sea-level anomaly at Galapagos. The model is a simple reduced gravity model with the thermocline depth being the nearest model analogue of observed sea level. The model thermocline depth anomaly is shown in Fig. 11.10(b). Although one cannot always relate thermocline depth variations to sea level variations, at low frequency this may be possible. Given the inadequacy of the wind data and the simplicity of the model, the agreement is surprisingly good.

11.4.5 Discussion

The models discussed in the previous section were all adiabatic. Further, changes in SST did not induce any variation in the atmospheric circulation. This is an obvious deficiency that can be rectified by using coupled models. That changes in the winds can occur following El Niño has been indicated both by observational studies (Bjerknes, 1969; Rasmusson and Carpenter, 1982) and atmospheric model studies (Rowntree, 1972; Keshavamurty, 1982;

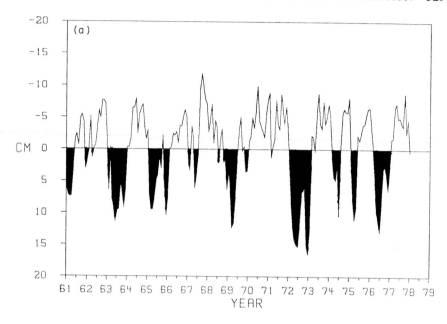

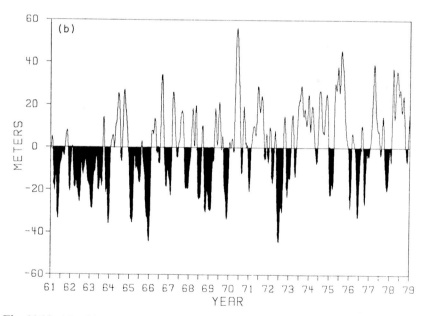

Fig. 11.10. (a) Observed sea-level anomaly at Galapagos (from Busalacchi O'Brien, 1982). (b) Model pycnocline height anomaly at Galapagos (from Busalacchi and O'Brien, 1982).

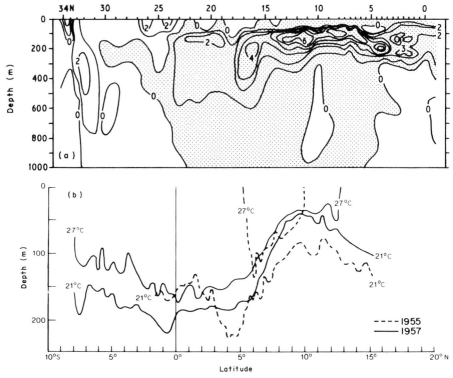

Fig. 11.11. (a) Latitudinal *v.* depth plot of temperature anomaly along 137°E in winter of 1973 (from Asai, 1979). (b) Meridional sections of temperature across the central equatorial Pacific during cold (1955) and warm (1957) events (from Barnett, 1978, based on data by Austin, 1960).

Shukla and Wallace, 1983). There is evidence too to suggest that such wind changes can then have a significant effect on the ocean circulation.

Attention has been drawn by Gill (1979) to the very large drop in sea level at Guam in October 1972. Meyers and White (1979) have shown that this drop in sea level corresponds to an uplifting of the thermocline by 100 m. The most likely way for this to occur is via Ekman pumping—increased vertical velocity induced by the stronger than normal wind stress curl. Recent calculations by Kutsuwada (personal communication) support this. To verify this requires additional wind data at a resolution good enough to define at least the wind stress curl and possibly the curvature. Finally, to emphasize the thermocline change, the difficulty of interpreting SST dynamically, and the merit of long temperature sections, Fig. 11.11(a) shows the subsurface anomalous temperature structure at this time. SST changes were virtually zero, yet at 150 m the temperature was 6° colder. The scale is non-equatorial but more comparable with the scale one might expect of the atmospheric wind field. However, it is quite likely that the anomalous winds induce a non-equatorial response. Figure 11.11(b) shows the depths of two isotherms in the Pacific during cold and warm events on a different occasion, indicating subsurface temperature changes of 6°C. Though not in phase over the whole range, the isotherm displacements to 15° of the equator, are again much larger than the equatorial radius of deformation.

The results of the model studies in Section 11.4.4 suggest that the Kelvin wave mechanism for eastward communication may be reasonably correctly modelled in linear and adiabatic models. It is not so clear that the mechanism for westward movement of the SST anomaly following El Niño can be represented in such models since it is not self evident that Rossby wave propagation need be the key component. Advection in the South equatorial current is possible. The currents are suitable for advecting warm anomalies from the eastern region westward, particularly in the eastern half basin. But it is also possible that the oceanic response is not a free wave but a forced response. A warm SST anomaly would lead to reduced winds and hence upwelling (west) of the anomaly and this can lead to westward progression. Only ocean models with all these processes modelled can help determine the relative importance of the various mechanisms which may vary from place to place.

Adiabatic models have been very useful in helping to clarify many of the relevant aspects of wave dynamics but the time is now ready for baroclinic effects and heat sources and sinks to be included. Only by using such models can we begin to quantify the relative importance of the various mechanisms responsible for the amplification and westward propagation of the thermal anomaly.

11.4.6 Annual and interannual variation in the Atlantic and Indian Oceans

It is when the anomaly is located in the central Pacific and at its most extensive, that the largest changes to the atmospheric circulation occur. This involves large changes to the convection over Indonesia and New Guinea. The anomalous atmospheric heating does not appear to be local to the SST anomaly, i.e., is not located over the maximum SST anomaly but rather seems to prefer the region of maximum SST presumably because it is here that the potential for latent heat release is greatest. (Wallace, personal communication). There are also indications that SST in the Pacific may also have a predicative capability for monsoon conditions over India. Rasmusson and Carpenter (1983) have shown a strong correlation between the occurrence of El Niño and the strength of the Indian Monsoon a few months later. El Niño conditions usually precede a weak monsoon although El Niño is not on every occasion associated with reduced precipitation. Other factors must also be able to modulate monsoon rainfall but the association with El Niño is

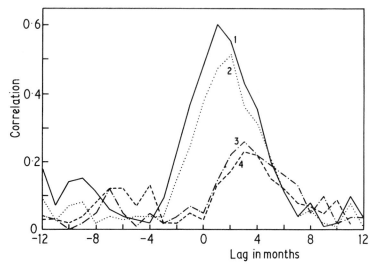

Fig. 11.12. Plot of the correlation between the wind stress near Brazil and (1) SST in the equatorial Gulf of Guinea, (2) SST along the Northern Guinea coast. Curves (3) and (4), which correspond to more local winds, show a smaller correlation. The correlation is maximum for curve (1) at about 1 month lag, which would be consistent with eastward Kelvin wave propagation at a speed of between 1 and 1.5 m s^{-1}. The lag is greater in the case of (2) since the wave has to propagate further. From Servain *et al.* (1982).

marked: 20% of El Niño's were accompanied by *catastrophic* drought (rainfall anomalies of two or more standard deviations) and 44% are associated with anomalies of above standard deviation. Exactly how this takes place is unclear but changes in the low-level jet, the Somali current, and Arabian Sea cooling are plausible contenders. Shukla and Misra (1977) have indicated a connection between Arabian Sea temperatures and Indian rainfall from model studies, but this link is not universally accepted. Space does not permit a fuller discussion of Indian Ocean dynamics. Reference may be made to a review of the low-level jet by Anderson (1980), articles on the Somali current by various authors in *Science* (1981) and *Ocean Modelling* (Delacluse and Philander, 1981; Anderson, 1981) and a review by Anderson (1979).

The idea of 'remote' forcing in low latitude oceans appears to have had its recent origins in the paper by Lighthill (1969), who proposed the Somali current was driven by winds along the equator. This may be partly so in the near equatorial region, but most of the Somali current response appears to be local. In fact, seasonal changes in wind stress suggest that the Atlantic is in principle a more likely ocean for remote forcing of western boundary currents. Model studies (Anderson, 1983) in fact suggest that 30% of the amplitude of the North Brazilian coastal current could be attributed to remote winds and this may also hold for the seasonal response. To date there are no data to support this contribution to the western boundary current from westward propagating Rossby waves, and the amplitude of the seasonal variation of the near surface flow of the North Brazilian coastal current is not well known either. Confirmation of Rossby wave propagation is thus lacking, though as noted in Section 11.4.2 there are changes in slope of the thermocline consistent with wind stress which could be established by Rossby waves. Gonella *et al.* (1981) interpret their buoy motions in the Indian Ocean in terms of Rossby waves effecting a change from Yoshida jet dynamics to thermocline slope.

There is support for equatorial Kelvin wave propagation in the Atlantic, however. Figure 11.12 from Servain *et al.* (1982) shows a correlation between the zonal wind stress in the western equatorial Atlantic and SST in the eastern equatorial region and along the Guinea coast, with a phase lag consistent with eastward propagation. This suggests that some of the upwelling in the eastern Atlantic and Gulf of Guinea is forced by changes in the wind in the Western equatorial Atlantic.

Not all of the eastern Atlantic upwelling need be remotely forced, however. There appears also to be a local response in the equatorial region (Voituriez, 1981; Anderson, 1983), and this bears a certain similarity to the temperature field in the Pacific upwelling in both occurring at times of strong cross-equatorial winds.

References

ANDERSON, D. L. T. (1979). Basin models: The general circulation of regions of the World ocean. *Dyn. Atmos. Oceans*, **3**, 345–371.

ANDERSON, D. L. T. (1980). Orographically controlled cross equatorial flow. In *Orographic Effects in Planetary Flows* (R. Hide and P. White, Eds), Global Atmospheric Research Program. Publ. 23.

ANDERSON, D. L. T. (1981). The Somali current. *Ocean Modelling*, **34**.

ANDERSON, D. L. T. (1982). Low latitude seasonal adjustment in the Atlantic (in press).

ANDERSON, D. L. T. and ROWLANDS, P. B. (1976). The role of inertia gravity and planetary waves in the response of a tropical ocean to the incidence of an equatorial Kelvin wave on a meridional boundary. *J. Mar. Res.*, **34**, 295–312.

ASAI, T. (1979). A report of existing and future activities relevant to the POMS in Japan. JOC/SCOR planning meeting for Pilot Ocean Monitoring Study, Miami 1979.

AUSTIN, T. (1960). In *Symposium on the Changing Pacific Ocean 1957 and 1958*. California Co-operative Fisheries Investigation, vol. 7, pp. 52–55.

BARNETT, T. P. (1978). The role of the oceans in the global climate system. In *Climatic Change* (J. Gribbin, Ed.), Cambridge University Press, 280 pp.

BARNETT, T. P. (1981). Statistical prediction of North American air temperatures from Pacific predictions. *Mon. Weath. Rev.*, **109**, 1021–1041.

BENNET, A. F. (1978). Poleward heat fluxes in the Southern Hemisphere oceans. *J. Phys. Ocean.*, **8**, 785–798.

BJERKNES, J. (1969). Atmospheric teleconnections from the equatorial Pacific. *Mon. Weath. Rev.*, **97**, 163–172.

BROECKER, W. S. (1981). Geochemical tracers and ocean circulation. In *Evolution of Physical Oceanography* (B. A. Warren and C. Wunsch, Eds), M.I.T. Press, 620 pp.

BRYAN, K. (1962). Measurements of meridional heat transport by ocean currents. *J. Geophys. Res.*, **67**, 3403–3413.

BRYAN, K. (1982). Seasonal variation in meridional overturning and poleward heat transport in the Atlantic and Pacific Oceans. *J. Mar. Res.*, **40** (supplement).

BRYAN, K. and LEWIS, I. J. (1979). A watermass model of the World ocean. *J. Geophys. Res.*, **84**, 2503–2517.

BRYAN, K. and RIPA, P. (1978). The vertical structure of North Pacific temperature anomalies. *J. Geophys. Res.*, **83**, C5, 2419–2429.

BRYAN, K., KOMRO, F. G., MANABE, S. and SPELMAN, M. J. (1982). Transient climate response to increasing atmospheric carbon dioxide. *Science*, **215**, 56–58.

BRYDEN, H. L. and HALL, M. M. (1980). Heat transport by ocean currents across 25°N latitude in the Atlantic Ocean. *Science*, **207**, 884–886.

BURRIDGE, D. (1982). In *The 'CAGE' Experiment. A Feasibility Study*. UNESCO JSC/CCCO Liaison Panel.

BUSALACCHI, A. J. and O'BRIEN, J. J. (1982). Interannual variability of the equatorial Pacific in the 1960s. *J. Geophys. Res.*, **86**, 10 901–10 907.

DAVIS, R. E. (1976). Predictability of sea level pressure and sea surface temperature over the North Pacific. *J. Phys. Ocean*, **6**, 249–266.

DAVIS, R. E. (1978). Predictability of sea level pressure anomalies over the North Pacific Ocean. *J. Phys. Ocean.*, **8**, 233–246.

DELACLUSE, P. and PHILANDER, S. G. H. (1981). The Somali current. *Ocean Modelling*, **36**.

ELLIS, H. (1751). A letter to the Rev. Dr. Hales, FRS. *Trans. R. Soc. Lond.*, **47**, 211–214.

ENFIELD, D. B. (1980). El Niño Pacific eastern boundary response to interannual forcing. Chapter 8 in *Resource Management and Environmental Uncertainty* (Michael H. Glantz, Ed.), John Wiley & Son.

ENFIELD, D. B. and ALLEN, J. S. (1980). On the structure and dynamics of monthly mean sea level anomalies along the Pacific coast of North and South America. *J. Phys. Ocean.*, **10**, 557–578.

ERIKSEN, C. C. (1979). An equatorial transect of the Indian Ocean. *J. Mar. Res.*, **37**, 215–232.

FLETCHER, J. O. (1979). An ocean climate research plan NOAA. Environmental Research Lab., Boulder, Colorado.

FRANKIGNOUL, C. and HASSELMANN, H. (1977). Stochastic climate models, Part II. *Tellus*, **29**, 289–305.

FU, L. L. (1981). General circulation and meridional heat transport of the subtropical South Atlantic determined by inverse methods. *J. Phys. Ocean.*, **11**, 1171–1193.

GILL, A. E. (1979). Why did the sea go out at Guam? *Ocean Modelling*, **20**.

GILL, A. E. and TURNER, J. S. (1976). A comparison of seasonal thermocline models with observations. *Deep Sea Res.*, **23**, 391–401.

GONELLA, J., FIEUX, M. and PHILANDER, S. G. H. (1981). Mise en évidence d'ondes de Rossby équatoriales dans l'Océan Indien au moyen de bonées dérivantes. *Comptes Rendus*, *T292*, Série II, 1397–1399.

HANEY, R. (1980). A numerical case study of the development of large-scale thermal anomalies in the central North Pacific Ocean. *J. Phys. Ocean.*, **10**, 541–556.

HASTENRATH, S. (1980). Heat budget of tropical ocean and atmosphere. *J. Phys. Ocean.*, **10**, 159–170.

HICKEY, B. (1975). The relationship between fluctuations in sea level, wind stress and sea surface temperature in the equatorial Pacific. *J. Phys. Ocean.*, **5**, 460–475.

HOLLAND, W. R. and RHINES, P. B. (1980). An example of eddy-induced ocean circulation. *J. Phys. Ocean.*, **10**, 1010–1031.

HOREL, J. D. and WALLACE, J. M. (1982). Planetary scale atmospheric phenomena associated with the interannual variability of sea surface temperature in the equatorial Pacific. *Mon. Weath. Rev.*, **109**, 813–829.

HOSKINS, B. J. and KAROLY, D. J. (1981). The steady linear response of a spherical atmosphere to thermal and orographic forcing. *J. atmos. Sci.*, **38**, 1179–1196.

HURLBURT, H. E., KINDLE, J. C. and O'BRIEN, J. J. (1976). A numerical simulation of the onset of El Niño. *J. Phys. Ocean.*, **6**, 621–631.

HUYER, A. (1980). The offshore structure and subsurface expression of sea level variations off Peru 1976–77. *J. Phys. Ocean.*, **10**, 1755–1768.

JENKINS, W. J. and RHINES, P. B. (1980). Tritium in the deep North Atlantic Ocean. *Nature*, **286**, 877–879.

KESHAVAMURTY, R. N. (1982). Response of the atmosphere to sea-surface temperature anomalies over the equatorial Pacific and teleconnections of the Southern Oscillation. *J. atmos. Sci.*, **39**, 1241–1259.

KILLWORTH, P. D. (1983). Deep convection in the World's oceans. *Rev. Geophys. and Space Physics* (in press).

KINDLE, J. C. (1979). Equatorial Pacific ocean variability—seasonal and El Niño time scales. Ph.D. Thesis, Florida State University.

LAU, N-C. (1981). A diagnostic study of recurrent meteorological anomalies appearing in a 15-year simulation with a GFDL General Circulation Model. *Mon. Weath. Rev.*, **109**, 2287–2311.

LIGHTHILL, M. J. (1969). Dynamic response of the Indian Ocean to the onset of the southwest monsoon. *Phil. Trans. R. Soc. Lond.*, **A265**, 45–92.

LUTHER, D. S. (1980). Observations of long period waves in the tropical oceans and atmosphere. Ph.D. Thesis, M.I.T./W.H.O.I., 210 pp.

MCCREARY, J. P. (1976). Eastern tropical ocean response to changing wind systems: with application to El Niño. *J. Phys. Ocean.*, **6**, 632–645.

MARSIGLI, L. M. (1681). Osservazioni intorno al Bosforo Tracio o vero. Canale di Constantinopli, rappresentate in lettera alla Sacra Real Maestà Cristina Regina di Svezia, Roma. Reprinted in *Bolletino di pesca, di piscicoltura e di idrobio logia*, **11** (1935), 734–758.

MAURY, M. F. (1844). Remarks on the Gulf Stream and currents of the sea. *Amer. J. Sci. Arts*, **47**, 161–181.

MERLE, J. (1980). Seasonal heat budget in the Equatorial Atlantic Ocean. *J. Phys. Ocean.*, **10**, 464–469.

MEYERS, G. and WHITE, W. (1979). El Niño in the West Pacific. *Ocean Modelling*, **23**.

MEYERS, G. (1979). Annual variation in the slope of the 14° isotherm along the equator in the Pacific Ocean. *J. Phys. Ocean.*, **9**, 885–891.

NAMIAS, J. (1976). Negative ocean-air feedback systems over the North Pacific in the transition from warm to cold seasons. *Mon. Weath. Rev.*, **104**, 1107–1121.

O'BRIEN, J. J. and HURLBURT, H. E. (1974). Equatorial jet in the Indian Ocean: Theory. *Science*, **184**, 1075–1077.

OORT, H. A. and VONDER HAAR, T. H. (1976). On the observed annual cycle in the ocean-atmosphere heat balance over the Northern Hemisphere. *J. Phys. Ocean.*, **6**, 781–800.

PHILANDER, S. G. H. and PACANOWSKI, R. C. (1980). The generation and decay of equatorial currents. *J. Geophys. Res.*, **85**, 1123–1131.

RASMUSSON, E. M. and CARPENTER, T. H. (1982). Variations in tropical sea surface temperature and surface wind fields associated with the southern oscillating El Niño. *Mon. Weath. Rev.*, **110**, 354–384.

RASMUSSON, E. M. and CARPENTER, T. H. (1983). The relationship between eastern equatorial Pacific sea-surface temperature and summer monsoon rainfall over India. (In press.)

RIPA, P. and HAYES, S. (1981). Equatorially trapped waves in the Galapagos Array. *J. Geophys. Rec.*, **86**, 6509–6516.

ROEMMICH, D. (1980). Estimation of meridional heat flux in the North Atlantic by inverse methods. *J. Phys. Ocean.*, **10**, 1972–1983.

ROEMMICH, D. (1981). Meridional transport of heat, fresh water, oxygen and silicate in the South Pacific Ocean (in press).

ROWNTREE, P. R. (1972). The influence of tropical East Pacific Ocean temperatures on the atmosphere. *Q. Jl R. met. Soc.*, **98**, 290–321.

RUMFORD, B., COUNT (1800). Essay VII, The propagation of heat in fluids. In *Collected Works of Count Rumford* (S. C. Brown, Ed.), vol. 1, *The Nature of Heat*, 1968, Harvard University Press, Cambridge, pp. 117–285.

SABINE, E. (1846). On the cause of remarkably mild winters which occasionally occur in England. *Phil. Mag.*, **28**, 317–324.

SERVAIN, T., PICAUT, J. and MERLE, J. (1982). Evidence of remote forcing in the equatorial Atlantic. *J. Phys. Ocean.*, **12**, 457–463.

SHUKLA, J. and MISRA, B. M. (1977). Relationships between sea-surface temperature and wind speed over the Arabian Sea and monsoon rainfall over India. *Mon. Weath. Rev.*, **105**, 998–1002.

SHUKLA, J. and WALLACE, J. M. (1983). Numerical simulation of the atmospheric response to equatorial Pacific sea-surface temperature anomalies (in preparation).

STOMMEL, H. (1958). The abyssal circulation. *Deep Sea Res.*, **5**, 80–82.

STOMMEL, H. and CSANADY, G. T. (1980). A relation between the TS curve and global heat and atmospheric water transports. *J. Geophys. Res.*, **85**, 495–501.

TSUCHIYA, M. (1972). A subsurface north equatorial countercurrent in the Eastern Pacific Ocean. *J. Geophys. Res.*, **77**, 5981–5986.

TSUCHIYA, M. (1979). Seasonal variation of the equatorial zonal geopotential gradient in the eastern Pacific Ocean. *J. Mar. Res.*, **37**, 399–407.

VOITURIEZ, B. (1981). *Equatorial Upwelling in the Eastern Equatorial Atlantic*. Report of final meeting of SCOR Working Group 47, Venice. Edited by McCreary, Moore and Witte. Nova University Press, 466 pp.

VONDER HAAR, T. H. and OORT, A. H. (1973). New estimate of annual poleward energy transports by Northern Hemisphere oceans. *J. Phys. Ocean.*, **3**, 169–172.

WARREN, B. A. (1981). Deep circulation of the World ocean. In *Evolution of Physical Oceanography* (B. A. Warren and C. Wunch, Eds), M.I.T. Press, 620 pp.

WEISBERG, R. H., HORIGAN, A. and COLIN, C. (1979). Equatorially trapped Rossby-gravity wave propagation in the Gulf of Guinea. *J. Mar. Res.*, **37**, 67–86.

WHITE, W., BERNSTEIN, R., MCNALLY, G., PAYAN, S. and DICKSON, R. (1980). The thermocline response to transient atmospheric forcing in the interior mid-latitude North Pacific 1976–78. *J. Phys. Ocean.*, **10**, 372–384.

WORTHINGTON, L. V. (1969). An attempt to measure the volume transport of Norwegian Sea overflow water through the Denmark Strait. *Deep Sea Res.*, **16** (supplement), 421–432.

WORTHINGTON, L. V. (1977). Intensification of the Gulf Stream after the winter of 1976–77. *Nature*, **270**, 415–417.

WYRTKI, K. (1973). An equatorial jet in the Indian Ocean. *Science*, **181**, 262.

WYRTKI, K. (1975). El Niño—the dynamic response of the equatorial Pacific Ocean to atmospheric forcing. *J. Phys. Ocean.*, **5**, 572–584.

WYRTKI, K. and MEYERS, G. (1976). The trade wind field over the Pacific Ocean. *J. appl. Met.*, **15**, 698–704.

— 12 —

Medium-range weather prediction— operational experience at ECMWF

L. BENGTSSON and A. J. SIMMONS

12.1 Introduction

In his introduction to the First Conference on Climate Modelling, held in Princeton, 1955, John von Neumann outlined an overall strategy in atmospheric modelling and prediction (Pfeffer, 1960). He considered that prediction problems could conveniently be divided into three different categories depending on the time scale of the forecast. In the first was the short-range prediction of motions that are determined mainly by the initial state of the atmosphere. The second comprised much longer-term predictions of characteristics of the motion that are largely independent of the initial conditions, and thus included the problem of climate simulation. Thirdly, between these two extremes there was another category of predictions for which it was necessary to consider the details both of the initial state and of the external forcings which determine the final equilibrium. The logical approach was to attack these problems in the order in which they were listed.

This strategy was already being followed at the time of von Neumann's talk. Short-range numerical forecasting had begun with the pioneering experiments reported by Charney *et al* (1950), and a description of the first numerical general circulation experiment was published 6 years later by Phillips (1956). Further progress in both these categories was, however, required before a serious attempt could be made at predictions falling within the third, intermediate category. This latter category will be seen to include the prediction of the medium range, which we here interpret as the period from a few days to a week or two ahead.

The initial experiments in medium-range forecasting came at the end of the 1960s when a series of 2-week hemispheric predictions was carried out at GFDL, Princeton (Miyakoda *et al.*, 1972). The results of these forecasts were encouraging, and served an important catalytic function. They stimulated the planning and later the successful implementation of the First GARP Global Experiment (FGGE) in 1978–1979, and they constituted a very important impetus for the setting up of the European Centre for Medium Range Weather Forecasts (ECMWF). Operational production of forecasts for a period up to 10 days ahead began at ECMWF on the first of August, 1979.

In this article we describe briefly the design of the forecasting system chosen for the initial phase of operational prediction at ECMWF, and present more extensively the results obtained over the first 2 years of its implementation. Concentration on the performance of just one forecasting centre may seem inappropriate in a volume such as this, but the forecasts to be discussed are at the time of writing unique in providing a regular record of atmospheric predictability on a time scale as long as 10 days. In addition, objective verification has shown that forecasts for the 1–3-day range produced by the ECMWF system are distinctly more accurate than those produced by other current operational systems (Bengtsson and Lange, 1982). The results described here thus serve to indicate present practical limits of predictability.

12.2 The ECMWF global forecasting system

A global forecasting system has two major components, namely the data assimilation and the prediction model. Although these two components can be set up independently, they are mutually dependent, and highly integrated in the ECMWF system. This system is outlined in Fig. 12.1. The rationale behind its design is as follows.

12.2.1 The data assimilation

Two major conditions influence the design of the scheme for data analysis. Firstly, the irregular (and in many places inadequate) distribution of observations necessitates the composition of the initial state from previous data (via the forecast model), from climatology, and from a variety of different observations, all of them having different accuracies and representativeness. Secondly, the initial state must be in balance with the forecast model. This includes not only the well known dynamical balance necessary to avoid the generation of spurious gravity-wave motion, but also the correct physical balance, for example, to avoid large deficiencies in the overall rate of precipitation in the early stages of the forecasts.

The irregularity of the observations and the requirement of dynamical balance lead to the use of a three-dimensional multi-variate analysis scheme that simultaneously analyses the mass and wind fields (Lorenc, 1981). An additional advantage of this scheme is that it provides a consistent way of checking the data. This has made it possible to establish an essentially automatic operational procedure at ECMWF, and no manual intervention during the analysis process is carried out. Analyses are performed every 6 hours with observations used in a 6-hour 'time window'. The model provides the necessary 'first guess' for the analysis, which in turn provides the initial state for the 6-hour forecasting step required to produce the first guess for the following analysis.

The resulting analysis is in a dynamical balance sufficiently good for any gravity wave motion to have little impact on the resulting forecast beyond a day or so. Gravity waves are, however, large enough to have a significant influence on the subsequent acceptance of data when a 6-hour forecast is used as the 'first guess' for the next analysis. This is overcome by use of a non-linear normal-mode initialization (Machenhauer, 1977; Temperton and Williamson, 1981; Williamson and Temperton, 1981). An initial physical balance has largely been achieved during the period of operational forecasting by ensuring a correct distinction between temperature and virtual temperature, and by an interpolation of analysed increments to the first guess (rather than complete fields) from the pressure levels of the analysis to the sigma levels of the forecast model. The latter procedure essentially preserves the model boundary-layer structure through a series of analysis cycles. A further improvement is expected to be achieved by including diabatic processes in the initialization procedure.

The data produced in this manner were used for the general circulation diagnostics presented in Chapter 1.

12.2.2 The forecast model

The numerical model chosen for the first phase of operational forecasting at ECMWF uses a finite-difference scheme based on a staggered grid of variables known as the C-grid (Arakawa and Lamb, 1977). Choice of this grid was made mainly because of its low computational noise and the ease of implementation of a semi-implicit time scheme. Following the work of Arakawa (1966) and Sadourny (1975), the finite-difference scheme was designed to conserve potential enstrophy during vorticity advection by the horizontal flow. Further detail has been given by Burridge and Haseler (1977), and Burridge (1979).

Vertical and horizontal resolutions were selected, within overall computational constraints, to provide a reasonable description of the fundamental large-scale instabilities, some representation of the stratosphere, and an explicit boundary-layer structure. The related parameterization scheme (Tiedtke et al., 1979) describes the interactions thought to be of importance in

ECMWF—Global forecasting system, 15-level grid point model

(Horizontal Resolution 1.875° Lat/Lon)

Analysis	Prediction
Φ, u, v, q	T, u, v, q
p(mb)	σ
10	$0.025(\sigma_1)$
20	0.077
30	0.132
50	0.193
70	0.260
100	0.334
150	0.415
200	0.500
250	0.589
300	0.678
400	0.765
500	0.845
700	0.914
850	0.967
1000	$0.996(\sigma_{15})$

Vertical and horizontal (latitude / longitude) grids and dispositions of variables in the analysis (left) and prediction (right) coordinate systems.

Analysis

Method	Three-dimensional multivariate (15 analysis levels, see above)
Independent variables	λ, ϕ, p, t
Dependent variables	Φ, u, v, q
Grid	Non-staggered, standard pressure levels
First guess	6-hour forecast (complete prediction model)
Data assimilation frequency	6-hour (±3-hour window)

Initialization
Method   Non-linear normal mode, five vertical modes, adiabatic

Prediction
Independent variables λ, ϕ, σ, t
Dependent variables T, u, v, q, p_s
Grid   Staggered in the horizontal (Arakawa C-grid). Uniform horizontal (regular lat./lon.). Non-uniform vertical spacing of levels (see above)
Finite difference scheme   Second order accurate
Time-integration   Leapfrog, semi-implicit ($\Delta t = 15$ min) (time filter $\nu = 0.05$)
Horizontal diffusion   Linear, fourth order (diffusion coefficient $= 4.5 \times 10^{15}$ m^4 s^{-1})
Earth surface   Albedo, roughness, soil moisture, snow and ice specified geographically. Albedo, soil moisture and snow time dependent
Orography   Averaged from high resolution ($10'$) data set
Physical parameterization
    (i) Boundary eddy fluxes dependent on roughness length and local stability (Monin–Obukov)
    (ii) Free-atmosphere turbulent fluxes dependent on mixing length and Richardson number
    (iii) Kuo convection scheme
    (iv) Full interaction between radiation and clouds
    (v) Full hydrological cycle
    (vi) Computed land temperature, no diurnal cycle
    (vii) Climatological sea-surface temperature

Fig. 12.1. The ECMWF Global Forecasting System as on 1 August 1981.

the medium range, including a full hydrological cycle, a relatively detailed stability-dependent representation of boundary and free-atmospheric turbulent fluxes, and an interaction between the radiation and model-generated clouds.

12.2.3 Operational timings

Operational medium-range prediction at ECMWF has involved the production of 10-day forecasts on 5 days of the week starting from 1 August 1979, and daily since 1 August 1980. Forecasts have been run from a final analysis for 12 GMT with a data cut-off that varied from 17.30 GMT in the early stages of operational activity (using an incomplete telecommunication system) to 20.45 GMT at the end of the second year. This data cut-off is substantially later than is of necessity used in other centres for the operational production of short-range forecasts. Its impact on the quality of the forecasts has not been ascertained definitively, although several forecasts run from 00 GMT using a 03 GMT cut-off show no indication that a significant fall in quality results from an earlier cut-off, at least for the Northern Hemisphere.

The computational efficiency of the forecasting system has been significantly improved over the 2 years of routine use, and the most recent timings on the ECMWF's CRAY-1 computer are about 1 h 50 min for a 24-hour data assimilation, and 3 h 40 min for the ensuing 10-day forecast.

12.2.4 Changes in the forecasting system

In assessing the results to be presented below it should be borne in mind that some aspects of the system have been subject to revision over the 2-year period in question. In particular, several refinements of the data analysis have been introduced in the light of operational experience. Mention has already been made of factors influencing the physical balance of the initial state, and other changes have generally been such as to produce a more internally consistent and robust system.

Model changes have been fewer. Minor adjustments of the physical parameterizations were introduced early in the period and more efficient horizontal diffusion formulations have also been incorporated. A change to a more realistic representation of the orography and coastlines was made in April, 1981.

12.3 Methods of assessment

The methods used to assess the forecasts may be divided into three categories. The first is assessment by objective measures, and for convenience in this paper we concentrate attention on just one such measure. The second is the subjective evaluation of the predictions, and in this respect the ECMWF is fortunate to receive regular reports from experienced meteorologists working in its Member States. The third is the physical consistency of the forecast. Isolated trial forecasts incorporating a change to some aspect of the system may exhibit little change in the accuracy of the weather prediction itself, but more sensitivity can often be found in examination of some of the physical balances, for example in the budget of heat or moisture. An extensive range of diagnostic programs has thus been developed as part of the overall forecasting system.

The objective measure for which results will be discussed here is the anomaly correlation of the height field, the correlation between observed and predicted deviations from climatology of the height field at one or more levels of the atmosphere, evaluated over a certain area of the globe. This was one of the measures used by Miyakoda *et al.* (1972) in their experimental study of medium range predictability. It has generally been found to give a reliable indication of forecast skill, although care must be taken not to place too much reliance on one such method of assessment, particularly when small samples of forecasts are involved.

A question that immediately arises with objective scoring of forecasts is what score constitutes a limit beyond which a forecast is no longer useful. This question can have no unique answer as the level of accuracy required for a forecast to be useful depends very much on the purpose for which the forecast is to be used. At the extreme, a positive anomaly correlation may be an indication that a forecast possesses some skill, but in general a more restrictive criterion, an anomaly correlation of 50 or 60%, is used to define the limit of useful predictability (e.g., Hollingsworth *et al.*, 1980).

Justification of the latter choice may be made by comparison with subjective assessments of the forecasts. For example, an evaluation by the Swedish Meteorological and Hydrological Institute of all forecasts for the neighbourhood of Scandinavia in the period January 1980 to March 1981 showed about half the 1000 mb height forecasts to be useful at day 5. Over the same period, the mean anomaly correlation of 1000 mb heights averaged over a (larger) European area ($12°W-42°E$, $36°N-72°N$) reached the 60% level at day 4.5 of the forecast, and 50% at day 5.3.

It should be noted, however, that such criteria assume a forecast for one particular instant to be judged against the analysis for precisely that time. A forecast that is poor in synoptic detail or timing may score badly while still giving some indication of an overall change in the weather type.

12.4 The predictability of the extratropical Northern Hemisphere

12.4.1 Spatial variability

Anomaly correlations of the 500 mb and 1000 mb height fields calculated for most of the extratropical Northern Hemisphere and averaged over all operational forecasts from December 1980 are presented in the upper plot of Fig. 12.2. This shows a generally more accurate forecast at 500 than at 1000 mb, a result that is in agreement with synoptic assessment. Thus in the medium range, a particular value of the correlation is typically reached between 12 h and 1 day later at 500 mb than at 1000 mb. Representing as they do an average over several forecasts, these results show a steady decline in skill with increasing forecast period. Fifty per cent values for the anomaly correlation are reached at about day 6.5 at 1000 mb and about day 7.2 at 500 mb in this particular (slightly above average) winter month.

An idea of the variation in predictability with spatial scale may be gained from the lower plot in Fig. 12.2, which shows anomaly correlations for three separate groups of zonal wavenumber. These correlations are calculated using heights at standard pressure levels from 1000 to 200 mb, but they are generally similar to correlations calculated for the 500 mb level alone. A point to note is that these results give no indication that medium scales are forecast better than planetary scales of motion, a finding in contrast to some earlier experience reviewed by Leith (1978) and Bengtsson (1981). Instead, the spectral decomposition shows the larger scales to be the more accurately forecast. There is also a very much poorer forecast of the shorter synoptic scales, here represented by zonal wavenumbers 10 to 20. This latter result may be reflected on occasions in erroneous timings or intensities of individual weather events within an overall weather situation which is better forecast, but examples may also be found in which the erroneous forecast of a small-scale feature is followed by a deterioration of the forecast over a much larger area.

The spatial variability of predictability has also been examined by comparing objective measures over more limited areas. For the month of December 1980, anomaly correlations of the 500 mb height over the European area reached the 80% level at day 4 and the 60% level at day 5.8. Corresponding values of 4.1 and 8.5 days are found for an East Asian area (102°E–150°E, 24°N–60°N), and 4.3 and 6.1 days for North America (120°W–72°W, 24°N–60°N). Such regional results show more variability from month to month than those for the extratropical hemisphere, but generally indicate a poorer verification over the European area during the winter months. Synoptic assessments of the forecasts outside this area have been limited, but part of the reason for the deficient performance over Europe may be the systematic model height error discussed later in this article.

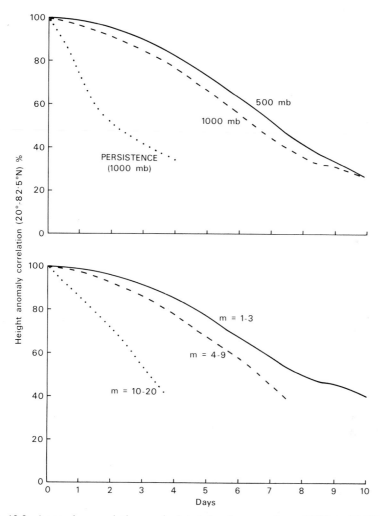

Fig. 12.2. Anomaly correlations calculated for the area from 20°N to 82.5°N and averaged for all forecasts from December, 1980. Values are plotted as functions of the length of the forecast. The upper graph shows results for the 1000 and 500 mb height fields, together with that of a persistence forecast for 1000 mb. The lower one is calculated using heights of standard levels between 1000 and 200 mb, but with different zonal wavenumber groups separately evaluated.

12.4.2 Temporal variability

Figure 12.3 indicates the temporal variability of the forecasts over the period from September 1979 (the first month for which results are available) to the end of the second year of operational forecasting. Illustrated is the time at

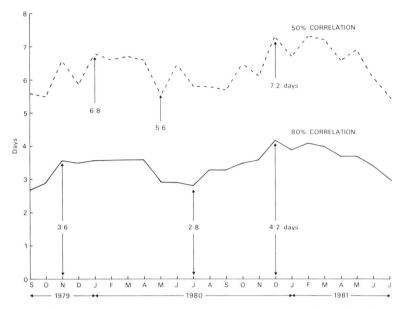

Fig. 12.3. The month-by-month variation of two levels of predictability. These are defined by the length of the forecast period for which monthly-mean anomaly correlations for the 1000–200 mb heights remain above either 80% or 50%.

which monthly mean anomaly correlations (again calculated over the extratropical Northern Hemisphere and from 1000 to 200 mb) reach the 80% and 50% levels. Three particular points of interest arise from examination of this figure.

The first is that there is a significant month-to-month variation in predictability. This is more marked for longer forecast periods, and changes from one month to the next in the time taken to reach the 50% level can be as large as one day.

The second point is that in addition to the month-to-month variability there is a clear seasonal variability in this (and other) objective measures of predictability. In particular, these hemispheric scores indicate a generally lower predictability in the summer months than in the winter months. Examination of the limited-area averages shows this result to be marked for the East Asian and North American regions. Conversely, over the European area somewhat better correlations are found for early summer than for winter. The reasons for these results remain to be determined.

The third feature of Fig. 12.3 is the generally higher level of correlations in the second year of forecasting than in the first. This is particularly marked in the shorter range, all months from September 1980 to July 1981 yielding better scores up to day 4 than obtained for the preceding year. As in other respects,

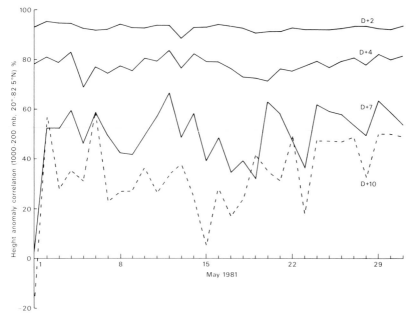

Fig. 12.4. Anomaly correlations for all 2, 4-, 7- and 10-day forecasts from initial dates in May 1981.

more variability is found further into the medium range, but improvements nevertheless occur in 8 of the 11 months in question. The regularity of the improvement in the short range suggests that it is not principally due to inherently more predictable synoptic situations during the second of the years in question, and the amount of data available actually declined during the period following the end of the FGGE observing year. Several comparisons using identical data suggest, however, that a distinct benefit has arisen from the revisions of the forecasting system introduced since the start of operational forecasting.

The day-to-day variability in the accuracy of the forecasts is illustrated (again using the anomaly correlation of height) in Fig. 12.4 for a particular operational month. Variability evidently increases as the forecast period increases, and such variation as does occur in the 2-day forecast is on a time scale of several days, only one forecast (from 13 May) appearing clearly out of sequence. Further into the forecast period quite large changes in correlation from day to day may be seen, although even at day 10 spells of above and below average predictability may be seen. In particular, 8 out of the last 10 forecasts of the month exhibit anomaly correlations above 40% at day 10. Only three such cases occurred over the preceding 21 days. Similar spells of relatively high predictability at the end of the forecast period have been found

Analysis 16 May 1981   **4-day Forecast**

Analysis 23 May 1981   **4-day Forecast**

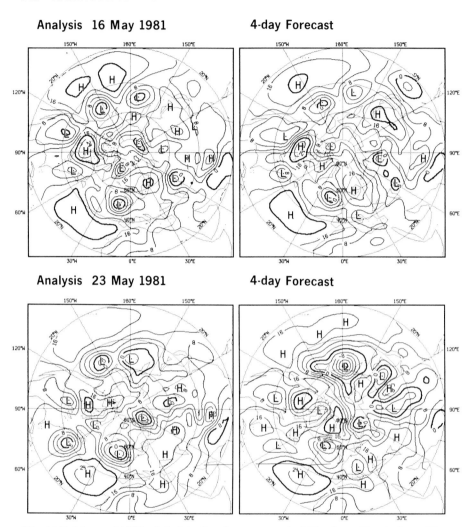

Fig. 12.5. Maps of 1000 mb height for the extratropical Northern Hemisphere. The left-hand plots are analyses for 16 May 1981 (upper) and 23 May 1981 (lower), and the right-hand plots are 4-day forecasts verifying on these two dates. The contour interval is 4 dam.

for other months, and we return to a discussion of these cases in the following section.

The results shown in Fig. 12.4 have been used to select examples of relatively good and bad forecasts at days 4 and 7. Figure 12.5 presents maps of the 1000 mb height field for the 4-day forecasts from 12th and 19th May, together with the verifying analyses. Corresponding 7-day forecasts for 500 mb are shown in Fig. 12.6. The forecast from the 12th is, according to our chosen

Analysis 19 May 1981

7-day Forecast

Analysis 26 May 1981

7-day Forecast

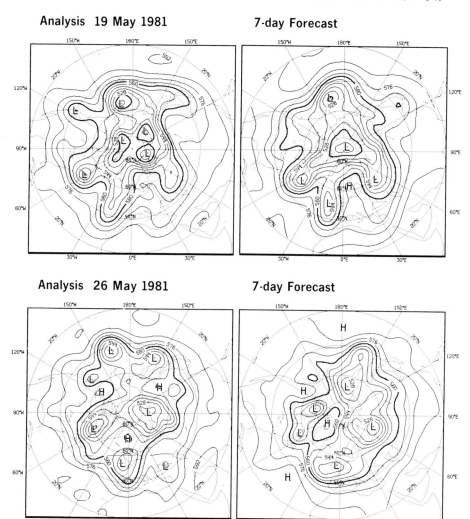

Fig. 12.6. As Fig. 12.5, but for 7-day forecasts and the 500 mb height. The contour interval is 8 dam.

objective measure, the best of the month at both day 4 and day 7, whereas that from the 19th is close to the worst of the month at both days. Anomaly correlations averaged over the extratropical Northern Hemisphere are 79% and 62% respectively for the day-4 forecasts at 1000 mb, and 62% and 32% for the two day-7 forecasts at 500 mb.

The surface synoptic situations were very similar on the 16th and 23rd May. Particular common features to note are the two lows in the Central and Eastern Pacific, the low to the west of Ireland and the developing low near the eastern coast of North America. Despite this similarity, the two day-4 forecasts are quite different. Although the better of the two has errors of detail, it has clearly forecast with reasonable accuracy the positions and intensities of the four lows in question.

The forecast from 19th May exhibits at day 4 several errors that may commonly be seen at later stages of the ECMWF forecasts. Particularly worthy of note is the Pacific sector, where the forecast has produced not two distinct lows but what appears closer to one, large-scale and large-amplitude depression. Conversely, the development of the low near 60°W has been substantially underestimated, and the position of the low to the west of Northern Europe has also been poorly forecast.

Overdevelopment of depressions, such as illustrated near 180°E in the above example, tends to occur most commonly over the Central and Eastern Pacific, and over the Eastern Atlantic and Northern Europe. In the mean this appears as an intensification and eastward shift of the Aleutian and Icelandic lows. This is discussed further elsewhere in this article. In contrast, late or inadequate development of new disturbances over the Western Atlantic occurs in a number of cases, and an underestimation of the phase speed of rapidly moving lows is frequently observed, although in the forecast from 19th May the low centred near 30°W had an origin significantly different from that of the true low centred further towards the east.

In view of the deficiencies of the 4-day forecast from 19th May it is not surprising that the 7-day forecast exhibits little skill. Figure 12.6 shows that although the analysed and forecast charts bear some overall resemblance (corresponding to the anomaly correlation of 32%) there is substantial error at most longitudes. In contrast, all main troughs exhibit a reasonably accurate position and amplitude in the 7-day forecast from a week earlier, although some detail is lost over the Pacific and the west of North America.

12.4.3 Cases of high predictability

Mention has already been made of a tendency for there to exist spells of relatively high predictability. This is illustrated by Fig. 12.7 which plots all cases between 1 December 1980 and the end of May 1981 in which the anomaly correlation at day 10 exceeded 40%. The distribution of these cases suggests that their high medium-range predictability is not purely due to chance, but rather that certain synoptic situations (perhaps related to the availability of crucial data) may be inherently more predictable over a time scale of a week or more.

Figure 12.7 shows that some 13% of forecasts exhibit a day-10 anomaly correlation of 50% or more. Such cases are clearly insufficient in number to be

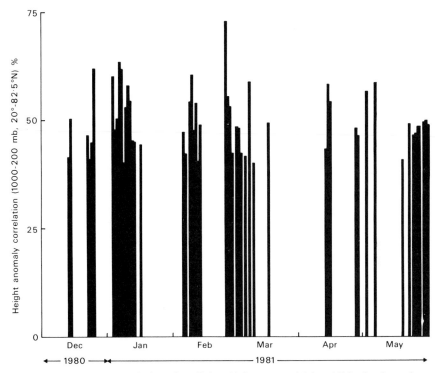

Fig. 12.7. Anomaly correlations for all day 10 forecasts which exhibited values above 40% during the period from December 1980 to May 1981.

of practical use unless a reliable indicator of the accuracy of forecasts can be obtained at the time of their production. A study of cases exhibiting skill towards the end of the forecast period is nevertheless of interest, and some reasons for this are discussed below.

In the first place, examination of these cases may improve understanding of their high predictability. This in turn may lead either to factors which may give guidance as to the expected accuracy of a particular forecast, or to ways of improving less accurate forecasts. The number of independent samples presented in Fig. 12.7 is too limited for definitive results, and indeed, no clear picture emerges from a first inspection of mean maps for the various periods of above average predictability. These periods include various synoptic situations, with both highly persistent patterns and cases in which significant change took place.

Two examples, in terms of 5-day mean maps of the 500 mb height field, are shown in Fig. 12.8. The left-hand maps show one of the forecasts from the first half of January 1981, during which time the forecasting system succeeded in reproducing a marked persistence of the atmospheric flow. Conversely, the

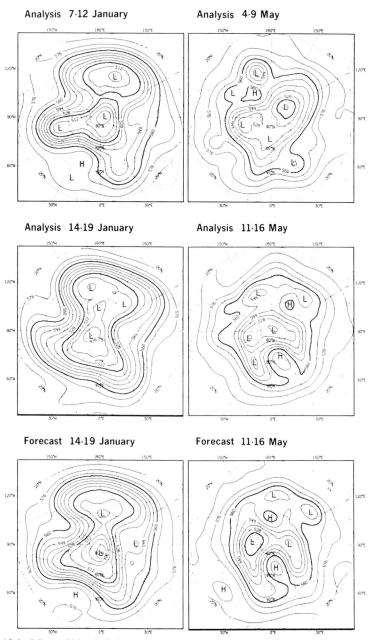

Fig. 12.8. Mean 500 mb height analyses for the periods 7–12 January (upper left), 14–19 January (middle left), 4–9 May (upper right) and 11–16 May (middle right), 1981. Corresponding means for 14–19 January and 11–16 May are also shown for the forecasts from 9 January and 6 May. The contour interval is 8 dam.

Observed anomaly   Forecast anomaly

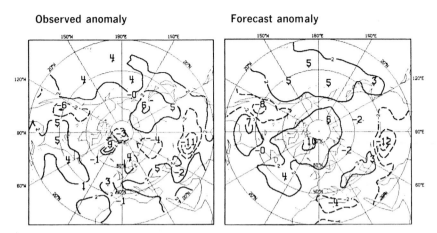

Fig. 12.9. Monthly mean deviations of the 850 mb temperature from climatology for the period from 11 July to 10 August 1980. The left hand map shows the analysed anomaly, and the right hand map the anomaly in the mean of the day 10 forecasts verifying in this period. The contour interval is 4 K and spot values in K are also shown. Apparent anomalies where temperature has been extrapolated under high ground should be disregarded.

right-hand plots illustrate a particularly successful forecast of major circulation changes. The accurately predicted development of the blocking high over Scandinavia was associated with a remarkable change from unusually cold to unusually warm spring weather over this region, and changes elsewhere over the hemisphere were largely captured.

Skillful performance of the model over the whole forecast period may also aid understanding of some of the mechanisms involved in the maintenance of anomalies over a longer time scale. Figure 12.9 shows the deviation from climatology of the mean of all day-10 forecasts of the 850 mb temperature from July 1980, together with the corresponding analysed mean anomaly. This month was characterized by a major heat-wave and drought over the eastern USA, and the maps clearly show the presence at day 10 of a realistic, if slightly strong, thermal anomaly over North America. Such an anomaly was, of course, present in the initial conditions for the forecasts, but the tendency for this model (which has fixed, climatological sea-surface temperatures) to somewhat strengthen, rather than decay, the anomaly over the forecast period suggests that the processes involved in the maintenance of the anomaly were amongst those included in the model.

A final point concerning the usefulness of the cases of high predictability concerns their role in the development of forecasting systems. A change to some aspect of a system may often result in little change in the forecast over the first half of a 10-day period, and in cases of average or below-average predictability it can be difficult to assess whether subsequent changes are such as to improve the prediction. In cases of high predictability, however, the forecasts and verifying analysis may be sufficiently close towards the end of the forecast period for one or other of the forecasts to be preferred. For reliable assessments, such experiments must be carried out for a number of such cases and a variety of synoptic situations.

12.5 Tropical and Southern Hemispheric forecasts

We now move from one end of the range of predictability to the other. Although the forecasts for the tropics have not been evaluated in as extensive an objective and subjective way as those for the extratropical Northern Hemisphere, there is no doubt that at present their accuracy and usefulness is very substantially less. Indeed, standard deviations of forecast winds show, in the mean, no improvement over persistence in the lower troposphere, and synoptic assessment reveals that some distinct errors in the low-level flow develop quite systematically in the earliest stages of the forecast. Maintenance of the quasi-stationary regional circulations of the tropical atmosphere is evidently a distinct modelling problem, and in the zonal-mean there is a partial suppression of the Hadley circulation and an underestimation of tropical precipitation by some 20%.

Despite these deficiencies, individual cases of quite accurate forecasts of transient behaviour can also be found. For the longer range there are arguments to suggest that the predictability of temporal and spatial means of the tropical atmosphere may be higher than that of middle latitudes (Shukla, 1981) but such studies have not as yet been carried out at ECMWF.

Objective verification indicates a very limited short-range predictive skill in the middle and upper troposphere. Thus for the belt from $18°N$ to $18°S$ the standard deviations of the 2-day forecast zonal and meridional velocity components averaged for 1980 are 3.7 and 3.4 m s^{-1} respectively at 500 mb, in comparison with values of 4.6 and 4.3 m s^{-1} for persistence. Corresponding forecast figures are 7 and 6.3 m s^{-1} at 200 mb, with persistence giving 7.7 m s^{-1} for both components. Seasonal variability has been examined for limited regions, and for a northern hemispheric area ($72°E–102°E, 6°N–33°N$) including India. Summer 2-day forecasts merely match persistence at both 500 and 200 mb, whereas winter forecasts appear slightly better, with standard deviations of 200 mb wind components of the order of 6 m s^{-1} in the 2-day forecast, persistence giving a corresponding value of about 8 m s^{-1}.

The indication from both subjective and objective assessments of the tropical forecasts is that there are serious deficiencies in the parameterization of convection, and a substantial effort to understand and correct these deficiencies is currently being made. As an example of the sensitivity that can

Analysis 11 June

Analysis 15 June

Forecasts 15 June

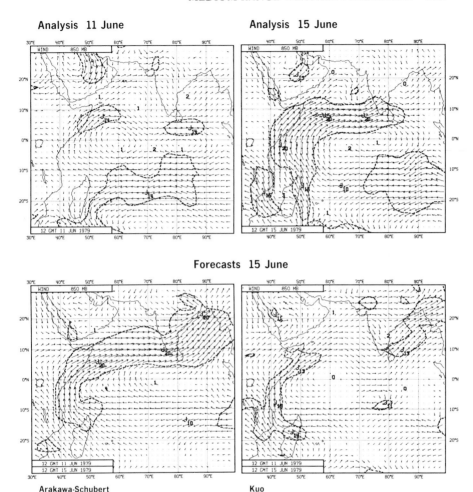

Arakawa-Schubert

Kuo

Fig. 12.10. Analyses of 850 mb wind for 11 June (upper left) and 15 June (upper right), and corresponding 4-day forecasts for 15 June using the Arakawa–Schubert (lower left) and Kuo (lower right) convection schemes. Flow maxima and minima are marked in m s^{-1}.

be found, we show in Fig. 12.10 two 4-day forecasts of the 850 mb wind over the Indian Ocean and bordering areas. These forecasts (which used FGGE rather than operational data) cover a period starting from 11 June 1979 which was marked by the onset, rather later than normal, of the south-west monsoon. The two differ only in their parameterization of convection, one using the scheme of Kuo (1974) adopted for operational forecasting and the other the Arakawa and Schubert (1974) scheme. The latter evidently produces a quite different forecast, and in fact a very much better representation of the development of the strong monsoon flow over the Arabian Sea. Just one experiment cannot of course be used to draw firm conclusions as to which of the convection schemes is the better (and indeed, the extratropical forecasts in this case were slightly better using the Kuo scheme), but the sensitivity of the forecast is worth noting. Other studies have in addition shown sensitivity to the prescription of soil moisture and orography, results in general agreement with those found elsewhere (Rowntree, 1978). Overall, it seems that the tropical forecasts respond more quickly and acutely to defects in the model than do forecasts at middle and high latitudes.

Assessments of the forecasts for the Southern Hemisphere have also been less detailed than for the Northern Hemisphere. Comparing anomaly correlations for $18°N–78°N$ with those for $18°S–78°S$ averaged over all 1980 forecasts shows the 60% value to be reached at days 4.5 and 5.4 at 1000 and 500 mb, respectively, for the Northern Hemisphere, and at 3.2 and 4 days for the Southern Hemisphere. The lower values found in the latter case are not surprising in view of the sparsity of southern hemispheric data, and generally more accurate forecasts for this hemisphere have been reported using the enhanced data coverage of the FGGE year (Bengtsson et al., 1982).

12.6 Systematic errors

An important part of the total model error is revealed by averaging forecast errors over a number of cases. These 'systematic errors' have been calculated routinely using, for convenience, monthly means. They characteristically grow in amplitude throughout the forecast period, and their general similarity to errors in the model climatology revealed by integration over extended periods indicates that these errors represent a gradual drift from the climate of the atmosphere towards that of the model. The rate of this drift is found to vary from case to case, but the overall error associated with it appears to be independent of the initial data.

The importance of this climatological component of the forecasts was recognized by Miyakoda et al. (1972) in their early series of medium-range forecasts. Two particular errors noted by them were a general cooling of the troposphere, and too low values of the 500 mb height, and to a lesser extent the 1000 mb height, over the North-Eastern Pacific and Atlantic Oceans. Similar, though larger, height errors were found in a trial series of forecasts using February cases carried out at ECMWF by Hollingsworth et al. (1980). These authors also estimated that correction of model deficiencies leading to this systematic error might directly lead to an increase of some 20% in objective estimates of the period of useful predictability. Related improvements in transient features might also be expected.

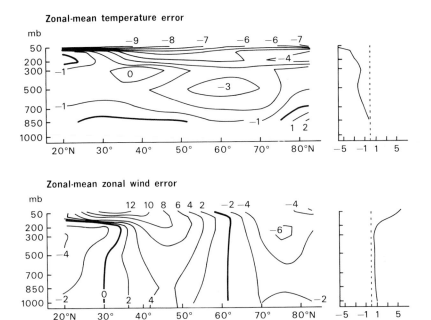

Fig. 12.11. Meridional cross-sections of zonal-mean temperature error (K) and zonal wind error (m s^{-1}) for the Northern Hemisphere calculated as averages from day 7 to day 10 of all forecasts from January 1981. Vertical profiles of the error averaged from 20° to 82.5°N are shown in the right-hand plots.

Operational experience has essentially confirmed the findings of Hollingsworth *et al.* (1980). We here restrict attention to the extratropical Northern Hemisphere, and present mean temperature and height errors towards the end of the forecast period for January 1981. This month was characterized by a persistent, relatively zonal circulation (see Fig. 12.8, for example), and the systematic error was particularly large.

Meridional cross-sections of errors in the zonal mean temperature and zonal mean zonal geostrophic wind, averaged from day 7 to day 10 of the forecasts, are shown in Fig. 12.11. An overall cooling of the troposphere, by a maximum of 3 K at 60°N and 500 mb is evident. The stratosphere was cooled by a substantially larger amount during this month, although this particular error has subsequently been reduced by ensuring an initial pressure-to-sigma interpolation of temperature which is more consistent with the formulation of the model's radiative parameterization. The tropospheric cooling shows little latitudinal variation below 500 mb, and consistent with this the zonal mean wind error is largely independent of height in this region, with the model flow generally stronger than in reality. At upper levels the zonal mean subtropical

500 mb temperature error   500 mb height error

850 mb temperature error   1000 mb height error

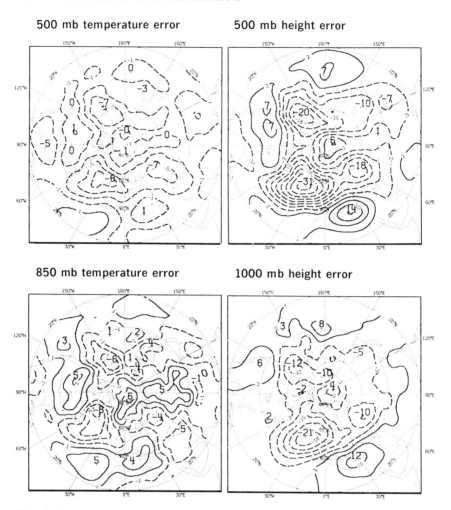

Fig. 12.12. Monthly mean error maps for day 10 forecasts of temperature at 500 mb (upper left) and 850 mb (lower left), and height of the 500 mb (upper right) and 1000 mb (lower right) pressure surfaces for January, 1981. Contour intervals are 2 K and 4 dam.

jet is displaced poleward, and its strength decreases less rapidly with height than is observed in the stratosphere.

The zonal mean temperature shows little error at 850 mb, but this disguises a substantially larger error which is revealed by study of the geographical distribution of temperature. Maps of the day 10 temperature error at 500 and 850 mb, and of the 500 and 1000 mb height error, are presented as Fig. 12.12.

Looking first at the height field we see very similar error patterns at 1000 and 500 mb, with distinct centres of low pressure over the North-Eastern Atlantic and Pacific Oceans, and a third centre at 60°E. The amplitude of the error increases with height, and consistent with this, areas of too low temperature tend to coincide with the areas of too low pressure, particularly at 500 mb. Elsewhere the 500 mb temperature error is small, but regions of substantially too warm 850 mb temperature are evident over North America, Siberia and Southern Europe. The temperature error over the latter region is atypical, but warm 850 mb temperatures over the other two areas occur commonly, and become even more pronounced over longer periods of integration. The general distribution of temperature error implies areas of quite erroneous static stability, and a serious impact on transient wave behaviour seems likely. At 850 mb the error may be related to the erroneous horizontal advection implied by the error in the height field.

Although large, these temperature errors for January 1981 should also be considered in the light of the observed and forecast deviations from climatology. An example of this has been shown in Fig. 12.9 for July 1980, and for the month of January 1981 the forecast model again succeeded in maintaining the principal anomaly pattern through the 10-day forecast period. Thus cold anomalies of up to -5 K were kept over much of the Pacific Ocean and near the eastern coast of the USA, and warm anomalies reached a maximum of 14 K over Western Canada in both analyses and 10-day forecasts. The errors shown in Fig. 12.12 result largely from a shift of some 10° of longitude in the positions of the anomalies, and from errors in their latitudinal extent.

The error map of the 1000 mb height field shown in Fig. 12.12 indicates the erroneously deep Aleutian and Icelandic surface lows noted previously. The phase change with height of the atmospheric standing wave pattern is, however, such that the negative centres at 500 mb correspond to an underestimate of the climatological ridges which occur in reality over the North-Eastern Atlantic and Pacific. The model thus exhibits a tendency to predict a too-zonal time mean flow at upper levels. Over Europe this is seen in a southward displacement of the jet stream, associated with which is a synoptically important southward displacement of cyclone tracks.

Examining maps corresponding to that shown in Fig. 12.12 shows the detailed pattern to vary from month to month, but centres of erroneously low pressure are almost invariably seen over the two ocean regions during the winter months. Maxima are generally less than the 31 dam illustrated for January 1981, but typically reach values around 20 dam at 500 mb. Error patterns appear more variable in summer and amplitudes are some 50% lower. 500 mb error maxima greater than 20 dam, and a general weakening of the monthly-mean wave pattern, may also be seen in day 10 forecasts for the Southern Hemisphere.

There are other points of interest concerning the systematic height error. In its equivalent barotropic structure and location of amplitude maxima it tends to resemble actual anomalies of the atmospheric circulation such as discussed by Wallace and Gutzler (1981) (see Chapter 3). In many respects it is by no means unique to the forecast model discussed here (Bengtsson and Lange, 1982; Wallace and Woessner, 1981), and experience at ECMWF confirms that of Manabe *et al.* (1979) who found this error to increase with increasing horizontal resolution in climate simulations. Solution of this particular model problem is thus of importance both for medium-range prediction and for the numerical simulation of climate. Some sensitivity of the error to the parameterization of turbulent fluxes and to the representation of orography has been found, and further investigations are being actively pursued.

12.7 Concluding remarks

Miyakoda *et al.* (1972) suggested that perhaps the results of their first comprehensive trial of 2-week predictions might be taken as a benchmark for future comparisons. Their ensemble mean anomaly correlations for the extratropical Northern Hemisphere, based on 12 January cases taken from the years 1964 to 1969, reached values of 80% after about 2 days of the forecast period for the 500 mb height and 1 day for the 1000 mb height. Corresponding times for the 50% correlation were about $4\frac{1}{2}$ and 4 days, respectively. Taking the average of the operational ECMWF forecasts from December 1980 to February 1981 gives values of 4 and $3\frac{1}{2}$ days for the 80% correlation at 500 and 1000 mb, and of 7 and $6\frac{1}{2}$ days for the 50% level. Anomaly correlations remained better than persistence throughout the ten-day period in these latter forecasts, whereas a 1000 mb forecast value equal to that of persistence was reported at day 6 by Miyakoda and his co-workers.

These comparisons indicate that a substantial improvement has taken place over the past 10–15 years in our ability to predict at least the larger scales of motion over the extratropical Northern Hemisphere. Considerable experimentation would be required to determine quantitatively the extent to which this is due to various aspects of the present forecasting system or to the availability of different types or quantities of data, but isolated case studies, such as that by Bengtsson (1981), suggest that many different factors may contribute to the better forecasts. It is of interest to note that much of the improvement is seen at anomaly correlations above 80%, the time taken for correlations to fall from 80% to 50% being largely similar in the two sets of forecasts. It is not clear if this is due to the existence of an inevitable error growth of the type indicated by the theory of atmospheric predictability since, for example, systematic model errors that have developed by day 4 in the more recent forecasts might result in loss of predictability at a rate similar to that due largely to similar model errors at, say, day 2 of the earlier forecasts.

The overall level of current skill is such that forecast charts for the extratropical Northern Hemisphere are generally of good quality for 3 to 4 days ahead with overall indications of weather type usually given for an additional 2 to 3 days. Cases in which major circulation changes are correctly forecast in the second half of the forecast period are not uncommon, and an important research task is to identify reasons for the spells of high predictability. It is more generally necessary to investigate the extent and type of useful information that can be extracted at different stages of the forecast. Although our discussion here has centred on traditional height, and to a lesser extent temperature, forecasts, various other fields, including precipitation, low-level wind and surface temperature, are predicted by the system, and the use of these remains a matter for more detailed experimentation and assessment.

In addition to their direct value as weather forecasts, the archived results of operational production, together with those of corresponding research experiments, comprise a source of information of potential value for the study of atmospheric dynamics, and predictability beyond the 10-day time range. Cases of particularly accurate prediction of, say, blocking (for example, as shown in Fig. 12.8) or stratospheric warming events (Bengtsson et al., 1982; Simmons and Strüfing, 1983) may be used to further our understanding of the mechanisms involved in these phenomena. This may be achieved both by diagnosis of the simulations themselves, which may compensate for any uncertainties in the diagnosis of the corresponding analyses, and by controlled numerical experimentation designed to identify aspects of the model or initial state crucial for the representation of the observed behaviour. The results over North America for July 1980 shown in Fig. 12.9 provide another example. Assuming the forecast anomaly at day 10 not to represent merely a systematic error which happened to coincide with the real anomaly, it is fascinating to speculate at what point in time a series of longer-range forecasts from initial dates within this month might cease to show the presence of the anomaly.

Scope evidently exists for further improvements in forecast quality in the medium range. Small but not insignificant benefits have been shown to result from replacement of the forecast model's finite difference scheme by a spectral scheme (Girard and Jarraud, 1982), and other developments of the forecasting system currently under test are expected to maintain the slow trend of increasing skill suggested by objective verification of the forecasts over the first two operational years. In the longer term it is reasonable to anticipate that improvements will result from correction of systematic model errors, use of higher resolution, and further refinement of the data assimilation, a marked sensitivity to which is an aspect of research experience at ECMWF which we have not discussed in this article. It will be of interest to see in the years ahead if the quantitative impact of such developments will be such as to maintain the rate of advance in numerical weather prediction which has occurred over the three decades since the original forecasts reported by Charney *et al.* (1950).

Acknowledgement

The work discussed here represents the efforts of many staff members and visiting scientists of ECMWF, to whom we express our gratitude.

References

ARAKAWA, A. (1966). Computational design for long-term numerical integration of the equations of fluid motion: Two-dimensional incompressible flow. Part 1. *J. Comp. Phys.*, **1**, 119–143.

ARAKAWA, A. and LAMB, V. R. (1977). Computational design of the basic dynamical processes of the UCLA general circulation model. *Methods in Computational Physics*, Vol. 17 (J. Chang, Ed.), Academic Press, 337 pp.

ARAKAWA, A. and SCHUBERT, W. H. (1974). Interaction of a cumulus cloud ensemble with the large-scale environment. *J. atmos. Sci.*, **31**, 674–701.

BENGTSSON, L. (1981). Numerical prediction of atmospheric blocking—A case study. *Tellus*, **33**, 19–42.

BENGTSSON, L. and LANGE, A. (1982). Results of the WMO/CAS NWP data study and intercomparison project for forecasts for the Northern Hemisphere in 1979–80. WMO, Geneva.

BENGTSSON, L., KANAMITSU, M., KÅLLBERG, P. and UPPALA, S. (1982). FGGE research activities at ECMWF. *Bull. Amer. meteor. Soc.*, **63**, 277–303.

BURRIDGE, D. M. (1979). Some aspects of large scale numerical modelling of the atmosphere, in *Proceedings of ECMWF Seminar on Dynamical Meteorology and Numerical Weather Prediction*, Vol. 2, 1–78.

BURRIDGE, D. M. and HASELER, J. (1977). A model for medium range weather forecasting—Adiabatic formulation. *ECMWF Technical Report No. 4*, 46 pp.

CHARNEY, J. G., FJÖRTOFT, R. and VON NEUMANN, J. (1950). Numerical integration of the barotropic vorticity equation. *Tellus*, **2**, 237–254.

GIRARD, C. and JARRAUD, M. (1982). Short and medium range forecast differences between a spectral and a grid-point model. An extensive quasi-operational comparison. ECMWF Technical Report No. 32, 178 pp.

HOLLINGSWORTH, A., ARPE, K., TIEDTKE, M., CAPALDO, M. and SAVIJÄRVI, H. (1980). The performance of a medium-range forecast model in winter—Impact of physical parameterizations. *Mon. Weath. Rev.*, **108**, 1736–1773.

KUO, H-L. (1974). Further studies of the parameterization of the influence of cumulus convection in large-scale flow. *J. atmos. Sci.*, **31**, 1232–1240.

LEITH, C. E. (1978). Objective methods for weather prediction. *Ann. Rev. Fluid Mech.*, **10**, 107–128.

LORENC, A. C. (1981). A global three-dimensional multivariate statistical interpolation scheme. *Mon. Weath. Rev.*, **109**, 701–721.

MACHENHAUER, B. (1977). On the dynamics of gravity oscillations in a shallow-water model with application to normal mode initialisation. *Beitr. Phys. Atmos.*, **50**, 253–271.

MANABE, S., HAHN, D. G. and HOLLOWAY, J. L. (1979). Climate simulation with GFDL spectral models of the atmosphere: Effect of spectral truncation, *GARP Publication Series No. 22*, 41–94.

MIYAKODA, K., HEMBREE, G. D., STRICKLER, R. F. and SHULMAN, I. (1972). Cumulative results of extended forecast experiments. I. Model performance for winter cases. *Mon. Weath. Rev.*, **100**, 836–855.

PFEFFER, R. L. (Ed.) (1960). *Dynamics of Climate*, Pergamon Press, 137 pp.

PHILLIPS, N. A. (1956). The general circulation of the atmosphere: A numerical experiment. *Q. Jl R. met. Soc.*, **82**, 123–164.

ROWNTREE, P. R. (1978). Numerical prediction and simulation of the tropical atmosphere. In *Meteorology over the Tropical Oceans*, Royal Meteorological Society, 278 pp.

SADOURNY, R. (1975). The dynamics of finite difference models of the shallow-water equations. *J. atmos. Sci.*, **32**, 680–689.

SHUKLA, J. (1981). Predictability of the tropical atmosphere. *Proceedings of ECMWF Workshop on Tropical Meteorology and its Effects on Medium Range Weather Prediction at Middle Latitudes*, 21–51.

SIMMONS, A. J. and STRÜFING, R. (1983). Numerical forecasts of stratospheric warming events using a model with a hybrid vertical coordinate. *Q. Jl R. met. Soc.*, **109**, 81–111.

TEMPERTON, C. and WILLIAMSON, D. L. (1981). Normal mode initialization for a multi-level grid-point model. Part I: Linear aspects. *Mon. Weath. Rev.*, **109**, 729–743.

TIEDTKE, M., GELEYN, J-F., HOLLINGSWORTH, A. and LOUIS, J-F. (1979). ECMWF model-parameterization of sub-grid scale processes. *ECMWF Technical Report No. 10*, 46 pp.

WALLACE, J. M. and GUTZLER, D. S. (1981). Teleconnections in the Geopotential Height Field during the Northern Hemisphere Winter. *Mon. Weath. Rev.*, **109**, 784–812.

WALLACE, J. M. and WOESSNER, J. K. (1981). An analysis of forecast error in the NMC hemispheric primitive equation model. *Mon. Weath. Rev.*, **109**, 2444–2449.

WILLIAMSON, D. L. and TEMPERTON, C. (1981). Normal mode initialization for a multilevel grid-point model. Part II: Nonlinear aspects. *Mon. Weath. Rev.*, **109**, 744–751.

— 13 —

Predictability in theory and practice

C. E. LEITH

13.1 Introduction

It is common knowledge that the weather is predictable with reliability for only a short time into the future. Weather prediction is the ultimate test of our understanding of the physics and dynamics of the atmosphere, and it is thus important to distinguish the various theoretical and practical limitations that we encounter.

This review will summarize some of these limitations and describe some research in progress that may extend them. The 'classical' theory of predictability will first be described. This treats the atmosphere as an unstable non-linear turbulent system in which any perturbation, no matter how small, will eventually grow to overwhelm the deterministic evolution of the system and destroy all but climatological knowledge of its state. The principal goal of the classical theory is to estimate the growth rate of small errors.

A problem has been posed for the classical theory by atmospheric blocking events which persist, and thus should be predictable, for times long compared with general predictability times. The atmosphere appears to shift between quasi-equilibrium states with and without blocking (Section 3.5.3 and Chapter 4). As discussed in Sections 6.2 and 7.44, recent studies with simple non-linear model systems show similar behaviour. Alternatively, an atmospheric block may be considered as a relatively stable structure embedded in an otherwise unstable turbulent flow. Solitons and modons have been studied recently as examples of such structures. In any case, it would be desirable to develop some easily computed measure of local predictability that could serve to identify relatively predictable situations.

Classical theory estimates the inherent internal growth rate of errors for a perfect prediction model, but predictability in practice depends on external error sources arising from both model imperfections and errors in the initial analysed state of the atmosphere. Numerical models of the atmosphere, whether in gridpoint or spectral representation, are limited in the spatial resolution that they can provide with available computing power. Although an attempt is made to treat the mean influence of unresolved dynamics and physics on resolved scales through an averaging process that has come to be known as parameterization, the fluctuating influence must remain as an error source. Estimates of subresolution error source rates for dynamics are available from turbulence theory, but for physical processes few quantitative error source estimates have yet been made.

The observations that are analysed to estimate the initial state of the atmosphere for a prediction are themselves of limited spatial resolution and are contaminated with errors. The widely used process of statistically optimal analysis provides information about associated error statistics. Optimal analysis is a linear Gaussian technique that must be merged with the recently developed non-linear initialization procedures. A Monte Carlo method is proposed for such a combined analysis and initialization system. Unbalanced gravitational mode errors have a smaller impact on predictability than do rotational mode errors, and it is necessary to distinguish these in any error budget analysis.

The significance of errors in the initial analysis and those arising from model imperfections can be assessed by fitting a simple error budget equation to prediction model error growth curves derived from experience. In current practice, analysis errors also depend on model errors since the result of a model prediction starting from a previous analysis is used as prior information in the statistical analysis procedure. The simple error budget equations can be extended to this more general data assimilation process to estimate the importance of model and observational errors.

The following sections contain a more detailed discussion of these aspects of atmospheric predictability and try to give a sense of the direction of current research. The main problems appear to be the better definition of local predictability, the optimal assimilation of observations, and the refinement of error budgets.

13.2 Classical predictability analysis

There are two basic objective methods for weather prediction (Leith, 1978). In the statistical method, regression prediction equations are developed to estimate a future state from knowledge of the present state and of climatologically observed time-lagged correlations. The advantage of the statistical method is that it is based on the observed behaviour of the real atmosphere. Its disadvantage is that it is feasible with the existing climate data set to develop only simple linear regression equations that are subject to significant sampling errors and that cannot reflect the atmosphere's complicated non-linear dynamics and physics. The predictability limit of the statistical method is imposed simply by the observed fact that time-lagged correlation functions approach zero and thus have dwindling predictive utility as time increases.

The second method, used now by all major weather services, is based on numerical weather prediction models in which the complexities of non-linear dynamics and physics are explicitly taken into account. In exchange for this advantage over the statistical method, numerical models have had to sacrifice their complete dependence on observed atmospheric behaviour. As Lorenz (1977) has pointed out, however, it is possible to combine the advantages of each method by treating the result of a numerical prediction as a non-linear predictor in a regression prediction equation.

A new predictability question was posed by the development of numerical models. If the predictability limit of linear methods was a consequence of non-linear effects, then the non-linear models might in principle remove any such limit. This proved not to be the case.

Predictability experiments have been carried out with many different numerical models. In these, two predictions with only slightly differing initial states diverge from each other until finally they differ by as much as any two states chosen randomly from the model climate ensemble. These model results suggest that the atmosphere is inherently unstable in the sense that any perturbation, no matter how small in scale or amplitude, would deflect the atmosphere from its original course and eventually lead to a completely unrelated sequence of weather events. The new sequence, however, would be consistent with climate statistics, and thus such small perturbations would change the weather but not the climate.

Classical predictability research concentrated on a quantitative determination of the average rate at which small errors would grow. The early model experiments, summarized by Smagorinsky (1969), found error growth rates with rms error doubling times of about 5 days. Jastrow and Halem (1970) and Williamson and Kasahara (1971) discovered, however, that this result depended on the spatial resolution of the model; with higher resolution the error doubling time decreased to about 3 days.

To avoid the problem of model dependence, Lorenz (1969a) attempted to estimate error growth rates from observations of the real atmosphere. He searched the past records to find pairs in which the two states resembled each other as closely as possible. The difference could then be interpreted as an error whose growth could be determined by observing the ensuing evolution of the two states. Unfortunately for this approach, the atmosphere has not repeated itself in the last few decades for which adequate data exist, and Lorenz showed that it was unlikely to do so over many centuries. He, therefore, had to rely on a plausible extrapolation procedure to estimate a small error doubling time of about $2\frac{1}{2}$ days.

Studies of predictability have also been carried out for purely turbulent flows free of the complicating physics of the atmosphere. These have shown that predictability depends on the shape and magnitude of the eddy kinetic energy spectrum. For example, Robinson (1967) estimated that if the large-scale atmospheric energy spectrum had the $-\frac{5}{3}$ power law of the Kolmogorov–Oboukhov inertial range of three-dimensional isotropic turbulence, then according to simple scaling laws the atmosphere would be predictable for only a day or so. Lorenz (1969b) made a simple turbulence model calculation for this case but also noted that if the atmospheric spectrum followed a -3 power law, predictability would be much improved. Such a power law is predicted for two-dimensional turbulence, which is likely to be more relevant to large-scale motions, and is indeed observed in the atmosphere (e.g., Wiin-Nielsen, 1967).

Calculations with the test-field model of turbulence (Leith and Kraichnan, 1972) confirmed and sharpened the earlier results for both three- and two-dimensional turbulence. These calculations showed explicitly how small-scale errors would contaminate successively larger scales leading to a back cascade of error and to error growth. For a two-dimensional eddy kinetic energy spectrum matching that observed in the atmosphere, the error doubling time was about 2 days. In light of the considerable oversimplification of turbulence models, this result is in reasonable agreement with a general consensus estimate of $2\frac{1}{2}$ days.

The turbulence theory results also accounted for the greater apparent predictability of low resolution models. As Miyakoda et al. (1971) showed, lack of resolution impeded baroclinic energy conversion processes in a model resulting in levels of eddy kinetic energy which were too low. All non-linear exchange processes were thus slowed down including the back cascade to larger scales and the error growth.

The moment closure techniques of statistical theories of homogeneous isotropic turbulence have also been applied to simple atmospheric models for which neither homogeneity nor isotropy is assumed. The resulting stochastic-dynamic prediction equations proposed by Epstein (1969) attempted to predict an error covariance matrix as well as a best estimate of the mean state vector. Such a procedure carries predictability information along with a prediction. Unfortunately, it requires an order N fold increase in arithmetic over a conventional prediction where N is the number of degrees of freedom or dimensionality of the model phase space. Since N can be of order 10^5 in a reasonable model, a stochastic dynamic prediction is not yet feasible. A Monte Carlo approximation with sample size M of order 10 is feasible, however, since it requires only an M fold increase in arithmetic (Leith, 1974).

All of these results of classical predictability analysis are based on the perfect model assumption. A theoretical error doubling time of about $2\frac{1}{2}$ days has been estimated for the atmosphere itself quite aside from any additional errors introduced in practice by discrepancies between the dynamics and physics of a model and that of the atmosphere. These will be discussed in a later section.

13.3 Quantum predictability analysis

In recent years, interest has been increasing in the possibility that the atmosphere may be found in distinguishable weather regimes between which it makes relatively rapid transitions. Perhaps the most noticeable example is an atmospheric blocking event that can persist for 2 or 3 weeks with a dramatic impact on weather patterns in the affected area (see Chapters 3 and 4). In such situations, it appears that the classical average predictability estimates may not be appropriate and that weather regimes can have a stability and persistence greater than that of individual weather events. What is needed then is what I shall call a quantum theory of atmospheric dynamics and predictability in which one studies transitions between quasi-stable states.

Objective statistical evidence is not yet overwhelming for the existence of bimodal or multimodal probability distributions in the dynamical phase space of the atmosphere. These would be the clearest indication of distinguishable weather regimes. White (1980), however, has found regions of low kurtosis of the 500 mb height field where blocking is most prevalent. Low kurtosis can be a consequence of a tendency toward the splitting of a probability distribution into two components.

There have been two approaches to the quantum dynamics of the atmosphere. One, which may be called the wave approach, developed from the study by Charney and colleagues of low-order non-linear systems in which only a few planetary waves have been kept. Charney and deVore (1979) showed that a simple non-linear barotropic model containing a few large-scale waves forced by Newtonian heating and by a single topographic wave could have two distinct stable states, one being of low zonal index characteristic of blocking events (see Section 6.2). Charney et al. (1981) extended this and a similar model by Hart (1979) to include more realistic topography and obtained results agreeing qualitatively with observations of blocking patterns. Charney and Straus (1980) extended the earlier study to a two-layer baroclinic model in which the role of orographic and baroclinic instability could be examined. They discovered that, although orography is necessary for the existence of the model blocking state, its maintenance appears to be by conversion of the potential energy of the mean flow. A refinement of this model is discussed in Section 7.4.4.

The main concern with these studies has been their severe truncation. As more and smaller modes are included, will they serve as random stochastic elements increasing the transition probability between states to the point of

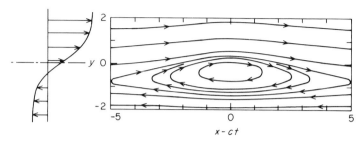

Fig. 13.1. Streamline pattern for a MKdV soliton in an unbounded asymmetric shear flow (from Redekopp, 1977).

smearing out the distinction between a quantum and classical view?

The other approach to the quantum dynamics of the atmosphere, which may be called the particle approach, has been through the study of stable structures, such as solitons or modons, which are special localized solutions of the non-linear dynamics equations. Each of the two aspects of these equations—non-linear interaction and linear dispersion—might destroy any local structure. Together, however, they can balance each other and preserve structures of certain forms. In atmospheric dynamics, linear dispersion arises from the Rossby wave effects of a mean vorticity gradient whereas non-linearity is a consequence of advection.

Long (1964) found soliton solutions in a β-channel and Benney (1966) carried through a detailed perturbation calculation to derive a Korteweg–deVries (KdV) equation appropriate for Rossby waves. Solitons are exact solutions of the KdV equation, but the KdV equation is only an approximation to, say, the barotropic vorticity equation that is formally valid for weak dispersion and weak non-linearity. The KdV (or the modified (MKdV)) equation describe dependence in only one space dimension taken as longitude in these applications. Latitudinal dependence is determined by the constraints of meridional boundary conditions. Redekopp (1977) has worked out the theory of Rossby solitons in considerable detail and Fig. 13.1, taken from his paper, illustrates one such case.

Linear dispersion can also be induced by effects of bottom topography in a shallow water model of the ocean. The corresponding solitons have been examined for stability and practical numerical simulation by Malanotte Rizzoli (1980). In her simulations, she finds the soliton-like solutions persist and appear to be robust for conditions that greatly exceed the formal requirements of the perturbation theory.

An alternate construction of a localized solution of the barotropic vorticity equation was provided by Stern (1975). His modon solution is a dipole confined within a circle that is stationary with respect to a uniform zonal flow (see Fig. 13.2). Modons are exact solutions rather than perturbation approxi-

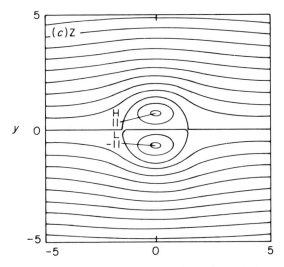

Fig. 13.2. Streamlines of an equivalent barotropic modon (relative to a frame moving with the modon) (from McWilliams, 1980).

mations but suffer from discontinuities at the circle boundary. Larichev and Reznik (1976) generalized Stern's modon and weakened the boundary discontinuity by attaching an exterior solution that decayed sufficiently rapidly to preserve the local nature of the modon. Such modons move with respect to a uniform zonal flow. Flierl *et al.* (1980) have generalized modons still further to equivalent barotropic and baroclinic cases and have shown that, having once constructed a dipole modon, monopole riders of great variety may be added. McWilliams (1980) has matched the parameters of an equivalent barotropic modon roughly to the observed characteristic of a dipole atmospheric blocking event observed over the North Atlantic Ocean in January 1963.

McWilliams *et al.* (1981) have carried out extensive gridpoint β-plane numerical studies of barotropic modons to determine the effects of limited resolution and dissipative processes and the resistance of modons to various levels and scale of perturbations. They find modons to be remarkably robust and not easily destroyed by perturbations. In the resolution experiments, even with only five grid intervals per modon diameter, a modon-like structure persisted although with a characteristic velocity about one-half of the theoretical value.

Should modons or modon-like structures be relevant to weather and climate simulation, then the question of required model resolution becomes of considerable interest. Gridpoint methods are notoriously poor in inducing erroneous linear dispersion with associated errors in group velocity propagation of wave packets (Grotjahn and O'Brien, 1976). The slowing down of modons at low resolution observed by McWilliams *et al.* (1981) is likely to be a consequence, in part, of this kind of error. Leith (unpublished) has repeated their resolution experiments with a spectral transform β-plane model in which the linear terms are treated exactly with a linear recursion operator. With five spectral transform grid intervals per modon diameter, the velocity is still diminished but only by 15%. At coarse resolution, it appears that the modon-like structure tends to enlarge and in accordance with modon dynamics to slow down. These results suggest that even low-resolution spectral transform global models, say with rhomboidal 15 truncation, are able to treat an atmospheric blocking structure of the sort studied by McWilliams (1980).

Many blocking events do not have the dipole structure of a simple modon and will require for their modelling either a soliton-like structure or a modon with rider. Further research is needed on the construction of such solutions and the investigation of their robustness. It will also be useful to examine the behaviour of solitons and modons in more general non-uniform mean flows to understand the preferred geographical location of blocking events and to bring together the results of the wave and particle approaches to the quantum dynamics of the atmosphere.

13.4 Local predictability

The possible existence of localized predictable structures makes it desirable to devise a practical measure of local predictability that would identify such situations. In principle, all aspects of the instantaneous behaviour of small errors about a given state can be determined by linearizing the dynamics equations about that state. In practice, the appropriate linear analysis is not feasible since the stability matrix in question is of dimensions $N \times N$, where N is the dimension of the phase space, perhaps 10^5. Even if it were possible to compute the eigenvalues and eigenfunctions of the stability matrix, it is not clear how these inherently non-local quantities could be interpreted in terms of local predictability. An effort was made to evaluate the diagonal elements of the instantaneous stability matrix in its configuration space representation by a Monte Carlo technique (Leith, 1980b). This would have determined at least the local response to local error, but it did not identify a significant response above the random noise of sampling error.

Stochastic dynamic prediction methods (Epstein, 1969) were designed to carry detailed error growth information along with a prediction but, as mentioned earlier, these too failed to be practical unless approximated by a Monte Carlo technique. This latter may be the only way to get at an indication of local predictability through calculation of local error growth during a prediction period of a few days. The Monte Carlo approach will be discussed more fully in a later section after the problems of model imperfections and observational errors that limit predictability in practice have been introduced.

13.5 Model imperfections

No numerical model can simulate perfectly the physics and dynamics of the atmosphere. Even if we knew accurately the details of all processes, the finite resolution of any model prohibits precise calculation. With a resolution of order 100 km for a global model, there must remain many unresolved physical and dynamical interactions influencing the larger resolved scales of motion. For example, the latent heat release associated with convective processes is on a spatial scale too fine to be resolved in a global model and must, at best, be estimated as an averaged influence on a larger resolved scale (convective parameterization). But statistical sampling theory ascribes to any average of a finite number of convective elements an uncertainty which must in this case represent a source of random error in the model. Other examples of such small-scale stochastic physical forcing effects include the influence of clouds on radiation, details of planetary boundary layer heat and momentum transport, and small-scale orographic interactions, each contributing an error source to the prediction process. Little has been done to make statistical estimates of the error source strengths of these various physical effects. To do so requires an understanding of the nature of the coupling of small-scale physical forcing with larger-scale dynamics.

In practice, a problem arises in achieving even the proper average effects of physical processes. This is reflected in the fact that a discrepancy exists between the mean climate of a prediction model and that of the atmosphere. Since prediction models expend computing power on as high a resolution as possible, it has been prohibitively expensive to make the long integrations necessary to tune them carefully to the proper climate. In carrying out an ensemble of predictions, the initial conditions will be drawn from the climate ensemble of the real atmosphere, but the predictions will tend toward the climate ensemble of the model. This climate mean drift represents a bias in each individual prediction. According to the fluctuation dissipation approximation (Leith, 1975; Bell, 1980), the climate mean adjustment occurs with a relaxation time of about 3 days and can therefore be an important source of error for practical prediction.

In considering the effects of limited resolution on the dynamical interactions between different scales of motion, it is possible to carry out simple experiments with purely dynamical models. There are two general concerns, one baroclinic and the other barotropic. As mentioned in an earlier section, Miyakoda et al. (1971) observed that inadequate resolution could interfere with the conversion of potential to kinetic energy. Quantitative estimation of this effect depends on knowledge of the conversion spectrum, and it is difficult to extract this from observations. Model studies (e.g., Gall, 1976) suggest that the conversion occurs on smaller scales than predicted by classical baroclinic instability theories (Charney, 1947; Eady, 1949) which linearize about an unrealistically smooth basic state.

The barotropic effects of truncation are more easily estimated since the energy spectrum is reasonably well known. Experiments of varying resolution in a barotropic spectral transform model by Puri and Bourke (1974) showed an N^{-2} dependence of truncation error where N is the truncation wavenumber. This is in agreement with theoretical expectation from turbulence scaling arguments for truncation in the -3 power-law spectral range of two-dimensional turbulence. Kraichnan (1976) has shown that an eddy viscosity in this range should scale as N^{-2} and thus so should the stochastic effects of unresolved scales.

The rapid diminishing of error with increasing resolution is a fortunate consequence of the -3 power-law spectrum in the atmosphere. Had the power law been $-\frac{5}{3}$, not only would the predictability of the atmosphere been much less but the resolution requirements of models and of observations would have been much greater.

13.6 Assimilation of observations

Predictability in practice is limited not only by errors arising from model imperfections and by the inherent growth of error but also by the errors in the initial state of a model. These are induced by instrumental errors in the observations and by analysis or interpolation errors in converting observations into initial state parameters. The problem of assimilation of observations into a prediction model is complicated by the fact that commonly used primitive equation models require that initial states satisfy certain balance requirements to avoid unrealistically rapid oscillations of gravitational modes during the prediction. The process of satisfying these model balance requirements is called initialization and the process of determining the best state of the atmosphere is called analysis; much research is in progress on combining the two optimally.

Many operational numerical weather prediction centres now use the non-linear normal mode initialization methods introduced by Machenhauer (1977) and Baer (1977). These methods lead to the selection, in the phase space of a primitive equation model, of an initial state vector out of a subset that has come to be known as the slow manifold (Leith, 1980a). In general, a state vector could be a set of grid-point values, an individual normal mode or a combination of such modes. Each state vector or 'point' in the slow manifold comprises only certain low-frequency modes (in middle latitudes effectively the quasi-geostrophic modes). State vectors in the non-linear slow manifold evolve during a model prediction without the rapid oscillations characteristic of unbalanced gravitational modes, i.e., they stay near the slow manifold subset.

Classical optimal analysis is based on multivariate linear regression methods (Gandin, 1963) that combine information on the statistical properties of meteorological fields known before an observation with the information newly acquired by the observation with its associated errors. The crudest source of prior information is the climate, but one does better to use the sharper estimate available from a previous prediction with associated errors. Prior knowledge will then be representable by the probability density distribution of an ensemble of states confined to the non-linear slow manifold. The non-linearity and lack of simple definition of the slow manifold lead, however, to practical computational difficulties in applying standard optimal analysis techniques to this problem.

Daley (1980) has carried out predictability experiments with pairs of primitive equation model predictions in which the initial small difference was either purely rotational or purely gravitational. He showed that initial rotational errors lead to the usual error growth rate but that initial gravitational errors had little influence on the succeeding evolution of the main rotational part of the prediction and therefore that error growth was greatly suppressed. These results show the importance of understanding how error sources may couple into the dynamics of a model if one is trying to construct an error budget. They also show that the principal task of analysis is to obtain a state with the proper rotational component; the associated balancing gravitational component is generated by the initialization process.

Figure 13.3 shows schematically the nature of the problem. The axes represent the *linear* rotational and gravitational manifolds; the slow manifold (*m*) is a non-linear deformation of the linear rotational manifold in which a balancing gravitational component is added. The point C represents a new observation; the surrounding ellipses are isolines of probability indicating the observational errors. Embedded in the slow manifold is a prior probability distribution (indicated by the number on either side of the point D on *m*). The problem is to determine the posterior probability distribution on the slow manifold that takes into account the new observation. The mean or maximum of this distribution will be the best least square or most likely posterior estimate of the initial state for a new prediction. According to Bayes' theorem, the posterior probability density is the product (renormalized) of the prior density and the density of the observational distribution shown by ellipses. The posterior probability density will thus again be confined to the slow manifold and may be indicated by the number on either side of point E.

This all seems to be simple enough in principle, but in practice the difficulty lies in characterizing the probability densities on the slow manifold for which it

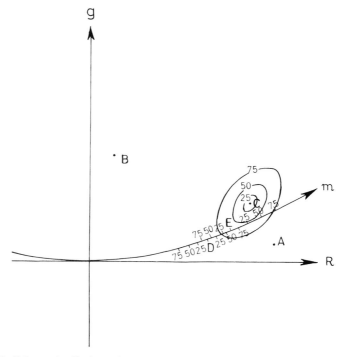

Fig. 13.3. Schematic display of the observational error distribution relative to the linear rotational and gravitational manifolds R and g and the non-linear slow manifold m. A point A in the diagram can be interpreted as representing a state dominated by rotational modes; point B would be dominated by divergent (gravitational) modes. The slow manifold line denoted by m has the property that points stay close to it as the flow evolves. The numbered contours surrounding point C indicate the probability (%) that the 'true' state to which the observation refers lies within the contour; the numbered positions on m indicate the probability that the best slow manifold representation of the true state lies within the given limits.

is not even computationally feasible to define a linear tangent manifold. The dimension of the slow manifold is about a third that of the whole phase space and precludes reasonable algebraic manipulations. In situations of this sort, one is forced to seek Monte Carlo approximations, a proposed procedure that extends earlier work (Leith, 1974) to this problem and which is now summarized.

The essential idea of a Monte Carlo approximation is that a moderately sized sample, say with $M = 10$ elements, is drawn from the infinite population. The probability distribution of this sample describes the extent of our knowledge about the state. In the present case, we assume that at the initial time for a prediction we have a sample of equally likely candidates to be the best or true representation of the state of the atmosphere. The M sample states are on the slow manifold (see Fig. 13.3) and can serve as the initial states for M predictions to the next observing time, say 12 hours later. At that time, the new set of M states is still on the slow manifold and is a sample of equally likely prior states, schematically representable as a cluster of points on the slow manifold of Fig. 13.3.

Now a new observation is made with associated error distribution as shown schematically in Fig. 13.3. The new observation, generally a set of grid point values, will in general not lie on the slow manifold. In accordance with Bayes' theorem, we next assign weights to the sample elements that are proportional to the probability of such an observational error having been made. Since the nature of observational errors is most simply described for meteorological variables at an observing point, the new weight calculation is likely to be most easily carried out by the evaluation of meteorological variables for the M sample states at all the observing points. The shift from an equally weighted to an unequally weighted sample reflects the shift from a prior to a posterior probability distribution based on the new observation.

The next step is to redefine a new equally weighted sample with the same mean and covariance as the former unequally weighted sample and with the new sample states being a linear combination of the old. It is assumed here that the sample cluster is small enough that for it the slow manifold is nearly linear and the new sample will be very close to the slow manifold. Initialization of each new sample member would correct any discrepancy. The condition that the means should agree is easily accomplished by translation. The condition that the covariances should agree requires some linear algebra with co-variances, and this would be prohibitive if it had to be carried out in the whole phase space. Fortunately, it need not be. The sample defines an $(M - 1)$-dimensional linear submanifold unchanged in the adjustment process and within which the algebraic manipulations are carried out.

We are now back at the beginning of the assimilation cycle with a sample of M equally likely states on the slow manifold to serve as initial states for M predictions.

A flaw in stochastic dynamic prediction methods and in Monte Carlo approximations to them is that model imperfections are not taken into account and thus sample clusters grow too slowly. As Pitcher (1977) recommended, this can be partly remedied by the introduction of random forcing terms to simulate the effects of model imperfections. Such a procedure is also appropriate in the present application, but the forcing terms should be confined to the sample submanifold and thus nearly to the slow manifold to avoid excitation of unbalanced gravitational modes.

The proposed Monte Carlo assimilation procedure may have a hidden virtue if the slow manifold has concealed in it attractor sheets (i.e., regions towards which solution states converge) of even lower dimension as in the low-order system of Lorenz (1980). The sample submanifold should be drawn into the attractor set after a few days of such assimilation cycles.

The price that must be paid is an M-fold increase in arithmetic over a single prediction and assimilation process. It will be necessary to determine rather carefully how large M must be for the sample submanifold to represent adequately the stochastic dynamic properties of the system.

13.7 Error budget analysis

Perhaps the best way to summarize the separate influences of internal error growth, initial observational error and model imperfections in limiting the predictability of the atmosphere is to refer to a simple error budget equation introduced for this purpose (Leith, 1978). Although the equation is based on rather crude assumptions, it seems to provide a consistent fit to observed error growths during the first day or so of prediction. It has recently been applied with some success by Bengtsson (1981) to the operational ECMWF prediction model described in Chapter 12.

The equation can also be extended to the calculation of the error budget for a data assimilation procedure. This extension provides an estimate of the relative impact of model errors on the equilibrium error level of the final analysis with assimilation.

In dealing with error budgets, it is far more natural to use mean square error or error variance E rather than the commonly used root mean square (rms) error. If, for example, a particular determination is afflicted by two independent errors with variances E_1 and E_2, then the resulting error variance is the simple sum $E = E_1 + E_2$. If, on the contrary, two independent determinations x_1 and x_2 with error variances E_1 and E_2 are combined into a better final determination $(E_2 x_1 + E_1 x_2)/(E_1 + E_2)$, then the inverses, which measure information content, are summed: $E^{-1} = E_1^{-1} + E_2^{-1}$. Both of these general statistical principles are used in this analysis.

The simple error growth equation (Leith, 1978) is:

$$\dot{E} = \alpha E + S. \tag{13.1}$$

The term αE describes the inherent tendency for error to grow owing to the unstable nature of atmospheric dynamics. The rms error doubling time of $2\frac{1}{2}$ days given by predictability theory translates into an error variance doubling time of 1.25 days and a value of $\alpha = 0.5545 \ \text{day}^{-1}$. The term S describes the model error source rate, which is model dependent and can be empirically determined by fitting observed error growth values.

Analysis error variance, which includes observation errors, will be denoted by E_0 and provides an initial value for the integration of Eqn. (13.1) with the result:

$$E(t) = E_0 + (E_0 + S/\alpha)[\exp{(\alpha t)} - 1].$$ (13.2)

It is convenient to replace the time variable t with the pseudo-time variable:

$$\tau = \frac{1}{\alpha}[\exp{(\alpha t)} - 1]$$ (13.3)

in terms of which the error growth is linear. The perceived error variance $E_p(\tau)$ involves a verification against a later analysis, and this contributes an additional term E_0 under the simple assumption that the verifying analysis has independent errors. Thus, for $\tau > 0$, we have:

$$E_p(\tau) = 2E_0 + (\alpha E_0 + S)\tau.$$ (13.4)

For a particular model a linear empirical fit to a plot of perceived values of E_p against values of $\tau > 0$ for a day or so determines first E_0 from the intercept and then S from the slope. Greater confidence is achieved by, at the same time, fitting values of $E_p(\tau)$ for a null model, namely, those of persistence forecasts. The intercept should be the same but the null model slope is greater and determines a value S_0. The ratio S/S_0 is a dimensionless figure of merit for a model. It must be remembered in making the linear empirical fit that Eqn. (13.1) includes no effects of saturation for large errors, thus that smaller errors at shorter times should be more heavily weighted.

It is assumed that the foregoing determination of E_0 is without any benefit of assimilation methods. Thus, E_0 is the error variance of an analysis which may use climate but not a model prediction as a source of prior information. The benefits of assimilation will be examined next.

The basic idea of data assimilation is to combine information from a new set of observations with the prior information about the state of the atmosphere available from a short-range prediction valid at the new observing time. In this way, information from earlier observations is carried forward, although somewhat degraded, to provide an independent source of information to be added to that newly acquired. It is straightforward to compute an error budget for the assimilation process by using Eqn. (13.1) between observation times and the general principle for compositing information at observation times.

Let now τ be the fixed pseudo-time interval of the assimilation cycle, and let E_n be the error variance after data assimilation at the nth cycle. According to Eqn. (13.1), prediction over the pseudo-time interval τ leads to a prediction error:

$$E_{\tau,n} = E_n(1 + \alpha\tau) + S\tau.$$ (13.5)

The introduction of new observations with error variance E_0 will lead to a new value E_{n+1} according to the general principle by which information is composited, thus:

$$E_{n+1}^{-1} = E_0^{-1} + E_{\tau,n}^{-1}$$ (13.6)

To cast the problem in dimensionless form, let $\varepsilon_n = E_n/E_0$, $\sigma = S/\alpha E_0$ and $\beta = 1 + \alpha\tau$. Then Eqns. (13.5) and (13.6) may be combined to give the iterative expression:

$$\varepsilon_{n+1} = [1 + \{\beta\varepsilon_n + \sigma\alpha\tau\}^{-1}]^{-1}. \tag{13.7}$$

As n increases, an equilibrium level:

$$\varepsilon = \lim_{n\to\infty} \varepsilon_n = \lim_{n\to\infty} \varepsilon_{n+1} \tag{13.8}$$

is reached which is the factor by which data assimilation reduces the observational error variance E_0. It is easy to deduce from Eqns. (13.7) and (13.8) that ε must satisfy the quadratic equation:

$$\varepsilon^2 + \eta(\sigma - 1)\varepsilon - \eta\sigma = 0 \tag{13.9}$$

where $\eta = \alpha\tau/\beta$ is a dimensionless parameter depending only on the assimilation time interval. The relevant root of Eqn. (13.9) is given by:

$$\varepsilon = [\eta\sigma + \{\eta(\sigma - 1)/2\}^2]^{1/2} - \{\eta(\sigma - 1)/2\} \tag{13.10}$$

and is displayed in Fig. 13.4 as a function of σ for values of η corresponding to assimilation time intervals of 0.25 day and 0.50 day and for $\alpha = 0.5545$ day^{-1}.

Error growth results from an early GISS research model (Druyan, 1974) were fitted by Eqn. (13.4) both for 500 mb height errors and velocity errors (Leith, 1978). For height errors, the resulting values were $E_0 = 200$ m^2, $S/\alpha = 1650$ m^2 and thus $\sigma = 8.25$. For velocity errors, the fitting parameters are $E_0 = 15$ m^2 s^{-2}, $S/\alpha = 31.3$ m^2 s^{-2} and thus $\sigma = 2.1$. More recently, Bengtsson (1981) reports for the ECMWF operational prediction model 500 mb height error values of $E_0 = 150$ m^2 and $S = 400$ m^2 with $\sigma = 4.8$. It is not clear, however, whether E_0 in this case reflects already the benefits of assimilation.

Figure 13.4 shows quantitatively how a decrease in model error sources leads to an improved equilibrium error variance. The greatest benefits appear to accrue when $\sigma = S/\alpha E_0$ is reduced to less than 1. It also appears that velocity errors may be decreased by assimilation more than are height errors.

13.8   Conclusion

In this review, an attempt has been made to establish a number of connections between studies of predictability and our general understanding of the physical and dynamical laws governing the atmosphere's behaviour. It is difficult to discuss all the relevant contributions to such an all-embracing subject; they have been limited to those with which the author is most familiar.

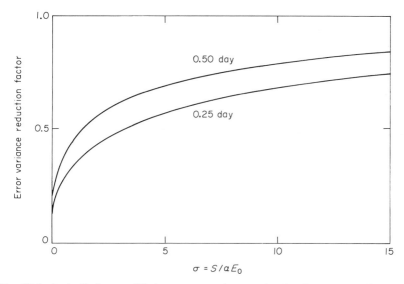

Fig. 13.4. Assimilation equilibrium error variance reduction factor ε as a function of dimensional model error source rate σ. The inherent error growth rate parameter $\alpha = 0.5545 \text{ day}^{-1}$.

What emerges most strongly as a general principle is that, since prediction is an ultimate test of our understanding, any effort to quantify the relative importance of different error sources will provide an implicit agenda for significant research in coming years. It is also clear that as analysis is becoming increasingly dependent on a prediction model for data assimilation, so also are diagnostic studies. Thus, the interpretation of observed truths about atmospheric behaviour must consider the assumptions built into our models.

References

BAER, F. (1977). Adjustment of initial conditions required to suppress gravity oscillations in non-linear flows. *Beitr. Phys. Atmos.*, **50**, 350–366.

BELL, T. L. (1980). Climate sensitivity from fluctuation dissipation: Some simple model tests. *J. atmos. Sci.*, **37**, 1700–1707.

BENGTSSON, L. (1981). In *Proceedings of the Symposium on Current Problems of Weather Prediction, Vienna, June 1981*. Extended Abstract.

BENNEY, D. J. (1966). Long non-linear waves in fluid flows. *J. Math. Phys. (Cambridge, Mass.)*, **45**, 52–63.

CHARNEY, J. G. (1947). The dynamics of long waves in a baroclinic westerly current. *J. Meteor.*, **4**, 135–162.

CHARNEY, J. G. and DEVORE, J. G. (1979). Multiple flow equilibria in the atmosphere and blocking. *J. atmos. Sci.*, **36**, 1205–1216.

CHARNEY, J. G. and STRAUS, D. M. (1980). Form-drag instability, multiple equilibria and propagating planetary waves in baroclinic, orographically forced, planetary wave systems. *J. atmos. Sci.*, **37**, 1157–1176.

CHARNEY, J. G., SHUKLA, J. and MO, K. C. (1981). Comparison of a barotropic blocking theory with observation. *J. atmos. Sci.*, **38**, 762–779.

DALEY, R. (1980). On the optimal specification of the initial state for deterministic forecasting. *Mon. Weath. Rev.*, **108**, 1719–1735.

DRUYAN, L. M. (1974). Short-range forecasts with the GISS model of the global atmosphere. *Mon. Weath. Rev.*, **102**, 269–279.

EADY, E. T. (1949). Long waves and cyclonic waves. *Tellus*, **1**, 258–277.

EPSTEIN, E. S. (1969). Stochastic dynamic prediction. *Tellus*, **21**, 739–759.

FLIERL, G. R., LARICHEV, V. D., MCWILLIAMS, J. C. and REZNIK, G. M. (1980). The dynamics of baroclinic and barotropic solitary eddies. *Dyn. Atmos. Oceans*, **5**, 1–41.

GALL, R. (1976). A comparison of linear baroclinic instability theory with the eddy statistics of a general circulation model. *J. atmos. Sci.*, **33**, 349–373.

GANDIN, L. S. (1963). In *Objective Analysis of Meteorological Fields*, Translation TT65-50007, Dept. of Commerce, National Technical Information Service, Springfield, Va., USA.

GROTJAHN, R. and O'BRIEN, J. J. (1976). Some inaccuracies in finite differencing hyperbolic equations. *Mon. Weath. Rev.*, **104**, 180–194, 989.

HART, J. E. (1979). Barotropic quasi-geostrophic flow over anisotropic mountains. *J. atmos. Sci.*, **36**, 1736–1746.

JASTROW, R. and HALEM, M. (1970). Simulation studies related to GARP. *Bull. Amer. met. Soc.*, **51**, 490–513.

KRAICHNAN, R. H. (1976). Eddy viscosity in two or three dimensions. *J. atmos. Sci.*, **33**, 1521–1536.

LARICHEV, V. and REZNIK, G. (1976). In *Rep. USSR Academy Sciences*, 321(5).

LEITH, C. E. (1974). Theoretical skill of Monte Carlo forecasts. *Mon. Weath. Rev.*, **102**, 409–418.

LEITH, C. E. (1975). Climate response and fluctuation dissipation. *J. atmos. Sci.*, **32**, 2022–2026.

LEITH, C. E. (1978). Objective methods for weather prediction. *Ann. Rev. Fluid Mech.*, **10**, 107–128.

LEITH, C. E. (1980a). Non-linear normal mode initialization and quasi-geostrophic theory. *J. atmos. Sci.*, **37**, 958–968.

LEITH, C. E. (1980b). In *Workshop on Stochastic Dynamics Forecasting*, ECMWF, May 1980.

LEITH, C. E. and KRAICHNAN, R. H. (1972). Predictability of turbulent flows. *J. atmos. Sci.*, **29**, 1041–1058.

LONG, R. R. (1964). Solitary waves in the westerlies. *J. atmos. Sci.*, **21**, 197–200.

LORENZ, E. N. (1969a). Atmospheric predictability as revealed by naturally occurring analogues. *J. atmos. Sci.*, **26**, 636–646.

LORENZ, E. N. (1969b). The predictability of a flow which possesses many scales of motion. *Tellus*, **21**, 289–307.

LORENZ, E. N. (1977). An experiment in non-linear statistical weather forecasting. *Mon. Weath. Rev.*, **105**, 590–602.

LORENZ, E. N. (1980). Attractor sets and quasi-geostrophic equilibrium. *J. atmos. Sci.*, **37**, 1685–1699.

MACHENHAUER, B. (1977). On the dynamics of gravity oscillations in a shallow water model with applications to normal mode initialization. *Beitr. Phys. Atmos.*, **50**, 253–271.

MALANOTTE RIZZOLI, P. (1980). Solitary Rossby waves over variable relief and their stability. Part II: Numerical experiments. *Dyn. Atmos. Oceans*, **4**, 261–294.

MCWILLIAMS, J. C. (1980). An application of equivalent modons to atmospheric blocking. *Dyn. Atmos. Oceans*, **5**, 43–66.

MCWILLIAMS, J. C., FLIERL, G. R., LARICHEV, V. D. and REZNIK, G. M. (1981). Numerical studies of barotropic modons. *Dyn. Atmos. Oceans*, **5**, 219–238.

MIYAKODA, K., STRICKLER, R. F., NAPPO, C. I., BAKER, P. L. and HEMBREE, G. D. (1971). The effect of horizontal grid resolution in an atmospheric circulation model. *J. atmos. Sci.*, **28**, 481–499.

PITCHER, E. J. (1977). Application of stochastic dynamic prediction to real data. *J. atmos. Sci.*, **34**, 3–21.

PURI, K. and BOURKE, W. (1974). Implications of horizontal resolution in spectral model integrations. *Mon. Weath. Rev.*, **102**, 333–347.

REDEKOPP, L. G. (1977). On the theory of solitary Rossby waves. *J. Fluid Mech.*, **82**, 725–745.

ROBINSON, G. D. (1967). Some current projects for global meteorological observation and experiment. *Q. Jl R. met. Soc.*, **93**, 409–418.

SMAGORINSKY, J. (1969). Problems and promises of deterministic extended range forecasting. *Bull. Amer. met. Sod.*, **50**, 286–311.

STERN, M. E. (1975). Minimal properties of planetary eddies. *J. Mar. Res.*, **33**, 1–13.

WHITE, G. H. (1980). Skewness, kurtosis and extreme values of Northern Hemisphere geopotential heights. *Mon. Weath. Rev.*, **108**, 1446–1455.

WIIN-NIELSEN, A. (1967). On the annual variation and spectral distribution of atmospheric energy. *Tellus*, **19**, 540–559.

WILLIAMSON, D. L. and KASAHARA, A. (1971). Adaptation of meteorological variables forced by updating. *J. atmos. Sci.*, **28**, 1313–1324.

— Appendix —

Potential vorticity and quasi-geostrophic theory

These subjects are dealt with in standard texts (e.g., Holton, 1979; Pedlosky, 1979) but, for convenience, we present here a summary of the theory.

A.1 Vorticity and potential vorticity

Using pressure as vertical coordinate, the adiabatic, frictionless, hydrostatic equations for the motion of a fluid on a β-plane may be written:

$$\frac{D\mathbf{v}}{Dt} + f\mathbf{k}\mathbf{x}\mathbf{v} + \nabla_h\phi = 0, \tag{A1}$$

$$\frac{\partial \Phi}{\partial p} = -\frac{1}{\rho}, \tag{A2}$$

$$\frac{D\theta}{Dt} = 0, \tag{A3}$$

where $f (= f_0 + \beta y)$ denotes the Coriolis parameter, $\mathbf{v}$ is the horizontal velocity vector, $D/Dt = \partial/\partial t + \mathbf{u} \cdot \nabla$ where $\mathbf{u} = \mathbf{v} + \omega\mathbf{k}$ ($\mathbf{k}$ = unit vertical vector), $\omega = Dp/Dt$, $\nabla \cdot \mathbf{u} = 0$, Φ is the geopotential, ρ the density and θ the potential temperature. Using the equation of state and the definition of potential temperature, and defining $\hat{R} = (p/p_0)^\kappa R/p$ where R is the gas constant and p_0 a standard surface pressure, the hydrostatic equation (A2) may be written:

$$\partial\Phi/\partial p = -\hat{R}\theta. \tag{A4}$$

Taking the curl of the horizontal momentum equation (A1) and using (A4)

385

gives the three-dimensional vorticity equation:

$$\frac{D\zeta_a}{Dt} = (\zeta_a \cdot \nabla)\mathbf{u} + \hat{R}\mathbf{k} \wedge \nabla\theta, \tag{A5}$$

where the absolute vorticity $\zeta_a = \left(-\dfrac{\partial v}{\partial p}, \dfrac{\partial u}{\partial p}, f + \dfrac{\partial v}{\partial x} - \dfrac{\partial u}{\partial y}\right)$. Since the coordinates x, y and p have different dimensions, so do the components of ζ_a. If preferred, one can use as vertical coordinate $z = H(1 - p/p_0)$, where H is a height scale. Then the equations are unchanged except that $\partial/\partial p$ is replaced by $\partial/\partial z$, ω by $w = Dz/Dt$, and $\hat{R}$ by $-\hat{R}p_0/H$. The first term on the right-hand side of (A5) is the stretching, twisting term and the second the 'solenoidal' term.

Taking the dot product of (A5) with $\nabla\theta$ and adding to it the result of taking the dot product of ζ_a with the gradient of (A3) gives, after some analysis:

$$\frac{D}{Dt} q_E = 0, \tag{A6}$$

where

$$q_E = \zeta_a \cdot \nabla\theta \tag{A7}$$

is the *Ertel potential vorticity*. For the medium- and large-scale atmosphere in middle latitudes, scale analysis shows that the term $\zeta_a \, \partial\theta/\partial p$ dominates in q_E, where $\zeta_a = \mathbf{k} \cdot \zeta_a$ is the vertical component of vorticity.

A.2 Quasi-geostrophic theory

The basic assumption of quasi-geostrophic theory is the smallness of the Rossby number $Ro = V/f_0 L$, where V and L are typical horizontal velocity and length scales. Assuming also that $\partial/\partial t \sim \mathbf{u} \cdot \nabla$, then the acceleration term in (A1) is of a smaller order than the Coriolis and pressure gradient terms. To lowest order f may be replaced by f_0, so that $\mathbf{v}$ is approximately equal to the geostrophic velocity $\mathbf{v}_g$ which is non-divergent and determined by the streamfunction Φ/f_0.

To obtain predictive capability one must proceed to a higher order in Ro, which is most conveniently done with reference to the vorticity equation. Because it is the vertical component of vorticity that is most important in q_E, it is this component of the vorticity equation (A5) that must be considered. Consistent with the small Ro, it is found that the vertical advection, the nonlinear stretching and the twisting terms may be neglected, giving:

$$(\partial/\partial t + \mathbf{v}_g \cdot \nabla_h)(f + \zeta_g) = f_0 \partial\omega/\partial p \tag{A8}$$

where ζ_g is the geostrophic vertical component of relative vorticity. The thermodynamic equation is modified by letting the vertical advection act only on a standard potential temperature distribution (Θ) which is a function of p only:

$$(\partial/\partial t + \mathbf{v}_g \cdot \nabla_h)\theta = -\omega \, d\Theta/dp. \qquad \text{(A9)}$$

This specification of a static stability independent of x, y and t is probably the most restrictive assumption of the theory.

Eliminating ω from (A8) and (A9) gives:

$$(\partial/\partial t + \mathbf{v}_g \cdot \nabla_h)q = 0, \qquad \text{(A10)}$$

where now the *quasi-geostrophic (quasi-) potential vorticity* is:

$$q = f + \zeta_g + f_0 \frac{\partial}{\partial p}\left(\theta' \bigg/ \frac{d\Theta}{dp}\right), \qquad \text{(A11)}$$

and:

$$\mathbf{v}_g = \mathbf{k} \wedge \nabla\psi . \qquad \text{(A12)}$$

Here:

$$\psi = [\Phi - \Phi_0(p)]/f_0, \qquad \text{(A13)}$$

and:

$$\theta' = \theta - \Theta = -f_0 \hat{R}^{-1} \, \partial\psi/\partial p. \qquad \text{(A14)}$$

Therefore we have:

$$q = f_0 + \beta y + \nabla_h^2\psi + f_0^2 \frac{\partial}{\partial p}\left[\frac{\partial\psi}{\partial p}\bigg/\left(-\hat{R}\frac{d\Theta}{dp}\right)\right]. \qquad \text{(A15)}$$

This elliptic equation for ψ is soluble if the distribution of q is given by (A10) and boundary conditions are specified. Since $w = g^{-1}D\phi/Dt$, (A12), (A13) and (A4) imply that:

$$w = (f_0/g) \, \partial\psi/\partial t - (\hat{R}\Theta/g)\omega, \qquad \text{(A16)}$$

where Θ has been substituted for θ in the last term. On a flat boundary $w = 0$, and ω from (A16) may be substituted into the thermodynamic equation (A9) to give:

$$\left(\frac{\partial}{\partial t} + \mathbf{v}_g \cdot \nabla_h\right)\theta' = -\frac{f_0}{\hat{R}\Theta}\frac{d\Theta}{dp}\frac{\partial\psi}{\partial t}. \qquad \text{(A17)}$$

Substituting for θ' from (A14) and rearranging gives:

$$\left(\frac{\partial}{\partial t} + \mathbf{v}_g\cdot\nabla_h\right)\left(\frac{\partial\psi}{\partial p} - \frac{1}{\Theta}\frac{d\Theta}{dp}\psi\right) = 0. \qquad (A18)$$

Because of the small variation of surface pressure, this equation may be applied on a surface p = constant. Since the vertical scale for change of ψ is generally much smaller than that for Θ, the right-hand side of (A17) and the second term in the second bracket in (A18) are usually neglected which is equivalent to replacing the condition $w = 0$ by $\omega = 0$. (See White, 1983, for further comment.)

It should be noted that q is not an approximation to the Ertel potential velocity q_E. With large-scale mid-latitude scalings, the first and second approximations to q_E defined in (A5) are:

$$q_{E_1} = f_0\,d\Theta/dp, \qquad (A19)$$

$$q_{E_2} = (f_0 + \beta y + \zeta_g)\,d\Theta/dp + f_0\,\partial\theta'/\partial p. \qquad (A20)$$

Since q_{E_1} is a function of p only, at the level of quasi-geostrophic theory, (A6) is replaced by:

$$(\partial/\partial t + \mathbf{v}_g\cdot\nabla_h)q_{E_2} + \omega\,dq_{E_1}/dp = 0. \qquad (A21)$$

Substituting for ω from (A9), (A21) may be reduced to:

$$(\partial/\partial t + \mathbf{v}_g\cdot\nabla_h)\,\frac{d\Theta}{dp}\,q = 0, \qquad (A22)$$

and hence to the conservation of the quasi-geostrophic potential vorticity q moving with the horizontal geostrophic velocity (A10).

In situations where the lower boundary is not level and there is Ekman pumping, the boundary condition $w=0$ is replaced by:

$$w = \mathbf{v}_g\cdot\nabla_h h_T + \alpha\zeta_g, \qquad (A23)$$

where h_T is the height of the topography and $\alpha = (K/2f)^{1/2}$, K being the eddy viscosity in the Ekman layer. From (A16) we then have at $p = p_0$:

$$-(\hat{R}\Theta/g)\omega = -(f_0/g)\,\partial\psi/\partial t + \mathbf{v}_g\cdot\nabla_h h_T + \alpha\zeta_g, \qquad (A24)$$

and the boundary condition (A18) becomes

$$\left(\frac{\partial}{\partial t} + \mathbf{v}_g\cdot\nabla_h\right)\left(\frac{\partial\psi}{\partial p} - \frac{1}{\Theta}\frac{d\Theta}{dp}\psi\right) = -\frac{g}{f_0\Theta}\frac{d\Theta}{dp}(\mathbf{v}_g\cdot\nabla_h h_T + \alpha\nabla_h^2\psi) \quad (A25)$$

A.3 Equivalent barotropic motion

A quasi-geostrophic flow is said to be equivalent barotropic if $\mathbf{v}_g = A(p)\hat{\mathbf{v}}(x,y)$. Thus there are no phase tilts in the vertical and the thermal advection is zero. The theory for equivalent barotropic motion is discussed in some detail in, for example, Thompson (1961) but here, for simplicity, the discussion is restricted to the steady version of (A8) linearised about a zonal flow $[u]$ which is a function of p only:

$$[u]\frac{\partial}{\partial x}\zeta^* + \beta v^* = f_0 \partial \omega^*/\partial p. \tag{A26}$$

Defining a vertical average $\{A\} = p_0^{-1}\int_0^{p_0} A\,dp$, and performing this average on (A24) gives:

$$\left\{[u]\frac{\partial}{\partial x}\nabla_h^2\psi^*\right\} + \beta\,\partial\{\psi^*\}/\partial x = f_0 p_0^{-1}\omega|_{p=p_0} \tag{A27}$$

An equivalent barotropic perturbation has the same horizontal structure at each height:

$$\psi^* = A(p)\hat{\psi}(x,y) \tag{A28}$$

where $\{A\}$ may be taken to be unity, so that $\{\psi^*\} = \hat{\psi}$. In this case the first term in (A25) may be written $U\,\partial/\partial x\,\nabla_h^2\hat{\psi}$ where

$$U = \{A[u]\}. \tag{A29}$$

Using the linearized, steady state form of (A24) for $\omega|_{p=p_0}$, (A27) may then be written:

$$U\frac{\partial}{\partial x}\nabla_h^2\hat{\psi} + \beta\frac{\partial}{\partial x}\hat{\psi} = -\frac{f_0}{H}\left([u](0)\frac{\partial h_T}{\partial x} + \alpha\nabla_h^2\psi^*(0)\right)$$

$$= -\frac{f_0}{H}\left([u](0)\frac{\partial h_T}{\partial x} + \alpha A(0)\nabla_h^2\hat{\psi}\right), \tag{A30}$$

where $H = R\Theta(0)/g$. Thus the horizontal structure of an equivalent barotropic perturbation is given by an equation with the same form as the barotropic vorticity equation with topographic forcing and linear damping. The zonal flow U is a weighted average of the actual flow $[u]$.

A.4 Log-pressure coordinates

Finally, we record the forms of the quasi-geostrophic equations when log-pressure coordinates are used. Defining:

$$z = H \ln (p_0/p), \tag{A31}$$

then (A10) is unchanged, but the quasi-geostrophic potential vorticity becomes:

$$q = f_0 + \beta y + \nabla_h^2 \psi + \frac{1}{\rho_0} \frac{\partial}{\partial z} \left(\frac{f_0^2}{N^2} \rho_0 \frac{\partial \psi}{\partial z} \right), \tag{A32}$$

where

$$N^2 = \frac{g}{\tilde{\Theta}} \frac{d\Theta}{dz}, \quad \tilde{\Theta} = \frac{gH}{R} \exp \left(\frac{\kappa z}{H} \right) \quad \text{and} \quad \rho_0 \propto \exp \left(-\frac{z}{H} \right).$$

The boundary condition (A18) transforms to:

$$\left(\frac{\partial}{\partial t} + \mathbf{v}_g \cdot \nabla_h \right) \left(\frac{\partial \psi}{\partial z} - \frac{N^2}{g} \psi \right) = 0 \tag{A33}$$

on $z = 0$, provided that $H = R\Theta(0)/g$.

References

Holton, J. R., 1979: *An Introduction to Dynamic Meteorology*. Second edition. Academic Press, 391 pp.

Pedlosky, J., 1979: *Geophysical Fluid Dynamics*. Springer-Verlag, 624 pp.

Thompson, P. D., 1961: *Numerical weather analysis and prediction*. Macmillan, 170 pp.

White, A. A., 1983: Some theoretical aspects of the non-Doppler quasi-geostrophic formulation. (in preparation).

Epilogue

No doubt the perspective of history will show that some of the topics discussed in some detail in this book are not of great relevance whereas, because of our present lack of knowledge, others that get a passing mention turn out to be of great significance. For instance, at the time of going to press tremendous interest is being shown in the possibility of the atmosphere being a system with multiple equilibria for the same external parameters. This theory is discussed only in a few paragraphs in Chapters 3, 6 and 7. Stable finite amplitude structures such as modons and solitons are referenced only in Chapter 13. This indicates that the authors are not yet convinced of the lasting importance of these phenomena for the atmosphere, yet time could easily prove them wrong. The subject of critical lines discussed in Chapter 6 is another, the significance of which is not clear. Is it just an artifact of linear thinking, or is it of lasting importance?

There are many areas of research covered by this volume in which advances need to be made. To pick out just one example, the way in which the tropical troposphere interacts with and responds to diabatic heating and the reality and precise nature of the so-called 'cumulus friction' needs clarification. Our hope is that this volume, by indicating the present state of the art and stressing the philosophy of using observations and an hierarchy of models to illuminate the processes responsible for particular phenomena, will promote these advances.

BJH
RPP

Subject index